Saas-Fee Advanced Course   42

More information about this at http://www.springer.com/series/4284

Cathie J. Clarke · Robert D. Mathieu
I. Neill Reid

Dynamics of Young Star Clusters and Associations

Saas-Fee Advanced Course 42

Swiss Society for Astrophysics and Astronomy
Edited by Cameron P.M. Bell, Laurent Eyer
and Michael R. Meyer

 Springer

Cathie J. Clarke
Institute for Astronomy
University of Cambridge
Cambridge
UK

Robert D. Mathieu
Department of Astronomy
University of Wisconsin
Madison, WI
USA

I. Neill Reid
Space Telescope Science Institute
Baltimore, MD
USA

Volume Editors
Cameron P.M. Bell
Department of Physics and Astronomy
University of Rochester
Rochester, NY
USA

Laurent Eyer
Department of Astronomy
University of Geneva
Sauverny
Switzerland

Michael R. Meyer
Institute for Astronomy
ETH Zürich
Zürich
Switzerland

This Series is edited on behalf of the Swiss Society for Astrophysics and Astronomy: Société Suisse d'Astrophysique et d'Astronomie Observatoire de Genève, ch. des Maillettes 51, CH-1290 Sauverny, Switzerland

Cover illustration: The Orion Nebula Cluster, hosting several thousand stars and substellar objects, is the richest aggregate of young stars within 1500 light years (about 500 parsecs), and can be found within the 'sword' region of its eponymous constellation. CREDIT: NASA, ESA, M. Robberto (Space Telescope Science Institute/ESA) and the Hubble Space Telescope Orion Treasury Project Team.

ISSN 1861-7980   ISSN 1861-8227   (electronic)
Saas-Fee Advanced Course
ISBN 978-3-662-47289-7   ISBN 978-3-662-47290-3   (eBook)
DOI 10.1007/978-3-662-47290-3

Library of Congress Control Number: 2015946088

Springer Heidelberg New York Dordrecht London

Printed on acid-free paper

Springer-Verlag GmbH Berlin Heidelberg is part of Springer Science+Business Media
(www.springer.com)

Foreword

The Saas-Fee schools are legendary and I vividly remember the one I attended as a second-year graduate student, exactly 30 years before this one, entitled: "Morphology and Dynamics of Galaxies". This year's topic is very timely, as the study of the dynamics of young stellar clusters and associations will get an enormous observational boost from the *Gaia* mission and the many ongoing spectroscopic programmes, including those on ESO's telescopes. Adriaan Blaauw (1914–2010) pioneered this field in the 1940s and 1950s, and young star clusters and associations continued to have his interest throughout his entire life. I was privileged to carry out a small student project with him which much later, but by now more than 15 years ago, led to the *Hipparcos* census of the nearby associations and follow-up work on the run-away OB stars, in both of which Blaauw had an active role. Since then the field has moved forward again, and the lectures by Cathie Clarke, Bob Mathieu and I. Neill Reid provide an excellent overview of the state of the field, with an exciting glimpse of the future.

Tim de Zeeuw

Preface

Where do most stars (and the planetary systems that surround them) in the Milky Way form? What determines whether a young star cluster remains bound (such as an open or globular cluster), or disperses to join the field stars in the disc of the galaxy? These questions not only impact understanding of the origins of stars and planetary systems like our own (and the potential for life to emerge that they represent), but also galaxy formation and evolution, and ultimately the story of star formation over cosmic time in the Universe.

To help young (and older) scientists understand our current views concerning the answers to these questions as well as frame new questions that will be answered by the European Space Agency's *Gaia* satellite that was launched in late 2013, we proposed the 42nd Saas-Fee Advanced Course "Dynamics of Young Star Clusters and Associations" to the Swiss Society of Astronomy and Astrophysics in October, 2010. The course was approved and we began to organise the school. The lectures were held in the alpine village of Villars-sur-Ollon in March 2012. We were very fortunate to have such world renowned experts agree to participate as lecturers, including Cathie Clarke (University of Cambridge) who presents the theory of star formation and dynamical evolution of stellar systems, Robert Mathieu (University of Wisconsin) who discusses the kinematics of star clusters and associations, and I. Neill Reid (Space Telescope Science Institute) who provides an overview of the stellar populations in the Milky Way and speculates on from whence came the Sun. We also benefitted from the participation of Dr. Timo Prusti (ESA) who presented a special lecture on the expected performance and impact of the *Gaia* satellite (material presented in that lecture can be found at https://cast.switch.ch/vod/clips/29ksi0s1o8/link_box and is not part of this book). Although Prof. Tim de Zeeuw was not able to participate in the school, we are grateful for his thoughtful words that grace the preceding page of this volume.

In March 2012, over 60 Ph.D. students, post-doctoral fellows and senior scientists of every career stage came to hear these lectures, think about the formation and evolution of star clusters, argue, discuss and learn (there was probably some skiing involved as well). We are grateful to each of the attendees for their active and enthusiastic participation in the school. As is often the case, the organisers and lecturers learnt as much from them as the students did from us. Each lecturer presented seven individual lectures, and a few combined lecture/discussions were also held. We also provided electronic transcripts of the lectures to each author as an aid in preparing this written version. In this volume we attempt to capture most of the material presented. Each of the lecturers of course has a unique style, as well as associated strengths and weaknesses in the presentations. We have tried to preserve those, while also injecting some uniformity of content and format. If you wish to review our work, you may view the lectures online at http://www.astro. phys.ethz.ch/sf2012/index.php?id=videos-slides.

Of course with any undertaking such as this, there are many people to thank. Ms. Marianne Chiesi (ETH) is chief among them. She was crucial to the organisation of the school, was on-site during the week of the event, and played a critical role afterwards in helping us get organised to produce this volume. We also thank Ms. Myriam Burgener from the University of Geneva for providing very helpful organisational advice in the lead up to the school. The local organising committee consisted of Dr. Richard Parker, Ms. Maddalena Reggiani, Mr. Michiel Cottaar, Dr. Richard I. Anderson and Mr. Lovro Palaversa. In particular, Dr. Richard Anderson and Lovro Palaversa were instrumental in helping to record the lectures for online access and transcription. The staff of the Eurotel Victoria Villars Hotel was very helpful, friendly, and accommodating to our needs. We also thank the Community of Villars-sur-Ollon for helping us to arrange some social events (at discount) for the participants of the school. Finally, we would also like to thank our colleagues at Springer, in particular Mr. Ramon Khanna and Ms. Charlotte Fladt, for their patience and support throughout this process.

One of us (MRM) would like to also express his gratitude to his colleagues, Dr. Cameron Bell and Dr. Laurent Eyer, for their dedicated effort, and unwavering support throughout this editorial process. In particular, Cameron Bell, who joined our efforts after the school was completed, deserves all possible recognition and accolade for his hard work in completing this volume as lead editor. Without him, this book would never have been published. However, for any errors that remain, despite our best efforts to catch them, we take full responsibility.

ETH Zürich   Michael R. Meyer
Observatoire de Genève   Laurent Eyer

Contents

List of Figures

List of Tables

Part I
Theory of Star Formation and Dynamical Evolution of Stellar Systems

Chapter 1
The Raw Material of Cluster Formation: Observational Constraints

Cathie J. Clarke

Star clusters form from reservoirs of dense cold gas ('Giant Molecular Clouds', henceforth GMCs) and in the following chapters we explore the wealth of recent simulations that follow this process, together with simulations that model the later (essentially gas-free) evolution of clusters.

A first step in any simulation is to decide on the initial conditions and for the cluster formation problem we need to specify the properties of GMCs (their typical densities and temperatures, levels of internal motions, homogeneity, etc.). We will mainly base these parameter choices on observational data and hence this chapter provides an overview of GMCs' observed properties. We will also use insights from larger scale (galaxy-wide) calculations in which GMCs emerge from simulations of the large scale interstellar medium (henceforth ISM). This brief overview is angled towards the kinds of issues that are relevant to understanding cluster formation and is no substitute for the kind of broader review of molecular clouds that can be found elsewhere: see for example Blitz (1991), Williams et al. (2000), McKee and Ostriker (2007), Fukui and Kawamura (2010), and Tan et al. (2013).

1.1 Overview of Molecular Cloud Observations

The total inventory of molecular gas in the Galaxy is estimated to be around 2.5×10^9 $M_\odot$, with about a third of the gas mass inward of the solar circle believed to be in molecular form (Wolfire et al. 2003); this number is somewhat uncertain because of the difficulty in detecting a possibly significant component in very cold gas (Loinard and Allen 1998). It is well known that molecular clouds are associated with spiral arms, both in the Milky Way (Heyer et al. 1998; Stark and Lee 2006) and in external galaxies (Helfer et al. 2003). This association is to be expected since spiral arms are

C.J. Clarke (✉)
Institute for Astronomy, University of Cambridge, Cambridge, UK
e-mail: cclarke@ast.cam.ac.uk

© Springer-Verlag Berlin Heidelberg 2015
C.P.M. Bell et al. (eds.), *Dynamics of Young Star Clusters and Associations*,
Saas-Fee Advanced Course 42, DOI 10.1007/978-3-662-47290-3_1

conspicuous in the blue light associated with young stars; since these stars have not had time to migrate far from their birth locations one would expect their natal gas to trace a similar pattern. The origin of the spiral pattern in the gas is believed to be the formation of shocks as the gas flow responds to the spiral pattern in the underlying mass distribution (Roberts 1969).

Stars form from molecular gas because the associated *Jeans mass* is low. The Jeans mass is the minimum mass required for gravitational collapse against support by pressure gradients and is given by:

$$M_{\mathrm{J}} = 0.2\,\mathrm{M}_{\odot} \left(\frac{T_{10}^3}{n_5} \right)^{1/2}, \tag{1.1}$$

where T_{10} is the temperature in units of 10 K and n_5 is the number density of hydrogen normalised to 10^5 cm^{-3} (which is typical of the densest regions within GMCs). The corresponding length scale (r_{J}) is obtained by equating M_{J} with the mass contained within a sphere of radius r_{J} so that we obtain:

$$r_{\mathrm{J}} = 0.06\,\mathrm{pc} \left(\frac{T_{10}}{n_5} \right)^{1/2}. \tag{1.2}$$

A simple heuristic way of arriving at these scales is obtained by equating the timescales for free-fall collapse with the sound crossing timescale across a region. It is then unsurprising that cold and dense conditions (as found in GMCs) are associated with a low Jeans mass and a tendency towards gravitational collapse on small scales.

It is hard to assign meaningful 'average' properties to molecular clouds because of the hierarchical organisation of the ISM, consisting of nested structures on a large dynamic range of scales (see e.g. Scalo 1990; Elmegreen 2002). A plethora of terminology is used to describe local over-densities in terms of 'clumps' or 'cores' (Williams et al. 2000; Bergin and Tafalla 2007). We will discuss methods of characterising this hierarchy in Chap. 3 but for now there are a few numbers that are worth noting: molecular gas is organised into GMCs with typical masses in the range of a few $\times 10^5\,\mathrm{M}_{\odot}$ (these often being surrounded by atomic envelopes of similar mass; Blitz 1991). The mass distribution of GMCs is describable as a power-law with index -1.6 (Blitz et al. 2007), i.e. the fraction of clouds by number in a given mass range scales with cloud mass, M_{cl}, as $M_{\mathrm{cl}}^{-1.6}$. This distribution is shallower (more mass on large scales) than the corresponding distribution for massive stars where the power-law index is -2.35 (Salpeter 1955). In the case of GMCs the power-law is however only defined over about an order of magnitude in mass since the largest GMCs in the Milky Way have masses slightly in excess of $10^6\,\mathrm{M}_{\odot}$ and the distribution is limited by completeness at the low mass end.

The mean column densities and mean volume densities of molecular clouds are of particular interest, being a little less than $\sim 10^{22}$ cm^{-2} and ~ 300 cm^{-3} respectively. We discuss below how the former quantity depends on how the cloud boundary is defined (see Lombardi et al. 2010). The typical column density is at least partly set by the requirement that clouds are dense enough to be self-shielded against photodissociation by the Galaxy's ambient ultraviolet radiation field (van Dishoeck and Black 1988). The mean density $\tilde{\rho}$ (which does not necessarily relate to a typical density of structures within clouds, given their clumpy structure, but is simply derived from the ratio of total mass to total volume) can be used to estimate a characteristic free-fall timescale through $t_{ff} \sim (G\tilde{\rho})^{-1/2}$: this turns out to be about a Myr. This number will be relevant to our later discussions about whether GMCs collapse and form stars on a free-fall timescale.

Before proceeding further with a description of the empirical 'laws' that are applied to the internal structure of GMCs we now set out a brief guide to the techniques that are used to measure the properties of molecular clouds.

1.2 Observational Techniques Applied to GMCs

GMCs are predominantly composed of molecular hydrogen: it is therefore highly inconvenient that this molecule has no permanent dipole moment since this limits the transitions corresponding to observable lines. Indeed the lowest pure rotational level of H_2 has an excitation temperature of 510 K, which is far higher than the temperatures of molecular clouds (typically 10s of K away from regions of massive star formation). This problem has led to a number of other diagnostics being used as a *proxy* for H_2. Below we summarise the complementary information that can be gleaned from line emission, dust emission and dust absorption, and discuss the advantages and disadvantages of each technique.

1.2.1 Molecular Line Emission

Line emission from a variety of abundant molecules is used to study cloud structure and kinematics. Early surveys (Solomon et al. 1987) used the second most abundant molecule in GMCs (^{12}CO); molecular clouds are generally optically thick in this emission so that it does not provide a good measure of cloud *mass*. It is thus preferable to use lower abundance molecules that are optically thin up to higher overall column densities. One of the most commonly used tracers is ^{13}CO (see Heyer et al. 2009); other commonly used molecules instead trace the densest gas within molecular clouds (e.g. NH_3: Bergin and Tafalla 2007; Juvela et al. 2012; HCN: Gao and Solomon 2004; Wu et al. 2005 and CS: Plume et al. 1997; Shirley et al. 2003).

The most obvious benefit of using molecular line data in the present context is that it provides a unique diagnostic of cloud *kinematics* via Doppler shifted emission. It thus allows a determination of the dynamical state of GMCs and this will turn out to be very important information for initialising cluster formation simulations. Moreover, the use of transitions with different critical densities (i.e. densities at which collisional and radiative de-excitation rates are equal) provides information on *volume* densities, whereas dust emission/absorption only measures column densities. Finally, for those with an interest in the chemistry of molecular clouds, molecular emission spectra provide important diagnostic information (see the reviews of Bergin and Tafalla 2007; Caselli and Ceccarelli 2012), constraining for example the free electron abundance (e.g. Bergin et al. 1999; Caselli et al. 2002) and the ages of star-forming regions (Doty et al. 2006).

On the other hand, chemical considerations can be a complicating factor when it comes to deriving the column density of H_2 from the flux in a given spectral line. There is a considerable debate in the literature about whether one can use a global conversion factor between CO and H_2 (Solomon et al. 1997; Blitz et al. 2007; Tacconi et al. 2008; Liszt et al. 2010; Wolfire et al. 2010; Sandstrom et al. 2013); additionally at high densities there is the issue of depletion of molecular gas on to grains (Redman et al. 2002; Jørgensen et al. 2005) so that gas phase diagnostics do not necessarily relate straightforwardly to the total abundance levels.

1.2.2 Dust Emission

Another widely used diagnostic of molecular cloud structure is thermal emission from dust. In order to derive the column density of gas from the flux density of dust emission at a single wavelength (usually in the millimetre or sub-millimetre range) one needs to be confident in a number of assumptions. One needs to know the fractional abundance of dust grains (compared with hydrogen), the dust emissivity law and the temperature of the emitting material. In practice one does not usually know the temperature *a priori* and thus multi-wavelength data is used to constrain this. Mapping with the *Herschel* Far Infrared satellite at wavelengths of 70–500 μm has recently provided dust continuum measurements at shorter wavelengths and has proved valuable for improving temperature constraints (e.g. Könyves et al. 2010).

The great advantage of thermal dust emission measurements is that they can not only be used to survey entire clouds—since even relatively dense structures in molecular clouds are still optically thin at millimetre wavelengths—they also allow the mapping of the densest regions of GMCs known as 'dense cores' (e.g. Motte et al. 1998; Johnstone et al. 2000). The disadvantages (apart from the lack of kinematic information) relate to uncertainties in the relationship between dust emissivity and gas mass (deriving both from uncertainties in dust emission properties and the dust to gas ratio). Moreover, in the case where a telescope beam contains emission components at a range of temperatures the mapping between multiwavelength dust emission and the total dust column can be under-constrained by the data.

1.2.3 Dust Absorption

This last difficulty is circumvented in the case of dust absorption measurements. This is because the attenuation of background sources by intervening dust depends only on the dust opacity and absorption coefficients and not on the dust temperature. 'Extinction mapping' (e.g. Lombardi and Alves 2001; Lombardi et al. 2006) is based on measuring spatial variations in the distribution of infrared colours of background stars. By comparing this distribution with that in control fields 'off-cloud' such measurements can be used to deduce a column density map of the cloud (again with the above provisos about uncertainties in the dust opacities and dust to gas ratio). Deep near-infrared measurements mean that it is possible to penetrate large limiting column densities ($\sim 10^{23}$ cm^{-2}; Román-Zúñiga et al. 2010) and thus allow the mapping of dense cores; deep observations also improve the spatial resolution since they allow a denser sampling of the background stellar sources. Detailed comparison of maps obtained via continuum emission and via extinction mapping indicates fair agreement over all though with some differences (Bianchi et al. 2003; Goodman et al. 2009; Malinen et al. 2012).

1.3 Magnetic Support and the Star Formation Efficiency Problem

Following this summary of observational methods for measuring cloud masses and kinematics, we now consider the energy budget within clouds. It is well-established that the gravitational, kinetic and magnetic energies of GMCs are comparable in magnitude whereas their thermal energy is orders of magnitude smaller. (See below for a description of magnetic field measurements in molecular clouds). This hierarchy of energies immediately implies that GMCs are not (in contrast to stars) supported by thermal pressure and this has led to the view that clouds are supported by either turbulent motions or magnetic fields. There is however a problem with sustaining such support. As we have noted, clouds are highly clumped and since the kinetic energy densities are much higher than thermal energies this means that these clumps are in a state of highly supersonic motion. Collisions between clumps are expected to be highly dissipative and this should lead clouds to collapse on a free-fall time. At one time it was believed that this situation would be mitigated by magnetic fields (even if these fields were themselves insufficient to support a static cloud) since shocks are less dissipative if they are magnetically cushioned by fields in the plane of the shock. However, simulations of hydrodynamical and magneto-hydrodynamical (MHD) turbulence (Gammie and Ostriker 1996; Mac Low et al. 1998) demonstrated that magnetic fields do *not* increase the turbulent dissipation timescale (an effect that can be broadly understood from the fact that in a turbulent medium the fields are not always parallel to shock fronts).

On the other hand, magnetic fields of sufficient strength *can* impede cloud collapse even in the absence of internal cloud motions. The ability of magnetic fields to support a static cloud against gravitational collapse can be cast in terms of a critical mass-to-flux ratio (Mouschovias and Spitzer 1976). We can derive a heuristic estimate for this value (by analogy with our description of the Jeans mass above) by comparing the free-fall collapse time with the timescale for Alfven wave propagation: Alfven waves propagate through a magnetised medium at a speed of $(B^2/\mu_0\rho)^{1/2}$ (for magnetic flux density B, magnetic permeability μ_0 and density ρ) and represent an important dynamical communication mode in magnetised media. The result of this exercise is that the critical mass-to-flux ratio is simply given by a factor of order unity times $G^{-1/2}$. Note that the critical Jeans mass (see Eq. 1.1) depends on gas density and therefore this changes—if the initial mass exceeds the initial Jeans mass—as a cloud collapses; for magnetised clouds the critical mass-to-flux ratio is however constant. Thus—provided that the magnetic field remains 'frozen' to the gas (i.e. the mass-to-flux ratio is fixed)—the ratio of a cloud's mass-to-flux ratio to the critical value is itself constant. The extent to which a cloud is either subcritical or supercritical thus does not change during collapse. It was at one time widely assumed that magnetic fields are indeed sub-critical and thus non-ideal MHD effects (specifically ambipolar diffusion: Mestel and Spitzer 1956; Galli and Shu 1993; McKee et al. 1993) were invoked as a means to slowly increase the mass-to-flux ratio and hence modulate the rate of cloud collapse (and star formation).

Subsequently there has been considerable observational effort devoted to the measurement of magnetic fields in star-forming clouds. Although the morphology of the magnetic field in the plane of the sky can be inferred from dust polarisation measurements (e.g. Heiles 2000), its magnitude can only be estimated through Zeeman polarimetry on Zeeman sensitive lines such as OH, CN and HI. Such measurements however only measure the component of the magnetic field along the line-of-sight and thus needs to be assessed in a statistical sense from a large ensemble of measurements. Early studies (Crutcher 1999) indicated that the mass-to-flux ratios in molecular clouds were close to critical (i.e. confirming that the magnetic energy density was of comparable magnitude to the gravitational potential energy). A decade of further observations and analysis has led to the conclusion that the mass-to-flux ratio is roughly twice critical: Crutcher et al. (2010) noted that the Zeeman data was 'inconsistent with magnetic support against gravity' and noted that the observed scaling of magnetic field strength with density ($B \propto \rho^{2/3}$) was as expected if the magnetic field was being passively advected in a gravitationally dominated flow. This situation is in contrast to that in the diffuse (atomic) interstellar medium where magnetic fields are instead sub-critical (Heiles and Troland 2004) and the lack of correlation between magnetic field strength and density is indicative of magnetically dominated conditions.

Clearly, therefore, magnetic fields must be important in the process of GMC formation from the diffuse medium (Kim and Ostriker 2006; Mouschovias et al. 2009). Even on the scale of GMC interiors (which will form the subject of much of these chapters), the only mildly super-critical conditions mean that magnetic fields should *not* be ignored. It is worth emphasising that most of the simulations described below omit magnetic fields for purely practical reasons.

The above results have an important implication for what is often described as the 'star formation efficiency problem'. Given that hydrodynamical and MHD turbulence both dissipate on a free-fall time and given that magnetic fields are insufficient to prevent collapse, we are left to conclude that clouds should collapse on a free-fall timescale, unless there are mechanisms that re-inject energy into the turbulence. We might therefor expect—unless the formation of stars itself disperses the remaining gas—that the timescale on which a GMC is converted into stars is its free-fall time ($\sim$1 Myr). There are however a number of observational indications that this is *not* the case. If we divide the entire mass of molecular gas in the Milky Way ($\sim$$10^9\,\mathrm{M_\odot}$) by a typical cloud free-fall timescale, we would expect that the Galactic star formation rate would be $\sim$$10^3\,\mathrm{M_\odot\,yr^{-1}}$, which exceeds the observed rate by more than two orders of magnitude. We therefore conclude that the star formation rate associated with GMCs (averaged over the time that gas is within GMCs) is much less than the total mass in GMCs divided by the free-fall time.

This conclusion, based on galaxy-wide scales, has been confirmed by recent studies within individual GMCs. Probably the most comprehensive study to date is that of (Evans et al. 2009) which used the *Spitzer* 'Cores to Disks' Legacy Survey to compare the census of young stars with the magnitude of the available mass reservoir. The results of this exercise confirmed that star formation is indeed inefficient with around 3–6% of the cloud mass being converted into stars per free-fall time.

1.4 Scaling Relations

Following the first large-scale surveys of the structure and kinematics of molecular clouds, several correlations ('scaling relations') were noted by Larson (1981). These are now known widely as 'Larson's Laws' and concern the inter-relationship between mass, linear size and velocity width for structures within molecular clouds:

1. The velocity dispersion σ across structures of different size (R) scales as $\sigma \propto R^{0.5}$ (see Solomon et al. 1987). Since this relation was derived from radio line observations it is often termed the 'size linewidth' relation: see Fig. 1.1.
2. The mass (M), R and σ are related by $M \sim R\sigma^2/G$.
3. The mean density varies inversely with R or (equivalently) different structures within clouds share a roughly constant column density: $M \propto R^2$. Note that a situation of constant column density cannot be true in detail or else there would be no contrast between different structures within clouds. Lombardi et al. (2010) have shown that extinction mapping (see Sect. 1.2.3) within several GMCs demonstrates that the distribution of column densities within a cloud is describable as a log normal. The mean column density within a cloud depends on the level at which the data is thresholded (i.e. what is the lower limit on extinction used to define the cloud boundary). Since the distribution of column densities appears to be rather similar from cloud to cloud, the mean column density is indeed similar in different clouds, provided the clouds are analysed above the same extinction contour.

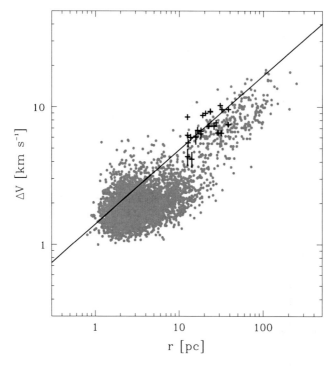

Fig. 1.1 Size-line width radius relationship for molecular clouds in M33 (*crosses*). The *grey dots* represent Milky Way molecular cloud data from Solomon et al. (1987) and Heyer et al. (2001). The power-law fit gives $\Delta V \propto r^{0.45 \pm 0.02}$. Figure from Rosolowsky et al. (2003)

It can immediately be seen that any two of Larson's laws imply the third and so one would like to know which two of the laws are 'fundamental' and which one is just a consequence of the other two. We will consider (1) and (2) in a little more detail.

It is often said that molecular clouds exhibit supersonic turbulence: supersonic motions are of course immediately implied by the high ratio of kinetic to thermal energy in GMCs that was noted in Sect. 1.3. It is debatable whether these strong internal motions can strictly be described as turbulence (where this is understood to represent a *steady state* cascade of energy from a large [driving] scale to the small scale at which it is dissipated). Larson however pointed out that the first law was roughly consistent with such a scenario. If one considers a power spectrum $P(k) \propto k^{-a}$ (where k is related to the wavelength, λ via $k = 2\pi/\lambda$) then the kinetic energy per unit mass associated with wave vectors in the range k, to $k + dk$ is given by $P(k)k^2 dk$. From this one can deduce that the mean square velocity associated with size scale R should scale as $\sigma^2 \propto R^{(a-3)}$ and thus the Larson Law would suggest a power spectrum with $a = 4$. This value is suggestively close to several well-studied categories of turbulence (e.g. incompressible 'Kolmogorov' turbulence has $a = 11/3$, compressible 'Burgers' turbulence has $a = 4$ while MHD turbulence has $a = 3.5$). Myers and Gammie (1999) explored this possibility further, examining the case of injection at finite driving scales. They pointed out that the asymptotic relationships described above should flatten out at size scales above the 'driving' scale: the fact that this is *not* observed within GMCs (see Fig. 1.1) then implies that the driving scale is large (of order 100 pc or above), and would suggest that energy is injected into the clouds from the larger scale galactic environment.

A large body of work has been devoted to modelling GMCs as turbulent systems but it is worth repeating that we do not know whether GMCs have had time to achieve the steady state turbulent cascade that is observed in situations of laboratory turbulence. If they are indeed in such a steady state, then the turbulent structure depends only on the physical conditions in the medium (compressibility, presence of magnetic fields, etc.) and not on the initial conditions. If they are not in a steady state (as is the case for a large category of the simulations that we will discuss in forthcoming chapters, where clouds fragment into stars on a free-fall time) then the statistics describing kinematic and density structures are constantly evolving. In this case the power spectrum partly reflects the formation history of the cloud and it becomes particularly important to understand the nature of the relationship between the internal kinematics of GMCs and their interaction with the wider environment.

Turning now to Larson's second law, we note that there are several different interpretations. One extreme interpretation would be to say that Larson's 1st and 3rd laws are 'fundamental' for some reason and that therefore the 2nd law ($\sigma^2 \propto M/R$) is just a mathematical consequence of the other two laws. In this extreme interpretation, this scaling has nothing to do with the role of gravity in molecular clouds. Another interpretation is to note that the constant of proportionality in this relation is of order G (the gravitational constant), suggesting that self-gravity plays an important role in determining cloud structure. Note that this is a weaker statement than another extreme version which maintains that clouds are in a state of virial equilibrium (which has led to this assumption being used in order to *determine* cloud masses, e.g. Blitz et al. 2007; Bolatto et al. 2008).

Much observational data has been assembled on the masses and kinematics of a range of clouds, both in the Galaxy (Heyer et al. 2009) and in extragalactic environments (Rosolowsky 2007; Bolatto et al. 2008). These studies express the degree of gravitational boundedness of clouds in terms of a parameter α_{vir} which is proportional to the ratio of kinetic energy to potential energy and which would be unity in the case of a spherical cloud in virial equilibrium (and equal to 2 in the case of a marginally unbound spherical cloud). A large scatter in α_{vir} values is found at all masses, with values ranging from somewhat less than 1 to around 10 (see Fig. 1.2). It is not clear what fraction of this scatter can be attributed to observational uncertainties. The mean is close enough to unity to discourage the idea that gravity is irrelevant to this relation. Nevertheless the large scatter means that it is still arguable whether the bulk of clouds are gravitationally bound or unbound; as we shall see later (see Chap. 4, Fig. 4.1), rather small differences in α_{vir} in the region of marginal boundedness can have dramatic effects on the star formation rate.

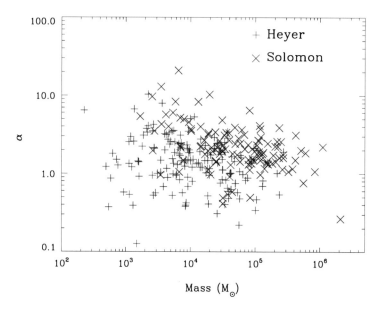

Fig. 1.2 Plot of the gravitational parameter (α; see text) as a function of molecular cloud mass from Solomon et al. (1987) and Heyer et al. (2009). Figure from Dobbs et al. (2011b)

1.5 GMCs and the Large-Scale ISM

At this point we can start to see that the properties of GMCs are quite well characterised observationally but that we do not understand some important empirical facts (such as why the fraction of clouds that is converted into stars per crossing time is so low, nor what drives the inter-relationship between the Larson scaling relations). In both cases it is likely that the answers are related to processes that originate beyond the GMCs themselves in the wider galactic environment.

In recent years, advances in computational power have enabled some ambitious galaxy-wide simulations that have started to shed some light on the relationship between GMCs and the wider ISM: for example, the grid based calculations of Tasker and Tan (2009) and the complementary smoothed particle hydrodynamics (SPH) calculations of Dobbs et al. (2011a). From the perspective of our present discussion, the main questions of interest are whether such galaxy-wide simulations give rise to GMC-like structures that, for example, obey the scaling relations discussed above (it should be stressed that the 'clouds' formed in this simulation are in themselves insufficiently resolved for one to follow star formation within them directly). It is found that the simulations do a reasonable job at reproducing the size-line width relation, although the dynamic range of the simulated clouds is small compared with that covered by observations. It is however hard to identify exactly what physical processes contribute to the form of the relationship in the simulated clouds (indeed

Dobbs and Bonnell 2007 have shown that such a relationship is readily obtained from a variety of situations where clouds form in clumpy shocks, even in the absence of self-gravity).

The kinetic and potential energy contents of clouds formed in galaxy-wide simulations have also been analysed and demonstrate that a number of factors (including numerical resolution; Tasker and Tan 2009) affect whether clouds are predominantly bound or unbound. For example, Dobbs et al. (2011a) found that clouds were predominantly unbound in the case of magnetised simulations on account of the role of magnetic fields in inhibiting collapse. In the absence of magnetic fields or (parametrised) supernova feedback, cloud collapse produces a roughly virialised (hence bound) state; however even in the absence of magnetic fields the number of unbound clouds increases strongly as the strength of the feedback is increased. Dobbs et al. (2011a) argued that the latter is more realistic since the unbound clouds in the simulation are rather aspherical (similar to those observed; Koda et al. 2006) whereas gravitationally bound clouds collapse to more spherical configurations.

A further cloud diagnostic that may relate to the mode of cloud assembly is whether the net cloud rotation is prograde or retrograde (with respect to the rotation of the host galaxy). Rosolowsky et al. (2003) noted that a surprisingly large number of clouds in M33 were counter-rotating and that the magnitude of rotation in the prograde population was too small to be consistent with angular momentum conserving collapse associated with gravitational instability. The simulations of Dobbs (2008) suggested that those (generally higher mass) clouds that are self-gravitating are indeed prograde, but that an important population of smaller clouds, which are formed mainly by agglomeration, display a mixture of prograde and retrograde spin directions.

A final point to emerge from these larger scale simulations is that they agree on the importance of encounters between clouds. This is a caveat that should be borne in mind when interpreting the simulations that we will be discussing later, which (for reasons of computational economy) consider cloud evolution *in isolation*. For example, Tasker and Tan (2009) emphasise that their cloud-cloud collision timescales are shorter than many estimates of GMC lifetimes and that '...an individual GMC is just as likely to have its properties dramatically altered by a merger than by a destructive mechanism such as supernova feedback or ionization feedback'. In a similar vein, Dobbs et al. (2011a) note that '...the constituent gas in GMCs is likely to change on timescales of Myr... A cloud seen after 30 Myr may not be a counterpart to any cloud present at the current time'.

Evidently such large-scale simulations are in their infancy and at this stage some of the insights are rather qualitative. It however appears 'rather easy' to produce clouds whose properties (mass, spins, morphology, internal kinematics, gravitational energy) roughly match those observed. In the simulations, both self-gravity and agglomeration play roles in cloud creation, with the former being increasingly important at larger cloud scales. Moreover, the simulations raise important doubts about the legitimacy of treating any GMC's evolution as being truly isolated from its environment and paint a picture in which clouds' individual identities are mutable on timescales of Myr.

1.6 Summary: Key Observational Constraints for Simulations

We will proceed in the following chapters to describe a variety of hydrodynamical simulations of star and cluster formation and so we end this chapter by listing the key factors that should inform the design and interpretation of such simulations: (i) It is necessary to model loosely bound and unbound clouds. (ii) Ideally such simulations should include magnetic fields, since the magnetic energy density in clouds is similar in magnitude to their kinetic and gravitational energies. (iii) Ideally such simulations should model interaction with the surroundings—this is particularly hard to model in a meaningful way without resorting to galaxy-scale simulations and thus sacrificing resolution within clouds. (iv) Clouds should not necessarily be regarded as examples of fully developed (steady state) turbulence, especially given the insights above about the transient lifetimes of clouds as distinct entities. (v) Finally, the results of such simulations need to be monitored with regard to the 'efficiency' (rate per free-fall time) of their resulting star formation, in order that they do not exceed the upper limits imposed by observations.

References

Bergin, E. A. & Tafalla, M. 2007, ARA&A, 45, 339
Bergin, E. A., Plume, R., Williams, J. P., & Myers, P. C. 1999, ApJ, 512, 724
Bianchi, S., Gonçalves, J., Albrecht, M., et al. 2003, A&A, 399, L43
Blitz, L. 1991, in NATO ASIC Proc. 342: The Physics of Star Formation and Early Stellar Evolution, ed. C. J. Lada & N. D. Kylafis, 3
Blitz, L., Fukui, Y., Kawamura, A., et al. 2007, in Protostars and Planets V, ed. B. Reipurth, D. Jewitt & K. Kell, University of Arizona Press, 81
Bolatto, A. D., Leroy, A. K., Rosolowsky, E., Walter, F., & Blitz, L. 2008, ApJ, 686, 948
Caselli, P. & Ceccarelli, C. 2012, A&A Rev., 20, 56
Caselli, P., Walmsley, C. M., Zucconi, A., et al. 2002, ApJ, 565, 331
Crutcher, R. M. 1999, ApJ, 520, 706
Crutcher, R. M., Wandelt, B., Heiles, C., Falgarone, E., & Troland, T. H. 2010, ApJ, 725, 466
Dobbs, C. L. 2008, MNRAS, 391, 844
Dobbs, C. L. & Bonnell, I. A. 2007, MNRAS, 374, 1115
Dobbs, C. L., Burkert, A., & Pringle, J. E. 2011a, MNRAS, 417, 1318
Dobbs, C. L., Burkert, A., & Pringle, J. E. 2011b, MNRAS, 413, 2935
Doty, S. D., van Dishoeck, E. F., & Tan, J. C. 2006, A&A, 454, L5
Elmegreen, B. G. 2002, ApJ, 564, 773
Enoch, M. L., Glenn, J., Evans, II, N. J., et al. 2007, ApJ, 666, 982
Evans, II, N. J., Dunham, M. M., Jørgensen, J. K., et al. 2009, ApJS, 181, 321
Fukui, Y. & Kawamura, A. 2010, ARA&A, 48, 547
Galli, D. & Shu, F. H. 1993, ApJ, 417, 220
Gammie, C. F. & Ostriker, E. C. 1996, ApJ, 466, 814
Gao, Y. & Solomon, P. M. 2004, ApJS, 152, 63
Goodman, A. A., Pineda, J. E., & Schnee, S. L. 2009, ApJ, 692, 91
Hatchell, J., Richer, J. S., Fuller, G. A., et al. 2005, A&A, 440, 151
Heiles, C. 2000, AJ, 119, 923
Heiles, C. & Troland, T. H. 2004, ApJS, 151, 271
Helfer, T. T., Thornley, M. D., Regan, M. W., et al. 2003, ApJS, 145, 259
Heyer, M. H., Brunt, C., Snell, R. L., et al. 1998, ApJS, 115, 241

Heyer, M. H., Carpenter, J. M., & Snell, R. L. 2001, ApJ, 551, 852
Heyer, M. H., Krawczyk, C., Duval, J., & Jackson, J. M. 2009, ApJ, 699, 1092
Johnstone, D., Wilson, C. D., Moriarty-Schieven, G., et al. 2000, ApJ, 545, 327
Johnstone, D., Di Francesco, J., & Kirk, H. 2004, ApJ, 611, L45
Jørgensen, J. K., Schöier, F. L., & van Dishoeck, E. F. 2005, A&A, 437, 501
Juvela, M., Harju, J., Ysard, N., & Lunttila, T. 2012, A&A, 538, 133
Kim, W.-T. & Ostriker, E. C. 2006, ApJ, 646, 213
Koda, J., Sawada, T., Hasegawa, T., & Scoville, N. Z. 2006, ApJ, 638, 191
Könyves, V., André, P., Men'shchikov, A., et al. 2010, A&A, 518, L106
Larson, R. B. 1981, MNRAS, 194, 809
Liszt, H. S., Pety, J., & Lucas, R. 2010, A&A, 518, 45
Loinard, L. & Allen, R. J. 1998, ApJ, 499, 227
Lombardi, M. & Alves, J. 2001, A&A, 377, 1023
Lombardi, M., Alves, J., & Lada, C. J. 2006, A&A, 454, 781
Lombardi, M., Alves, J., & Lada, C. J. 2010, A&A, 519, L7
Mac Low, M.-M., Klessen, R. S., Burkert, A., & Smith, M. D. 1998, Physical Review Letters, 80, 2754
Malinen, J., Juvela, M., Rawlings, M. G., et al. 2012, A&A, 544, 50
McKee, C. F. & Ostriker, E. C. 2007, ARA&A, 45, 565
McKee, C. F., Zweibel, E. G., Goodman, A. A., & Heiles, C. 1993, Protostars and Planets III, ed. E. H. Levy & J. I. Lunine, University of Arizona Press, 327
Mestel, L. & Spitzer, Jr., L. 1956, MNRAS, 116, 503
Motte, F., Andre, P., & Neri, R. 1998, A&A, 336, 150
Mouschovias, T. C. & Spitzer, Jr., L. 1976, ApJ, 210, 326
Mouschovias, T. C., Kunz, M. W., & Christie, D. A. 2009, MNRAS, 397, 14
Myers, P. C. & Gammie, C. F. 1999, ApJ, 522, L141
Plume, R., Jaffe, D. T., Evans, II, N. J., Martín-Pintado, J., & Gómez-González, J. 1997, ApJ, 476, 730
Redman, M. P., Rawlings, J. M. C., Nutter, D. J., Ward-Thompson, D., & Williams, D. A. 2002, MNRAS, 337, L17
Roberts, W. W. 1969, ApJ, 158, 123
Román-Zúñiga, C. G., Alves, J. F., Lada, C. J., & Lombardi, M. 2010, ApJ, 725, 2232
Rosolowsky, E. 2007, ApJ, 654, 240
Rosolowsky, E., Engargiola, G., Plambeck, R., & Blitz, L. 2003, ApJ, 599, 258
Salpeter, E. E. 1955, ApJ, 121, 161
Sandstrom, K. M., Leroy, A. K., Walter, F., et al. 2013, ApJ, 777, 5
Scalo, J. 1990, in Astrophysics and Space Science Library, Vol. 162, Physical Processes in Fragmentation and Star Formation, ed. R. Capuzzo-Dolcetta, C. Chiosi, & A. di Fazio, Kluwer Academic Publishers, 151
Shirley, Y. L., Evans, II, N. J., Young, K. E., Knez, C., & Jaffe, D. T. 2003, ApJS, 149, 375
Solomon, P. M., Rivolo, A. R., Barrett, J., & Yahil, A. 1987, ApJ, 319, 730
Solomon, P. M., Downes, D., Radford, S. J. E., & Barrett, J. W. 1997, ApJ, 478, 144
Stark, A. A. & Lee, Y. 2006, ApJ, 641, L113
Tacconi, L. J., Genzel, R., Smail, I., et al. 2008, ApJ, 680, 246
Tan, J. C., Shaske, S. N., & Van Loo, S. 2013, in IAU Symposium, Vol. 292, Molecular Gas, Dust and Star Formation in Galaxies, ed. T. Wong & J. Ott, Cambridge University Press, 19
Tasker, E. J. & Tan, J. C. 2009, ApJ, 700, 358
van Dishoeck, E. F. & Black, J. H. 1988, ApJ, 334, 771
Williams, J. P., Blitz, L., & McKee, C. F. 2000, Protostars and Planets IV, ed. V. Manning, A. P. Boss & S. S. Russell, University of Arizona Press, 97
Wolfire, M. G., McKee, C. F., Hollenbach, D., & Tielens, A. G. G. M. 2003, ApJ, 587, 278
Wolfire, M. G., Hollenbach, D., & McKee, C. F. 2010, ApJ, 716, 1191
Wu, J., Evans, II, N. J., Gao, Y., et al. 2005, ApJ, 635, L173

Chapter 2
The Numerical Tools for Star Cluster Formation Simulations

Cathie J. Clarke

GMCs collapse and form stars wherever gravitational forces can overwhelm the supportive effects provided by thermal pressure, magnetic fields and internal motions ('turbulence'). In practice, the bulk of star formation within molecular clouds is associated with dense gas where the Jeans mass and Jeans length are small (see Eqs. 1.1 and 1.2, Chap. 1). The characteristic scale of gravitational fragmentation (which gives rise to 'pre-stellar cores': see Chap. 1, Sect. 1.2) is somewhat less than a solar mass. Dense gas in GMCs is often organised into filamentary structures and cores and the protostars into which these evolve also trace a filamentary pattern (see for example Chap. 7, the right-hand panel of Fig. 7.1).

The evolution of dense self-gravitating gas from pre-stellar cores through to pre-main-sequence stars is usefully summarised in the famous cartoon of Shu et al. (1987) shown in Fig. 2.1. Roughly speaking the time spent in the pre-stellar/protostellar phase is of order 10^5 years which equates roughly with the free-fall time for the densest regions within GMCs; on the other hand the overall lifetime of star-disc systems (a few Myr; Haisch et al. 2001) is of order the dynamical time of the entire GMC. The transformation of objects along the path shown in Fig. 2.1 thus occurs on timescales on which there is ample time for dynamical evolution. Observations of the stars and gas within star-forming regions (see e.g. Mathieu Chap. 11) thus do not just capture conditions at birth but also reflect a history of significant dynamical evolution.

In order to interpret such observations it is essential to perform simulations. Ideally these simulations should start with a GMC (or even start with the formation of the GMC from the diffuse ISM) and self-consistently form stars, thereafter tracking both continued star formation and the dynamical evolution of the gas and stellar components. In order to be realistic, magnetic fields in the gas should be included and the temperature of the gas (given its irradiation by the stars that form) should be modelled through radiative transfer calculations. At some point the residual gas is

C.J. Clarke (✉)
Institute for Astronomy, University of Cambridge, Cambridge, UK
e-mail: cclarke@ast.cam.ac.uk

© Springer-Verlag Berlin Heidelberg 2015   17
C.P.M. Bell et al. (eds.), *Dynamics of Young Star Clusters and Associations*,
Saas-Fee Advanced Course 42, DOI 10.1007/978-3-662-47290-3_2

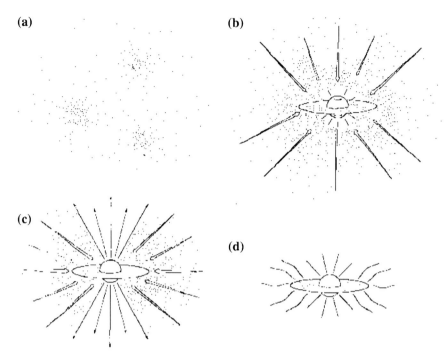

Fig. 2.1 Schematic of the evolutionary stages of young stars demonstrating the successive stages of collapse, infall, disc formation and ultimately disc dispersal. Figure from (Shu et al. 1987)

also likely to be impacted by so-called mechanical feedback (i.e. the momentum input from winds, jets or supernovae). All these effects modify the location of the gas and its gravitational effects on the groupings of stars formed in the simulations. At some point, the gas is expelled from the vicinity of the stars and then—and only then—is it legitimate to treat the stars with purely stellar dynamical simulations.

Simulations that include all the effects listed above are currently computationally unfeasible and we instead need to make progress in a piecemeal fashion, gathering insights from simulations that focus on different aspects of the problem and which treat a different subset of physical ingredients. In this chapter we present a brief overview of the numerical techniques that are involved in modelling the formation of star clusters. We will start with the simplest (pure gravitational problem) and then work towards codes that can handle a larger range of physical effects.

2.1 The Pure Gravitational Problem

One approach to solving for the evolution of an ensemble of gravitationally inter-acting point masses is to regard the system as a fluid in six-dimensional phase space. In this case the fluid evolves according to the Fokker-Planck equation which is the collisionless Boltzmann equation including also source terms representing the effect of gravitational encounters between individual stars. Monte Carlo codes (e.g. Hénon 1971; Freitag and Benz 2001) have been developed using this approach but are most appropriate to large-N systems where the dominant stellar motion is composed of collisionless trajectories in the smoothed cluster potential and where 'collisional effects' (which in the stellar dynamical community usually refer to gravitationally focused encounters between stars rather than actual physical collisions) are rare. It is not however a good approach to small-N systems where strong deflections are the norm. We will see later that the hierarchical nature of cluster assembly means that small-N dynamics are relevant in the early evolutionary stages of even rather populous clusters. Henceforth we will instead concentrate on the alternative method of simulating stellar systems, i.e. by direct N-body integration: the interested reader is directed to reviews by Aarseth (2003) and Dehnen and Read (2011).

The N-body problem splits into two components: determining the potential given the instantaneous distribution of stellar mass points and then integrating the orbits of each star in this time-dependent potential. The most obvious way to determine the gravitational acceleration experienced by each star is through direct evaluation of each pairwise force, an operation that scales with particle number N as N^2 and thus whose expense would prohibit the integration of populous systems. This difficulty is usually circumvented by the implementation of 'tree gravity' (Barnes and Hut 1986) in which particles are lumped together (for the purpose of evaluating the force on a given star) according to the angle that they subtend from the star (see Fig. 2.2). In this implementation, the contribution from close-by particles is evaluated directly whereas particles are increasingly grouped at larger distances. This computational economy is reflected in a more advantageous scaling with N (i.e. as $N \ln N$): 'tree gravity' forms the basis not only of many N-body codes but also for the gravitational component of SPH codes, as described below.

Close encounters pose a particular problem for N-body codes, which can be seen by considering the case of two stars in a highly elliptical but marginally bound orbit. The total energy should of course be a fixed quantity around the orbit, but when the stars are at pericentre, this fixed total is given by the difference between two large numbers (the gravitational and kinetic energy). A small fractional error in either of these quantities can then produce a very large fractional error in the total energy, resulting, for example, in a situation where numerical errors can unbind an initially bound pair. Since this is deeply undesirable, it is preferable to stop particles getting very close to each other and this is often achieved by 'softening' the potential (i.e. modifying the potential for each pairwise interaction so that it tends to a finite value at zero separation and is significantly modified for separations on the order of a prescribed 'softening' length). The penalty paid for such softening is that results are unreliable at small scales and that two-body relaxation effects are somewhat suppressed (although not entirely, since the cumulative effect of large numbers of more distant, unsoftened, encounters are also important for two-body relaxation: see Heggie and Hut 2003). The alternative option (which is much less intuitively obvious) is the expedient of 'Kustaanheimo-Stiefel (KS) regularisation' (Kustaanheimo and Stiefel 1965) which is a coordinate transformation that—by removing the singularity at the origin—allows the integration of the two-body problem down to arbitrarily

Fig. 2.2 *Upper panel*:
Computation of the force for
one of 100 particles
(*asterisks*) in two dimensions
(for graphical simplicity)
using direct summation:
every line corresponds to a
single particle-particle force
calculation. *Middle panel*:
Approximate calculation of
the force for the same
particle using the tree code.
Cells opened are shown as
black squares with their
centres z indicated by *solid
squares* and their sizes w by
dotted circles. Every *green
line* corresponds to a
cell-particle interaction.
Lower panel: Approximate
calculation of the force for
all 100 particles using the
tree code, requiring 902
cell-particle and 306
particle-particle interactions
($\theta = 1$ and $n_{\text{max}} = 1$),
instead of 4950
particle-particle interactions
with direct summation.
Caption and figure from
Dehnen and Read (2011)

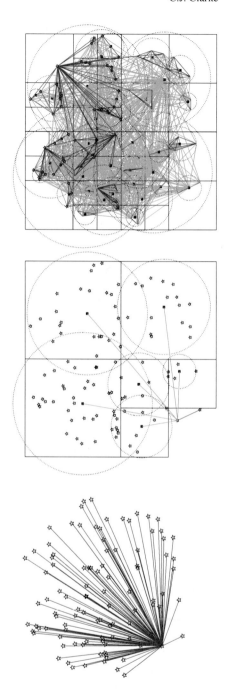

small separations. Regularised codes switch to this integration regime as required; 'chain regularisation' is the application of this approach to the case of nested multiple systems (see Mikkola and Aarseth 1990, 1993, 2002).

2.2 Hydrodynamical Problems: A Quick Guide to SPH

We start by noting that hydrodynamic equations can be written in either Lagrangian or Eulerian form, which respectively follow the evolution of individual fluid elements or else trace the evolution with respect to a spatially fixed grid. Each of these approaches is associated with a class of numerical techniques. We shall start with Lagrangian codes of which the most well developed is that of 'Smoothed Particle Hydrodynamics', a technique that is widely applied to star and cluster formation simulations. The interested reader is referred to reviews by Monaghan (1992, 2005) and Springel (2010b).

The fundamental problem to be addressed by a Lagrangian code is how—when modelling a fluid as a set of discrete mass points—one can evaluate quantities such as the density and pressure since these are needed for the evaluation of hydrodynamical forces. One might, most simply, just place a 'sampling volume' around each particle and calculate the density as the number of particles in this volume divided by the volume. This quantity would however fluctuate as particles left and entered the sampling volume, making the density estimate a noisy quantity and thus rendering accelerations deriving from pressure gradients inaccurate. In SPH, this problem is handled by weighting the contribution of each particle within the sampling volume according to its distance from the particle in question, the weighting being controlled by a so-called 'kernel function'. The spatial extent of the sampling volume is controlled by the SPH smoothing length, h; in modern implementations of SPH codes this is generally adaptive so as to ensure a roughly fixed number of particles within a 'neighbour sphere'.

As mentioned above, gravity within SPH simulations is often handled with a tree formulation; gravitational softening is generally set equal to the SPH smoothing length and is thus adaptive. It is worth noting that in hybrid systems consisting of both gas and star particles, the accuracy of the stellar dynamics is therefore affected by the resolution of the gaseous component.

As well as having to model pressure and gravitational forces, SPH codes need also to include artificial viscosity terms, i.e. components of the equation of motion that depend on the relative mutual velocity of particles (in the absence of viscosity, approaching supersonic particles would simply 'pass through' each other and viscosity is required in order to instead attain the desired hydrodynamic behaviour—i.e. the generation of a shock). For our purpose here, the most important thing to note is the possibly undesirable consequence of artificial viscosity when it comes to modelling centrifugally supported *discs* around young stars. In most implementations of viscosity within SPH, increased viscous dissipation is activated in the case that particles within a neighbour sphere (i.e. within a distance h) have a supersonic mutual relative velocity. However, in a Keplerian disc, the mutual velocity across a radial distance h is $h\Omega$ where Ω is the local Keplerian angular velocity. On the other hand, hydrostatic equilibrium normal to the disc plane requires that the sound speed in the disc is given by $\sim H\Omega$ where H is the disc vertical scale height. Thus supersonic relative motions within a neighbour sphere will be detected whenever $h > H$ and such a condition would, in default viscosity implementations, increase the levels of viscous dissipation (even though, in a shear flow, there is no shock and therefore such an enhancement is not necessary).

The reason why this enhancement of viscosity is undesirable is that it is associated with an increased (numerical) transport of angular momentum in the disc: mass will be transported radially through the disc and accrete on to the star by purely numerical effects. Such spurious accretion depletes circumstellar discs too rapidly and this then increases the accretion rate still further (because the declining surface density increases h further). Although the effect of unwanted artificial viscosity in pure shear flows has been mitigated by the 'Balsara switch' (Balsara 1995), there is no universally accepted solution to the problem, since it does not allow the treatment of problems (such as those encountered when material accretes on to a disc) which combine both shocks and shear flows (see Cullen and Dehnen 2010; Morris and Monaghan 1997) .

The reason why this point has been spelled out in some detail is that it represents a major shortcoming for the SPH simulations of star cluster formation that will be described later. When entire clusters are modelled, the resolution on the scale of individual discs is poor and numerical depletion of discs is likely. Disc evolution might not appear to be critical to the large-scale dynamics of the system, but it may in fact be important. For example the density and longevity of discs determines whether they are likely to fragment and produce low-mass companions; discs can also gravitationally influence neighbouring stars by gravitational drag. We will call attention several times to aspects of simulations which may be corrupted by this effect.

Despite this warning, SPH has many virtues. It obviously conserves total mass and—provided that interparticle forces are correctly symmetrised in the case of inter-actions between particles with different smoothing lengths (Price and Monaghan 2007)—it also conserves total momentum and total angular momentum to machine accuracy. In the absence of viscosity or cooling it also conserves total energy to machine accuracy. Its drawbacks (apart from the issue of angular momentum trans-port through artificial viscosity in shear flows highlighted above) concern its treat-ment of shocks and of fluid instabilities. In relation to shock modelling, it is often noted that Eulerian codes do a better job at producing sharp and narrow shock fea-tures. However, as noted by (Springel 2010b), the numerical shock width is *always* many orders of magnitude larger than the true width of the physical shock layer: what is more relevant is whether the properties of the post-shock flow are correct (which they are in SPH). Standard SPH however struggles with excessive mixing in the case of fluid instabilities involving interfaces between phases with a large con-trast in density and temperature. Agertz et al. (2007) published a detailed analysis of how such mixing suppresses Kelvin-Helmholtz instabilities in SPH simulations: see

Price (2008) and Read et al. (2010) for partial fixes. We will not discuss this issue further here since it is of little relevance to the kinds of (single phase) simulations that we shall mainly be considering.

2.3 Adding 'More Physics' to Hydrodynamical Codes

The description of molecular cloud parameters in Chap. 1 indicates that magnetic fields are an important ingredient and this has stimulated the development of MHD modules within SPH. Magnetic fields introduce extra terms in the momentum equation and need to be followed self-consistently in a manner that ensures that they satisfy the Maxwell equation specifying the 'no magnetic monopoles' requirement (i.e. $\nabla.B = 0$). At first sight, it would seem that this could be most readily achieved by adopting a vector potential (i.e. where B is set equal to $\nabla \times A$ for some vector field A) since this automatically ensures that $\nabla.B = 0$. Price (2010) found this approach to be inadvisable; an alternative approach is that of Euler potentials in which the magnetic field is set equal to the cross product of the gradients of two scalar fields (α and β: see Price and Bate 2007; Kotarba et al. 2009). What makes this approach so tractable is that in ideal MHD α and β are conserved for each fluid element; in practice, however, the technique does not perform well in simulations of hydromagnetic turbulence (Brandenburg 2010) which is an important requirement for modelling star-forming clouds. Alternatively, other codes periodically 'clean away' non-zero values of $\nabla.B$ (see Rosswog and Price 2007; Dolag and Stasyszyn 2009).

An important requirement of all the simulations reported here is that they should be able to handle the conversion of distributed gas into 'stars' without having to perform computationally prohibitive and irrelevant modelling of the internal structure of the stars so created. This has led to the implementation of 'sink particles' which excise regions of collapsed gas from the domain of detailed computation and replace them with point masses with the same mass and momentum. The condition for sink creation is generally a threshold density of bound gas and the spatial extent of this region sets the initial sink radius. Thereafter particles are accreted onto the sink if they fall within the sink radius, are also bound and have low enough angular momentum that they would circularise within the sink radius. The mass and momentum of accreted sink particles is simply added to the 'sink'.

Sink particles are relatively easy to incorporate in Lagrangian codes; the first implementation by Bate et al. (1995) opened up a generation of SPH star formation simulations on scales from binaries to entire clusters. In general they work well except for their influence on circumstellar discs: particles orbiting near the sink radius experience only spin-down torque from particles at larger radius (since the sink is devoid of SPH particles that interact hydrodynamically with gas beyond the sink) and hence the accretion of such particles is accelerated. This in turn accelerates the accretion of particles at larger radius. The net result of this is generally an unphysical depletion of gas in discs within a factor of a few times the sink radius (see Hubber et al. 2013c for details of an improved sink algorithm for disc applications).

So far, we have not specified how the gas temperature (which enters the hydro-dynamic equations via the pressure) is assigned in SPH calculations. This can be set by solving the thermal equation (see below) or instead by simply prescribing a barotropic equation of state where pressure is expressed as a function of density. This latter expedient is computationally cheap and, if motivated by detailed simulations that do solve the full thermal equation, can provide a useful way of exploring cloud evolution in a semi-realistic manner. A commonly employed prescription is based on the detailed frequency dependent radiative transfer calculations of Masunaga and Inutsuka (2000), wherein the gas is assumed to be isothermal at densities less than $\rho_{crit} = 10^{-13}\,\mathrm{g\,cm^{-3}}$ and 'adiabatic' ($p = K\rho^{7/5}$) at higher densities (where the gas becomes optically thick in the infrared and where cooling thus becomes inefficient). Note that 'adiabatic' is placed in inverted commas because this prescription is actually an isentropic equation of state (corresponding to *reversible* changes in the absence of cooling). In reality, gas that cannot cool should not be isentropic if it undergoes irreversible processes in shocks (i.e. the value of K should actually increase). This assumption is however found to have a rather minor effect on simulation results (see Bate 2011).

A more accurate alternative, which is however more costly than imposing an equation of state, is to solve equations following the evolution of the thermal energy in both the gas and dust and the radiation field. These equations must take account of pdV work, the flow of energy between matter and radiation (and between dust and gas by direct collisional interaction) and also the transport of radiation through the medium. This latter process should be modelled as a function of photon energy and thus becomes a particularly costly operation. A commonly used expedient is to apply a few simplifying assumptions: (a) the use of 'grey' opacities designed to model the propagation of radiative energy in a frequency averaged sense, (b) the imposition of thermal equilibrium between dust (the main opacity source at high densities and low temperatures) and gas (the main component by mass), and (c) the use of 'flux-limited diffusion' (Levermore and Pomraning 1981) to model radiation transport. This latter tends to the radiative diffusion approximation in regions of high optical depth (i.e. it models conditions where the radiation field is nearly isotropic and where the net flux results from gradients in temperature on length scales much larger than the photon mean free path). If however this formulation were (mis-)applied in regions of long photon mean free path, it could yield the unphysical result of radiative energy being advected at greater than the speed of light. The 'flux-limiter' prevents this and is formulated so that it caps the radiative flux as the product of the radiative energy density and the speed of light. 'Flux-limited diffusion' is thus a useful measure for modelling media that are largely optically thick and where one wants to avoid unphysically large energy fluxes in optically thin surface layers; it does *not* provide an accurate treatment of the cooling in these optically thin surface layers and should not be used in media that are largely optically thin (Kuiper et al. 2010; Owen 2012).

The implementation of flux-limited diffusion is straightforward in grid-based codes but the evaluation of double derivatives in Lagrangian codes like SPH requires some care (see Whitehouse and Bate 2004; Whitehouse et al. 2005). For this reason approximate cooling prescriptions have been developed in SPH which evaluate a cooling rate per particle based on the local temperature and an estimate of the local column density. This latter can be reasonably estimated from the local potential and density in the case of approximately spherical systems (Stamatellos et al. 2007; Forgan et al. 2009; Wilkins and Clarke 2012) or from the vertical component of the gravitational acceleration in the case of disc-like systems (Young et al. 2012).

In contrast to the Lagrangian codes discussed so far, Eulerian codes solve the hydrodynamic equations with respect to a grid. Fixed grid codes (such as ZEUS: Stone and Norman 1992) are however of limited utility in the case of star formation simulations due to the lack of predictable symmetry in the problem. It is generally impossible to predict at the outset of the simulation of a 'turbulent' cloud where and when high resolution will be required. Grid based codes have thus only become competitive with SPH codes in this field following the development of Adaptive Mesh Refinement (AMR) codes (Berger and Collela 1989; Bell et al. 1994). As the name suggests, these are able to map the simulation onto successively higher resolution meshes as locally required. As noted above, some problems (such as radiative transfer and the modelling of magnetic fields) are considerably easier in the case of Eulerian codes; moreover Eulerian codes perform better at modelling turbulent cascades over a significant dynamic range in size scales (Kritsuk et al. 2007; Lemaster and Stone 2008; Kitsionas et al. 2009, however see Price 2012 for a demonstration that SPH is also able to model the turbulent cascade provided that the artificial viscosity is appropriately reduced away from shocks). On the other hand, the use of Cartesian grids in AMR implies that spurious angular momentum transport can be problematical when modelling circumstellar discs. Moreover, the implementation of 'sink' particles is less straightforward than in the case of SPH.

Nevertheless the successful inclusion of moving sink particles in AMR (Krumholz et al. 2004: see Fig. 2.3) has allowed both SPH and AMR codes to tackle the same categories of problems. Despite well-publicised but unpublished claims from the AMR community that there were major numerical differences between star formation simulations conducted with AMR and SPH codes, subsequent calibration exercises have demonstrated generally fair agreement (Federrath et al. 2010; Junk et al. 2010; Price and Federrath 2010; Hubber et al. 2013b). We are now in the favourable situation, therefore, where simulation claims can be checked with two different classes of code. Within the last five years the debate within the numerical star formation community has therefore moved on from the purely technical issues of code comparison to a greater interest in the effect of varying the input physics.

An additional physical effect which is generally important in the case of high-mass star formation simulations is that of ionising radiation. This is most easily modelled in the case where densities are high enough for the 'on-the-spot' approximation to be valid: i.e. where it can be assumed that—whenever an ionising photon from a star is absorbed and then re-emitted due to recombination to the electronic ground state—it is then re-absorbed 'on-the-spot'. If this is true then it is not necessary to follow this process in detail: eventually a recombination to an excited electronic state, followed by a further cascade to the ground state, will lead to the re-emission of non-ionising photons. The net effect of this series of events is thus that the ionising photon is finally 'destroyed' close to the point where it is first absorbed. In equilibrium one

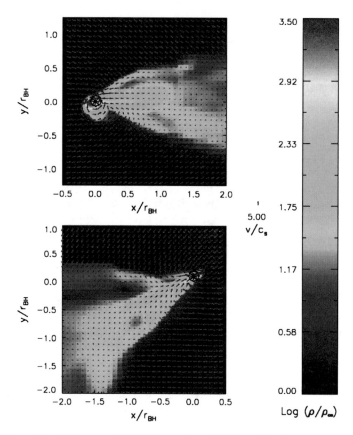

Fig. 2.3 Illustration of the successful implementation of sink particles in AMR for the case of a particle moving through a uniform background. In both panels the results are plotted in the rest frame of the particle but the two panels refer to calculations where the sink particle (*upper*) and the fluid (*lower*) are at rest with respect to the grid. Note the similarity in densities and opening angle of the Mach cones in both cases. Figure from Krumholz et al. (2004)

can then balance the emission of ionising photons from the star into a given solid angle with the integrated rate of recombinations to excited electronic states (so-called 'Case B' recombinations). This means that for any snapshot of a hydrodynamical simulation one can define an ionised ('Strömgren') volume around an ionising source and set the temperature within this region to an appropriate value ($\sim 10^4$ K) accordingly. The necessity of computing recombination integrals makes this easier in grid codes, though Strömgren volume techniques have been successfully developed in SPH also (Kessel-Deynet and Burkert 2000; Dale et al. 2005). Comparison with Monte Carlo radiative transfer codes (which do not *assume* the validity of the 'on-the-spot' approximation) indicates fair agreement in the high-density environments surrounding newly formed massive stars (Dale et al. 2007). In lower density environments it is instead necessary to follow the propagation of 'diffuse' ionising

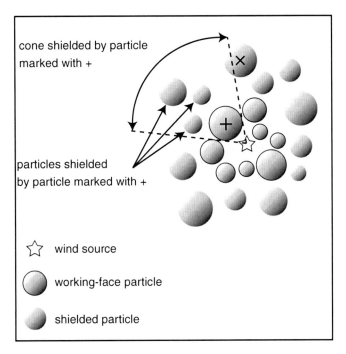

cone shielded by particle
marked with +

particles shielded
by particle marked with +

wind source

working-face particle

shielded particle

Fig. 2.4 Schematic depiction of momentum injection in SPH simulations of feedback from stellar winds, indicating which particles ('working-face' particles) receive momentum from the wind. Figure from Dale and Bonnell (2008)

photons re-emitted by recombinations to the ground state. This has led to the development of other numerical techniques (see e.g. Petkova and Springel 2009, 2011 for the implementation of a variable Eddington factor approach in SPH simulations).

Another physical effect associated with massive star formation is that due to powerful stellar winds. This has been incorporated into SPH simulations by Dale and Bonnell (2008) through the injection of momentum into those particles that define a wind 'working surface' around the star (see Fig. 2.4), while Rogers and Pittard (2013) have recently followed wind feedback from clusters using AMR. Algorithms dealing with each of these additional physical effects are naturally tested against analytic solutions where available (for example the case of steady momentum injection into a uniform medium; Ostriker and McKee 1988). Although agreement in these situations is encouraging, it is regrettable that there are no analytic solution treating the highly inhomogeneous conditions encountered in star formation simulations (and in real molecular clouds). This means that code intercomparison becomes of particular importance for these problems.

We end this summary of the numerical tools for star formation simulations with a brief look to the future. Recently, Springel (2010a) has developed the hybrid Eulerian-Lagrangian code 'AREPO' which solves the fluid equations on a moving Voronoi mesh. The superiority of this approach with respect to conventional SPH simulations has been demonstrated for a range of cosmological problems (Sijacki et al. 2012): in particular AREPO suppresses the spurious mixing between hot and cold phases that be-devils cosmological simulations modelled with conventional SPH. It remains to be seen whether there are commensurate advantages to such an approach when applied to the very different regimes encountered in star formation simulations; these differences are both physical and numerical, since cosmological simulations are—by necessity—typically less well resolved than the star formation simulations we discuss here. (Note that despite the advantages of AREPO mentioned above, there are some downsides—for example it does not strictly conserve total angular momentum, in contrast to conventional SPH.)

A development of more obvious relevance to star formation simulations is the capacity to combine high-accuracy N-body dynamics with simulations that also model the gas phase (the use of low accuracy integrators and softened gravitational potentials in SPH means that the gravitational dynamics of 'sinks' are not treated with the same degree of accuracy as in conventional N-body codes). On the other hand, N-body codes model the effect of gas only via the influence of prescribed gravitational potentials. This task—of combining high accuracy N-body dynamics with hydrodynamic simulations in which the gas is a 'live' component—is currently under way in several groups (see Hubber et al. 2013a).

More broadly, the 'AMUSE' initiative (Astrophysical Multipurpose Software Environment) seeks to combine suites of 'community software' (e.g. N-body codes, hydrodynamics codes, radiative transfer codes, stellar evolution codes) linked by a Python user script (see http://amusecode.org/). It will remain to be seen in the coming years whether the sort of complex hybrid problems encountered in star formation simulations are best served by such generic linkage of well-tested codes or whether it is more efficient to develop hybrid codes that are optimised for specific applications.

2.4 Summary

In this chapter we have reviewed the main numerical techniques employed in the simulations that will be discussed in following chapters, providing a brief overview of N-body codes and both grid-based and Lagrangian hydrodynamical codes. We have placed particular emphasis on those aspects of the numerical implementation that are important in the context of cluster formation simulations, emphasising in particular the problematical feature of disc evolution that is accelerated by numerical viscosity. We have also discussed the range of ways that so-called 'additional physics' (e.g. magnetic fields in addition to thermal and mechanical feedback) have been incorporated in such codes. This chapter is designed to give the non-specialist an overview of the state-of-the-art and an awareness of how numerical issues may influence the outcome of cluster formation simulations.

References

Aarseth, S. J. 2003, Gravitational N-Body Simulations, ed. Aarseth, S. J., Cambridge University Press

Agertz, O., Moore, B., Stadel, J., et al. 2007, MNRAS, 380, 963

Balsara, D. S. 1995, Journal of Computational Physics, 121, 357

Barnes, J. & Hut, P. 1986, Nature, 324, 446

Bate, M. R. 2011, MNRAS, 418, 703

Bate, M. R., Bonnell, I. A., & Price, N. M. 1995, MNRAS, 277, 362

Bell, J., Berger, M., Saltzman, J., & Welcome, M. 1994, SIAM Journal on Scientific Computing, 15, 27

Berger, M. & Collela, P. J. 1989, Journal of Computational Physics, 82, 64

Brandenburg, A. 2010, MNRAS, 401, 347

Cullen, L. & Dehnen, W. 2010, MNRAS, 408, 669

Dale, J. E. & Bonnell, I. A. 2008, MNRAS, 391, 2

Dale, J. E., Bonnell, I. A., Clarke, C. J., & Bate, M. R. 2005, MNRAS, 358, 291

Dale, J. E., Ercolano, B., & Clarke, C. J. 2007, MNRAS, 382, 1759

Dehnen, W. & Read, J. I. 2011, European Physical Journal Plus, 126, 55

Dolag, K. & Stasyszyn, F. 2009, MNRAS, 398, 1678

Federrath, C., Banerjee, R., Clark, P. C., & Klessen, R. S. 2010, ApJ, 713, 269

Forgan, D., Rice, K., Stamatellos, D., & Whitworth, A. 2009, MNRAS, 394, 882

Freitag, M. & Benz, W. 2001, A&A, 375, 711

Haisch, Jr., K. E., Lada, E. A., & Lada, C. J. 2001, ApJ, 553, L153

Heggie, D. & Hut, P. 2003, The Gravitational Million-Body Problem: A Multidisciplinary Approach to Star Cluster Dynamics, ed. Heggie, D. and Hut, P., Cambridge University Press

Hénon, M. H. 1971, Ap&SS, 14, 151

Hubber, D. A., Allison, R. J., Smith, R., & Goodwin, S. P. 2013a, MNRAS, 430, 1599

Hubber, D. A., Falle, S. A. E. G., & Goodwin, S. P. 2013b, MNRAS, 432, 711

Hubber, D. A., Walch, S., & Whitworth, A. P. 2013c, MNRAS, 430, 3261

Junk, V., Walch, S., Heitsch, F., et al. 2010, MNRAS, 407, 1933

Kessel-Deynet, O. & Burkert, A. 2000, MNRAS, 315, 713

Kitsionas, S., Federrath, C., Klessen, R. S., et al. 2009, A&A, 508, 541

Kotarba, H., Lesch, H., Dolag, K., et al. 2009, MNRAS, 397, 733

Kritsuk, A. G., Norman, M. L., Padoan, P., & Wagner, R. 2007, ApJ, 665, 416

Krumholz, M. R., McKee, C. F., & Klein, R. I. 2004, ApJ, 611, 399

Kuiper, R., Klahr, H., Dullemond, C., Kley, W., & Henning, T. 2010, A&A, 511, A81

Kustaanheimo, P. & Stiefel, E. L. 1965, J. Reine Angew. Math., 218, 204

Lemaster, M. N. & Stone, J. M. 2008, ApJ, 682, L97

Levermore, C. D. & Pomraning, G. C. 1981, ApJ, 248, 321

Masunaga, H. & Inutsuka, S.-I. 2000, ApJ, 531, 350

Mikkola, S. & Aarseth, S. J. 1990, Celestial Mechanics and Dynamical Astronomy, 47, 375

Mikkola, S. & Aarseth, S. J. 1993, Celestial Mechanics and Dynamical Astronomy, 57, 439

Mikkola, S. & Aarseth, S. 2002, Celestial Mechanics and Dynamical Astronomy, 84, 343

Monaghan, J. J. 1992, ARA&A, 30, 543

Monaghan, J. J. 2005, Reports on Progress in Physics, 68, 1703

Morris, J. P. & Monaghan, J. J. 1997, Journal of Computational Physics, 136, 41

Ostriker, J. P. & McKee, C. F. 1988, Reviews of Modern Physics, 60, 1

Owen, J. E., Ercolano, B., & Clarke, C. J. 2012, in Astrophysics and Space Science Proceedings, Vol. 36, The Labyrinth of Star Formation, ed. D. Stamatellos, S. Goodwin, & D. Ward-Thompson, Springer International Publishing, 127

Petkova, M. & Springel, V. 2009, MNRAS, 396, 1383

Petkova, M. & Springel, V. 2011, MNRAS, 412, 935

Price, D. J. 2008, Journal of Computational Physics, 227, 10040

Price, D. J. 2010, MNRAS, 401, 1475
Price, D. J. 2012, MNRAS, 420, L33
Price, D. J. & Bate, M. R. 2007, MNRAS, 377, 77
Price, D. J. & Federrath, C. 2010, MNRAS, 406, 1659
Price, D. J. & Monaghan, J. J. 2007, MNRAS, 374, 1347
Read, J. I., Hayfield, T., & Agertz, O. 2010, MNRAS, 405, 1513
Rogers, H. & Pittard, J. M. 2013, MNRAS, 431, 1337
Rosswog, S. & Price, D. 2007, MNRAS, 379, 915
Shu, F. H., Adams, F. C., & Lizano, S. 1987, ARA&A, 25, 23
Sijacki, D., Vogelsberger, M., Kereš, D., Springel, V., & Hernquist, L. 2012, MNRAS, 424, 2999
Springel, V. 2010a, MNRAS, 401, 791
Springel, V. 2010b, ARA&A, 48, 391
Stamatellos, D., Whitworth, A. P., & Ward-Thompson, D. 2007, MNRAS, 379, 1390
Stone, J. M. & Norman, M. L. 1992, ApJS, 80, 753
Whitehouse, S. C. & Bate, M. R. 2004, MNRAS, 353, 1078
Whitehouse, S. C., Bate, M. R., & Monaghan, J. J. 2005, MNRAS, 364, 1367
Wilkins, D. R. & Clarke, C. J. 2012, MNRAS, 419, 3368
Young, M. D., Bertram, E., Moeckel, N., & Clarke, C. J. 2012, MNRAS, 426, 1061

Chapter 3
The Comparison of Observational and Simulation Data

Cathie J. Clarke

Before we describe the results of star formation simulations in detail, we need to consider how, in principle, one should decide whether the output of a simulation is a good match to reality. This issue is not entirely straightforward given the complex morphology and hierarchical nature of observed molecular clouds (and also of numerical simulations). For example, the fact that simulations generally produce filamentary and highly structured clouds with a mixture of clustered and more distributed star formation is at first sight encouraging, because these are broadly properties shared by observed clouds (Men'shchikov et al. 2010; Peretto et al. 2012; Schneider et al. 2012). One however needs a more refined measure of whether simulations and observations are indeed quantitatively consistent. We therefore conduct a brief survey of statistical descriptors that have been applied to simulations and observations. We follow this by applying some of these methods to the simplest class of star cluster formation simulations (termed 'vanilla' calculations in these chapters) which contain only the three most basic physical ingredients: gas pressure, turbulence and gravity.

3.1 The Characterisation of Observational and Simulated Data

3.1.1 Characterising Gaseous Structures

There are many alternative descriptors of the wealth of structures found in the density and velocity fields of molecular clouds (see Blitz and Stark 1986 for early analyses of the hierarchical nature of the interstellar medium). For example, Padoan et al. (2003) analysed ^{13}CO emission maps of Taurus and Perseus by computing *structure functions* as a function of r (i.e. expectation values of the pth power of the difference

C.J. Clarke (✉)
Institute for Astronomy, University of Cambridge, Cambridge, UK
e-mail: cclarke@ast.cam.ac.uk

© Springer-Verlag Berlin Heidelberg 2015
C.P.M. Bell et al. (eds.), *Dynamics of Young Star Clusters and Associations*,
Saas-Fee Advanced Course 42, DOI 10.1007/978-3-662-47290-3_3

in intensity between points in the map separated by distance r): the power-law dependence of the structure function on r indicates the scale-free nature of much of the structure within molecular clouds. Padoan et al. (2003) attempted to compare these results with the predictions of various turbulence models for the form of the structure function for velocity. The relationship between structure functions for intensity and those for velocity (as derived in the case of supersonic turbulence; Boldyrev 2002) is however unclear.

A more intuitive method of analysing molecular cloud structures is via the use of *dendrograms*. These can be visualised by considering a dataset (e.g. integrated intensity as a function of two-dimensional position) as a topographical surface which one then 'thresholds' at various levels, identifying the distinct peaks above each threshold and tracing how these merge as the threshold level is reduced. (A good analogy here is how the distribution of 'islands' changes as the water level around a flooded mountain range is reduced.) The structure can then be depicted in terms of a network of branches whose geometry reflects the hierarchical organisation of the medium (see Fig. 3.1). Rosolowsky et al. (2008) applied such an analysis to the L1448 region of the Perseus molecular cloud and compared this with simulations of MHD turbulence by Padoan et al. (2006); they noted that dendrogram analysis can in some cases identify discrepancies between simulations and observations that are not discernible through analysis of the power spectrum (the power spectrum simply counts entities on different scales without directly assessing the spatial relationship between structures on different scales).

By far the most widely used algorithms for analysing the structure of molecular clouds are those of a 'friend of friends' type, such as the CLUMPFIND algorithm (Williams et al. 1994) which identifies peaks and then works downwards in intensity, assigning neighbouring regions to their local intensity peak and otherwise creating a new clump which is treated in the same way. Such an approach can be used to identify distinct clumps either in positional data or else, in the case of line emission maps, in

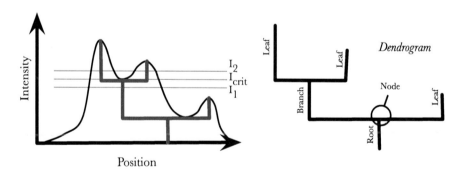

Fig. 3.1 A schematic depiction of how dendrogram analysis leads to the rendering of intensity positional data in terms of a root-branch-leaf structure. The *left panel* shows a one-dimensional emission profile with three distinct local maxima. The dendrogram of the region is illustrated in blue and shown in the *right panel* where the components of the dendrogram are labelled. Figure from Rosolowsky et al. (2008)

datacubes in joint positional and velocity space. Once such clumps have been identified, one can readily construct their mass spectrum (commonly termed the CMF: clump mass function) and compare with the corresponding quantity derived from simulations (see Klessen and Burkert 2000; Smith et al. 2008 for the application of such analyses to datacubes generated by SPH simulations). The extraction of a clump mass spectrum from either observations or simulations is however a non-unique procedure. As emphasised in their paper 'The perils of CLUMPFIND...', Pineda et al. (2009) conclude that the derived clump spectrum is highly dependent on the observational resolution and that, in particular, kinematic data is required to disentangle structures that are blended along the line of sight. In other studies, independent analyses of millimetre maps of ρ Ophiuchus (Motte et al. 1998; Johnstone et al. 2000) agree about the mass spectrum of the derived clumps but disagree about the masses and locations of individual clumps. Moreover, Smith et al. (2008) found (through applying the CLUMPFIND algorithm to simulation data) that the overall shape of the derived clump mass spectrum was fairly insensitive to the CLUMPFIND parameters employed but that the location of apparent breakpoints was rather strongly dependent on these algorithmic parameters. This result underlines the fact that such a method can only be used to compare observations and simulations if the algorithmic details, and the resolution, are well matched. It also raises obvious questions about the *physical* significance (if any) of such breakpoints in the derived CMF.

3.1.2 Characterising Stellar Distributions

We now turn to the issue of characterising stellar spatial distributions. One of the first measures to be applied to large-scale distributions of stellar positions was the Mean Surface Density of Companions (MSDC, Larson 1995; Simon 1997; Bate et al. 1998). This measure is computed by counting all the stars within an annulus of given radius (r) centred on each star, dividing by the area of the annulus, repeating this procedure with the annulus centred at every star in the region, and then averaging to obtain the mean surface density at that separation. The MSDC is thus closely related to the two-point correlation function which instead subtracts off the large-scale mean surface density: although appropriate to cosmological studies (where the distribution of galaxies is expected to be uniform on the largest scales) this is not useful in star-forming regions which are generally inhomogeneous on scales extending up to the size of the entire region analysed.

Analyses of nearby star-forming regions revealed a double power-law structure in the MSDC: the inner power-law is readily identified with size scales where stars have one companion on average—in other words it corresponds to the distribution of nearest neighbour (in reality bound—i.e. binary star—companion) distances. Since the binary separation distribution is rather flat in log separation, this translates into a MSDC of slope -2. At larger separations it transitions to a shallower slope which Larson (1995) interpreted as evidence of fractal clustering on larger scales (i.e. clustering with no characteristic size scale but a self-similar relationship between surface density and size). Bate et al. (1998) however argued that this interpretation was not unique and that global density gradients or non-fractal sub-clustering would also be consistent with the data over the limited dynamic range of size scales (2–3 orders of magnitude) between the binary regime and the total size of star forming regions.

Alternative measures of stellar distributions can be derived from the Minimum Spanning Tree (MST), which is the unique connection of points in a dataset so as to minimise the total length without involving any closed loops. Cartwright and Whitworth (2004) proposed a single number (the Cartwright Q parameter) that can be used to classify the nature of stellar distributions. Q is defined as the ratio of the value of $\bar{m}$ (the mean edge length normalised to the mean value for N random points in the area) to $\bar{s}$ (the 'normalised correlation length', i.e. mean separation divided by the cluster radius). The important distinction is that whereas the mean edge length refers to the mean separation of closest neighbours (i.e. those directly linked by the MST), the mean separation is the mean (over all stars) of the distance to *all the other stars* in the cluster.

We can start to understand how the Q parameter is able to distinguish qualitatively different stellar distributions by first considering a uniform distribution of points— empirically this yields a particular Q value (~ 0.7). Now we consider two different ways of driving the distribution away from the uniform: in both cases we move stars around so that there is now a range of densities, but whereas in one case the high-density regions are co-located (which we call the centrally concentrated case) in the other the islands of high-density are spatially dispersed (which we call the fractal case). As one moves away from the uniform distribution, both $\bar{m}$ and $\bar{s}$ are reduced as stars are brought closer together. However, the reduction in $\bar{s}$ is relatively small in the fractal case because when stars are moved to isolated high-density peaks, this affects the mean separation of relatively few stars. Consequently, Q falls as one proceeds from a uniform distribution to fractal distributions with decreasing fractal dimension (i.e. distributions that are more clumped); conversely Q increases as one proceeds from uniform distributions to distributions that are increasingly centrally concentrated. In general, real stellar distributions are neither necessarily centrally concentrated nor strictly fractal so the real utility of the Q parameter is that it provides a ready way to distinguish distributions in which high-density regions are co-located from those in which they are spatially dispersed. As such it is a useful tool when one compares the outcome of simulations with real observational data (see e.g. Schmeja et al. 2008).

Another use of the MST is that it can provide a simple empirical definition of a 'cluster': one can specify a 'cut-length' and sever all branches of the tree that exceed this length, thereby dividing a distribution of points into a set of distinct 'clusters'. Naturally, the numbers and identities of such 'clusters' are highly sensitive to the cut-length employed (see Fig. 3.2). Although the definition of clusters is thus arbitrary (and certainly does not correspond to entities that are necessarily gravitationally bound) it at least provides a consistent way to compare a simulation set with an observational dataset (provided, of course, that both datasets are analysed with the same cut-length; see Maschberger et al. 2010 for an analysis of the 'clustering' within the simulations of Bonnell et al. 2003, 2008 using this method).

Fig. 3.2 Illustration of the use of the minimum spanning tree in identifying 'clusters' (*numbered circles*) in the simulations of Bonnell et al. (2008). The three panels show how the 'clusters' identified depend on the value of the cut-length parameter d_{break} adopted (0.001, 0.025 and 0.05 from *top* to *bottom* respectively). Figure from Maschberger et al. (2010)

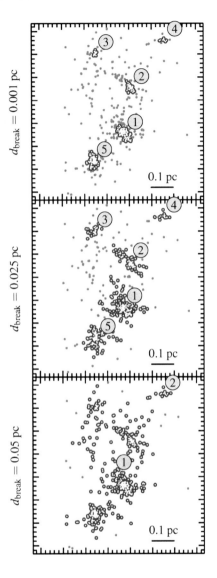

One of the most widely used applications of the MST is to characterise *mass segregation* within observational and simulation datasets. There is considerable interest in whether massive stars are preferentially located in dense regions and whether this is a consequence of two-body relaxation or instead reflects stellar birth sites. In the case of spherically symmetric clusters, mass segregation may be evaluated by comparing the radial distributions of stars as a function of mass (e.g. Bate 2009; Moeckel and Bonnell 2009). Such an approach is obviously not appropriate in the commonly encountered situation where stellar distributions lack any clear symmetry and this is where MST based techniques offer a clear advantage: Allison et al. (2009) proposed a method in which the mean edge length of the MST constructed from the i most massive stars is calculated and then compared with the corresponding quantity constructed from a MST based on random samples of i stars. The ratio of these quantities Λ is a measure of whether the i most massive stars are similarly distributed to the general population: its particular advantage is that it is self-calibrating because when one computes the mean edge length of i random stars through repeated sampling one obtains also the standard deviation of that quantity and thus can readily assess whether the value for the i most massive stars is significantly different. Results however need to be interpreted with care because the mean edge length is highly sensitive to the maximum edge length in the distribution. For example, in Taurus the mean edge length for massive stars is large because of a few massive stars lying at large distances from the remainder; this can produce an apparent signature of 'inverse mass segregation' even though the majority of massive stars in Taurus are actually more closely associated with each other than is the case for 'typical' (lower mass) stars in the region (Parker et al. 2011). The interpretation of the Λ statistic is thus improved if one also compares quantities (such as the median or geometric mean edge length, Maschberger and Clarke 2011; Olczak et al. 2011) which are less sensitive to the maximum edge length.

Finally it should be noted that MST-based methods can be applied to comparing the spatial distributions of stars as a function of age (as proxied by their possession of circumstellar disc diagnostics; see Ercolano et al. 2011).

3.1.3 Characterising the IMF

A widely used approach to describing the stellar mass function is to construct a log-log histogram such that—for a power-law IMF—the slope of the histogram gives the power-law index of the IMF. This is however problematical in several ways, as pointed out by Maíz Apellániz and Úbeda (2005). Firstly the derived slope is often sensitive to the binning. Secondly, Poissonian errorbars are largest in a log-log plot in the case of bins containing few objects (i.e. generally at high masses) and such errorbars are moreover asymmetric. This means that owing to Poisson noise, bins at high mass can frequently be sparsely populated. This introduces a bias which results in derived power laws being systematically too steep. The problem can be addressed by adopting bin sizes such that bins all contain the same number of objects. However this inevitably leads to large bin widths at the sparsely populated (high-mass) end of the IMF and thus reduces discrimination in this regime.

It is therefore preferable to use non-parametric tests (such as the Kolmogorov-Smirnov [KS] test) to test the consistency between observational data and a range of hypothesised functional forms. However, it needs to be borne in mind that the KS test is notoriously insensitive to deviations between distributions that occur near the extremes of the cumulative distribution: this is particularly problematical if one is trying to test, for example, whether data is consistent with an unbounded power-law or whether it requires a form that is truncated at high masses (this being an area of considerable debate: see discussion in Sect. 7.2.1 of Chap. 7). This can be remedied by applying a stabilising transformation to the variables (such that the test is uniformly sensitive at all centile values; see Maschberger and Kroupa 2009).

3.2 Simulation Results: Bonnell et al. (2008) as a Case Study

In what follows we term as 'vanilla' calculations all those that incorporate the minimum subset of physics that is required to produce a somewhat realistic star-forming complex. Such calculations incorporate gravity (obviously!), a supersonic velocity field and thermal properties prescribed according to a barotropic equation of state (see Masunaga and Inutsuka 2000). In order to mimic 'turbulent' velocity fields, a common expedient is to start with an unstructured cloud and then to impose a divergence-free random Gaussian velocity field with a power spectrum [$P(k) \propto k^{-4}$] that is designed to reproduce the Larson size-linewidth relations (Larson 1981; see Sect. 1.4 of Chap. 1). We will start with the simplest case of a one-off injection of 'turbulent' kinetic energy and will contrast this in the following chapter with the case of continually driven turbulence or cases where the turbulence is 'settled' prior to the switch-on of gravity.

There is now a large body of such 'vanilla' calculations, following the pioneering simulations of Bate et al. (2002a). These vary greatly in scale and numerical resolution and range from relatively cheap calculations (such as those of Delgado-Donate et al. 2004; Goodwin et al. 2004a,b) where the small cloud masses permit multiple realisations of a given parameter set, to very expensive 'one-off' calculations which push the limits either in scale (e.g. Bonnell et al. 2008) or in resolution (e.g. Bate 2009, 2012). Naturally, the introduction of additional physical effects involves sacrifices in terms of scale and/or resolution.

From the point of view of simulating cluster formation, perhaps the most instructive are the largest scale simulations since they permit the treatment of an entire complex of clusters and can trace the history of their hierarchical assembly. Accordingly we start with a discussion of the largest scale star formation simulation conducted to date, i.e. that of Bonnell et al. (2008) which models a cloud of 10^4 M$_\odot$. At the end of the simulation (at an age of 0.5 Myr) around 1500 M$_\odot$ of gas has been converted into stars (i.e. sink particles) and these are distributed in a number of 'clusters' comprising hundreds of stars as well as a distributed population.

The initial configuration of the simulation is a cylinder of radius 3 pc and length 10 pc with a mild axial density gradient which ensures that—following introduction of the initial injection of turbulent energy—the cloud is overall marginally bound (being mildly bound at one end and mildly unbound at the other).

The evolution follows a sequence that is characteristic of all similar calculations: supersonic turbulence creates a web of shocked layers of compressed gas which break up under the action of self-gravity to create a network of dense filaments (note that this feature is broadly consistent with the widespread observations of filaments in *Herschel* observations of molecular clouds (André et al. 2010; Juvela et al. 2012) although the simulations do not reproduce the invariant filament width that has has been reported in observations (Arzoumanian et al. 2011). In the simulations, small-scale inhomogeneities in the filaments are amplified by self-gravity and lead to fragmentation. The minimum spacing of such fragments is set by the characteristic (Jeans) length scale for collapse (see Eq. 1.2) for which the sound crossing and free-fall timescales along the length of the filament are similar. Stars (i.e. sinks) that form in a filament then follow large-scale self-gravitating flows along the filament and are thus transported towards the dense regions formed where filaments intersect (this itself also being a location of further fragmentation and star formation). It is worth noting at this stage that fragmentation often produces few body clusters and it is these (rather than single stars) that are conveyed along filaments.

This sequence of events leads to the formation of clusters via a bottom-up (hierarchical) process as demonstrated by the analysis of merger trees based on MST cluster identification (Maschberger et al. 2010). The ongoing merger sequence causes evolution of the Cartwrigth Q parameter which is low ('fractal') during the stage that (mini-)clusters are scattered along filaments but rises once mergers form a dominant centrally concentrated cluster. It should be stressed that these are not 'dry' (gas-free mergers) and that the gas (which remains the dominant mass component on large scales throughout the duration of the simulations, $\sim$0.5 Myr) plays an important role in channeling clusters together and facilitating the merger process.

If one looks at the properties of the clusters formed after 0.5 Myr (bearing in mind that the definition of a cluster depends on a particular choice of the 'cut-length' for the MST) one finds about 15 clusters containing more than 10 stars; obviously it is not sensible to define a formal cluster mass function from only 15 objects but it is clear that low-N clusters are more numerous and that the distribution is broadly compatible with the observed cluster mass function (where the fraction of clusters between N and $N + \mathrm{d}N$ scales as N^{-2}; Lada and Lada 2003). The most populous cluster in the simulations contains several hundred stars. These clusters are generally mildly aspherical (i.e. most frequently with projected axis ratios on the sky in the range of 1–2, though a few objects are at times more drastically aspherical). The cluster shape is sensitive to the history of mergers in the cluster, with highly aspherical shapes during ongoing mergers but with stellar two-body relaxation effects reducing the ellipticity between mergers. It is also found that these clusters are markedly mass segregated (especially the more populous clusters i.e. $N > 50$) at an age of 0.5 Myr; technically, this mass segregation is *not* primordial but is the result of rapid two-body relaxation within clusters that are assembled via mergers (see also McMillan et al. 2007; Allison et al. 2009, 2010). From an observational perspective, however, the system is so young that any observer would probably classify this situation as one of primordial mass segregation. It is worth noting that—as in the case of the ellipticity—the state of mass segregation changes during mergers: evidently while a merger is ongoing, there are two nuclei containing the most massive stars within a given cluster and this temporarily removes the mass segregation signature.

Before leaving this thumbnail portrait of cluster assembly in the Bonnell et al. (2008) simulation it is worth noting that the demographics of clustering within the simulation is quite sensitive to modest variations in the degree of gravitational

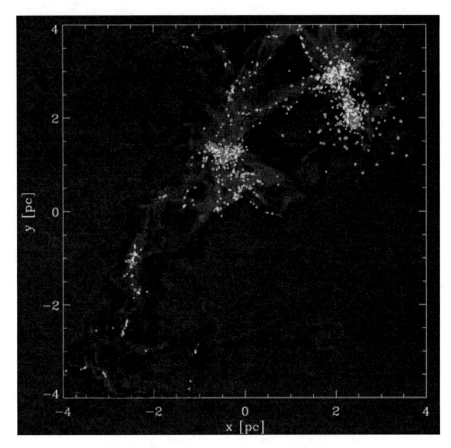

Fig. 3.3 The gas and stellar distribution in the simulation of Bonnell et al. (2008) at an age of ∼0.5 Myr. Note that the initial condition was cylindrical: the upper regions (where the populous clusters have formed) were initially mildly bound, whereas the lower half (where star formation is less intense) was initially mildly unbound. Figure from Bonnell et al. (2008)

boundedness in different regions of the simulation (see Fig. 3.3). As noted above, the mild density gradient along the axis of the initial gas cylinder means that the gas at one end is mildly unbound while it is mildly bound at the other end. Star formation proceeds more rapidly in the bound end of the cloud and the converging flows that develop along filaments lead to the formation of several populous clusters. At the unbound end, by contrast, much of the gas avoids significant compression and expands without forming stars: locally convergent flows do produce some stars even here but the large-scale flows are not conducive to significant merging and the star formation remains dispersed in rather small-N groupings.

We now proceed to a more general discussion of how various simulations of this 'vanilla' variety have contributed to our understanding of a range of issues in star and cluster formation.

3.3 The Relationship Between Gas, Cores and Stars in Simulations

Amongst the wealth of structures observed in molecular clouds there is a class of dense regions (identified in dust emission or absorption or else in molecular lines; see Chap. 1, Sect. 1.2) that are termed 'cores'. These are characterised by low internal velocity dispersions (comparable with the sound speed) and contain around a Jeans mass of gas. Such cores are widely regarded as being stellar progenitors and it is often claimed that the stellar mass function (IMF) is simply inherited from the core mass function (CMF, Motte et al. 1998; Johnstone and Bally 2006). We have already discussed 'the perils of clumpfind' (Pineda et al. 2009) and the difficulty in unambiguously identifying cores in observational data and this introduces some uncertainty about the reliability of observed CMFs. Nevertheless, from an observational perspective, comparison between the CMF and IMF is the only way to test the hypothesis that cores can be mapped directly onto resulting stars. Such a comparison has been claimed to indicate a systematic offset in logarithmic mass between the CMF and IMF (Lada et al. 2008), which can be interpreted as a universal 'efficiency' factor as cores turn into stars (see Goodwin et al. 2008 for an analysis of how this mapping is affected by the formation of multiple stars).

In the case of simulations, one has the luxury of being able to trace the fates of individual gas particles and of determining whether cores indeed turn directly into stars (bearing in mind the caveat that of course this does not necessarily indicate that the same evolutionary sequence is followed in reality!). Smith et al. (2009) identified gas cores in the simulations as local potential wells and showed that, in the simulations, the CMF and resulting IMF were indeed of similar functional form. However, they found that this situation represents a rather weak association between the masses of individual cores and the masses of the stars they produced—there is no more than a general tendency for more massive stars to form from more massive cores as the correspondence is blurred by the effect of subsequent accretion. Bonnell et al. (2004) also traced the assembly history of individual stars and showed that whereas low-mass stars form from rather local collapse, higher mass stars have mass contributions from a much larger volume. This is because stars that end up with high mass in the simulations are those that arrive early in cluster cores and are then able to accrete vigorously from a mass reservoir that is fed by material flowing in along filaments.

3.4 The Origin of the Stellar IMF in 'Vanilla' Calculations

Simulations of this kind routinely produce stellar (i.e. sink) IMFs which can be represented as a broken power-law—at high masses (above the so-called 'knee' of the IMF), the fraction of stars with masses between m and $m + dm$ is $\propto m^{-\alpha}dm$ where $\alpha \sim 2$ (comparable with the observed 'Salpeter' value of 2.35). Below the knee, the mass function is flatter (i.e. with $\alpha \sim 1.5$) and this functional form imparts the distribution with a 'characteristic' mass that is similar to the knee mass. Early simulations were conspicuously successful in creating an IMF that is well matched to the observed form (Bate et al. 2002a, b, 2003) with knee values of around a solar mass. This is however entirely fortuitous (with regard to the value of the knee) since in the case of an isothermal equation of state (as employed in all these simple 'vanilla' calculations) the knee value is simply related to the mean Jeans mass in the cloud at the onset of the simulation (Bonnell et al. 2006). A simple dependence on mean Jeans mass (and an insensitivity to the Mach number of the turbulence) is in fact in contrast to the conclusions based on studies of non-self-gravitating turbulence (Padoan and Nordlund 2002; Hennebelle and Chabrier 2008) in which an IMF is constructed by constructing nominally Jeans unstable peaks in the turbulent density field. In this case, the mean stellar mass decreases at large Mach number, since this increases the density of shocked layers and thus lowers the nominal Jeans mass associated with such layers. It is interesting that whereas the self-gravitating simulations also produce denser structures at high Mach number, this does not translate into lower mass stars as would be suggested by application of this simple Jeans criterion. The reason for this discrepancy is probably the inapplicability of a simple (density-based) Jeans criterion in slab geometry: compression in such geometry—which changes neither the lateral sound crossing time nor lateral free-fall time—has little effect on the Jeans mass (see Lubow and Pringle 1993; Whitworth et al. 1994; Bonnell et al. 2004).

Although there is some interest in studying the differences between the IMFs inferred from non-self-gravitating density fields and those produced in simulations which allow collapse and subsequent accretion, it should not detract from a much more fundamental problem with all isothermal calculations. It is strongly undesirable to have a situation where the 'knee' of the IMF can be simply shifted around by a change in mean cloud density and temperature (see Fig. 3.4). This is because the observed IMF appears to be remarkably invariant in all well studied regions (see Bastian et al. 2010 for a recent review of this issue) whereas star-forming clouds

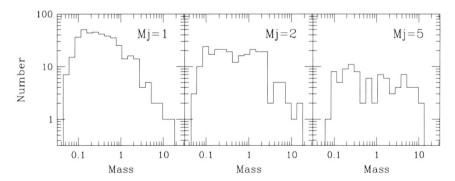

Fig. 3.4 Illustration of the IMFs produced in three different isothermal calculations in which the mean Jeans mass at the onset of the simulation is 1, 2 and 5 $M_\odot$ (*left* to *right* respectively). The 'knee' of the IMF then simply tracks the initial Jeans mass. Figure from Bonnell et al. (2006)

have a range of densities and temperatures which should—in this picture—cause corresponding variations in the IMF. It is however found that the situation may be remedied by using a rather modest departure from an isothermal equation of state. Larson (2005) proposed a barotropic equation of state in which the temperature falls mildly with increasing density ($T \propto \rho^{-0.25}$) in the regime dominated by line cooling (at number densities less than $10^6 \, \mathrm{cm}^{-3}$) but rises mildly with density ($T \propto \rho^{0.1}$) at higher density where dust cooling becomes important (see also Whitworth et al. 1998 for a similar proposal that the conditions associated with the onset of dust cooling imprint a characteristic Jeans mass on the IMF). Certainly, Bonnell et al. (2006) found that this modest revision of the equation of state had a remarkably stabilising effect on the IMF produced in simulations. Whereas in previous isothermal simulations, the IMF 'knee' had simply followed variations in the initial cloud Jeans mass, it was found that the modified equation of state produced similar IMFs for a range of cloud initial conditions. These authors argued that the IMF is imprinted at this mass scale because such an equation of state implies that the Jeans mass changes from being respectively more (less) density dependent than the free-fall time for densities below (above) this threshold. At higher density, the Jeans mass is less responsive to density changes on the free-fall time and this tends to suppress further fragmentation. Moreover, Larson (2005) argued that the relevant equation of state should not be very sensitive to the metallicity either, so that this adjustment might provide a good route to producing a near universal IMF. The recent hydrodynamic simulations of Dopcke et al. (2013) confirm that the metallicity dependence of the effect of dust cooling on the IMF is indeed rather mild, even down to extremely low ($<10^{-4} Z_\odot$) metallicities.

Whatever the details of the cooling physics invoked, it is encouraging that physically motivated modifications of the thermal physics can indeed stabilise the IMF. It should however be stressed that this is only one of the currently discussed ways in which 'additional physics' can achieve this stabilisation and we discuss other ideas (such as those relating this stabilisation to radiative feedback) in the following chapter.

Finally, we turn to the upper power-law of the IMF in simulations, i.e. in the regime above the 'knee'. Here the -2 power-law is generally ascribed to the role of Bondi-Hoyle accretion along the lines of the analysis first proposed by Zinnecker (1982). Such accretion gives rise to an accretion rate that scales quadratically with stellar mass, and, for a given initial mass, M_{in}, one can write an expression for the stellar mass at time t. One can then map a given range of initial masses dM_{in} into the corresponding range of masses dM at time t, which yields the relationship $dM_{\mathrm{in}} = M_{\mathrm{in}}^2 dM/M^2$. This then demonstrates that if one starts with a given small range of initial masses, these are transformed by accretion into a power-law probability density function for stellar mass with slope -2. Bonnell et al. (2001) examined this picture for the build-up of stellar mass through idealised simulations which placed stellar sinks in smooth collapsing parent gas distributions; they argued that, while the stars are more or less co-moving with the collapsing gas, the accretion cross-section associated with Bondi-Hoyle accretion (which scales as the relative star-gas velocity raised to the power of -4) is unphysically large and that instead the relevant cross-section is the smaller tidal radius: they showed that in this case the IMF slope should scale as mass to the power of -1.5 and that this is well matched to the simulation results at low mass. However, once stars dominate the potential in the core of the cloud they form a virialised sub-system in which stellar velocity directions are randomised with respect to radially inflowing gas. The increased random velocity then reduces the Bondi-Hoyle accretion cross-section to less than the tidal limit; hence Bondi-Hoyle accretion becomes the dominant process, thus explaining the power-law tail of the mass function with index -2.

It is not clear how much this explanation based on 'toy' models applies to the large range of subsequent turbulent fragmentation simulations which all show a similar IMF morphology (i.e. slope changing from ~ -1.5 to ~ -2 at an IMF 'knee'). At first sight, these simulations—such as the Bonnell et al. (2008) simulation described in detail above—bear little resemblance to the 'toy' model of a smooth radially collapsing gaseous background, since the turbulence generates a complex velocity and density field in the gas. Nevertheless, there may be more resemblance between the two situations than is visually apparent: Offner et al. (2009) demonstrated that when stars are formed in turbulent calculations their initial velocities with respect to the local gas is indeed low. On the other hand Kruijssen et al. (2012) showed that once clusters start to form via the hierarchical assembly process described above they form sub-systems which are in rough virial equilibrium and for which the increased relative velocity between stars and gas would make Bondi-Hoyle accretion a relevant process. This is consistent with the result mentioned above in which low-mass stars form from accretion of rather localised gas whereas more massive stars can attain a large fraction of their mass via accretion in cluster cores. Although this picture might still have some relevance to how the simulations build up stellar mass (and we emphasise that this is of course not the same as demonstrating its relevance to stellar mass acquisition in *real* systems), it should be noted that simulations apparently do *not* obey the quadratic relationship between stellar mass and accretion rate that under-pins the Bondi-Hoyle argument (Maschberger et al. 2014).

Before we leave such 'vanilla' calculations, it is worth dwelling further on the result that we have just noted, i.e. that the stellar motions within forming clusters appear to be in rough virial equilibrium with the potential produced by the stars alone and thus that the clusters within the simulations are internally gas-poor. This result does not appear to be a numerical issue with sink particle accretion inasmuch as the rapid accretion of gas within the region of the clusters dominated by the stars is insensitive to sink particle radius and resolution (Kruijssen et al. 2012) and also to numerical method (i.e. a similar result is found in the AMR based simulations of Girichidis et al. 2012). This result—if true also in the case of real protoclusters—would have profound consequences for the issue of cluster survival which we will discuss further in Chap. 6: it is often assumed that many star clusters become unbound (so-called cluster 'infant mortality') when gas—which was previously assumed to be the dominant mass component within embedded stellar clusters—was expelled. If in fact the clusters are already gas-poor on the scale of the stars (i.e. if the mass reservoir for further star formation is mainly located outside the stellar cluster) then gas-loss becomes irrelevant for star cluster survival.

3.5 Summary

We have surveyed the range of statistical descriptors that are used in the analysis of both observational data and the output of simulations, considering such issues as the spatial distribution of gas and stars, the stellar IMF, clustering and stellar mass segregation. We have then proceeded to a thumb-nail portrait of the largest scale simulation of cluster formation yet conducted—that of Bonnell et al. (2008) which models a cloud of mass $10^4 \, M_\odot$ which forms around 15 star clusters over a timescale of ~ 0.5 Myr. Although this simulation is considerably less sophisticated in terms of the physical processes modelled than are some of the simulations described in Chap. 4, it has nevertheless introduced some of the generic properties of cluster formation simulations. In particular we have drawn attention to the hierarchical nature of cluster assembly: the basic unit of cluster assembly on all scales is the small N (<10) cluster and large-scale clustering proceeds through a process of successive gas-mediated cluster mergers.

References

Allison, R. J., Goodwin, S. P., Parker, R. J., Portegies Zwart, S. F., & de Grijs, R. 2010, MNRAS, 407, 1098
Allison, R. J., Goodwin, S. P., Parker, R. J., et al. 2009, MNRAS, 395, 1449
André, P., Men'shchikov, A., Bontemps, S., et al. 2010, A&A, 518, L102
Arzoumanian, D., André, P., Didelon, P., et al. 2011, A&A, 529, L6
Bastian, N., Covey, K. R., & Meyer, M. R. 2010, ARA&A, 48, 339
Bate, M. R. 2009, MNRAS, 392, 590
Bate, M. R. 2012, MNRAS, 419, 3115
Bate, M. R., Bonnell, I. A., & Bromm, V. 2002a, MNRAS, 332, L65
Bate, M. R., Bonnell, I. A., & Bromm, V. 2002b, MNRAS, 336, 705
Bate, M. R., Bonnell, I. A., & Bromm, V. 2003, MNRAS, 339, 577
Bate, M. R., Clarke, C. J., & McCaughrean, M. J. 1998, MNRAS, 297, 1163
Blitz, L. & Stark, A. A. 1986, ApJ, 300, L89
Boldyrev, S. 2002, ApJ, 569, 841
Bonnell, I. A., Bate, M. R., & Vine, S. G. 2003, MNRAS, 343, 413
Bonnell, I. A., Clark, P., & Bate, M. R. 2008, MNRAS, 389, 1556
Bonnell, I. A., Clarke, C. J., & Bate, M. R. 2006, MNRAS, 368, 1296
Bonnell, I. A., Clarke, C. J., Bate, M. R., & Pringle, J. E. 2001, MNRAS, 324, 573
Bonnell, I. A., Vine, S. G., & Bate, M. R. 2004, MNRAS, 349, 735
Cartwright, A. & Whitworth, A. P. 2004, MNRAS, 348, 589
Delgado-Donate, E. J., Clarke, C. J., Bate, M. R., & Hodgkin, S. T. 2004, MNRAS, 351, 617
Dopcke, G., Glover, S. C. O., Clark, P. C., & Klessen, R. S. 2013, ApJ, 766, 103
Ercolano, B., Bastian, N., Spezzi, L., & Owen, J. 2011, MNRAS, 416, 439
Girichidis, P., Federrath, C., Banerjee, R., & Klessen, R. S. 2012, MNRAS, 420, 613
Goodwin, S. P., Nutter, D., Kroupa, P., Ward-Thompson, D., & Whitworth, A. P. 2008, A&A, 477, 823
Goodwin, S. P., Whitworth, A. P., & Ward-Thompson, D. 2004a, A&A, 414, 633
Goodwin, S. P., Whitworth, A. P., & Ward-Thompson, D. 2004b, A&A, 423, 169
Hennebelle, P. & Chabrier, G. 2008, ApJ, 684, 395

Johnstone, D. & Bally, J. 2006, ApJ, 653, 383

Johnstone, D., Wilson, C. D., Moriarty-Schieven, G., et al. 2000, ApJ, 545, 327

Juvela, M., Ristorcelli, I., Pagani, L., et al. 2012, A&A, 541, A12

Klessen, R. S. & Burkert, A. 2000, ApJS, 128, 287

Kruijssen, J. M. D., Maschberger, T., Moeckel, N., et al. 2012, MNRAS, 419, 841

Lada, C. J. & Lada, E. A. 2003, ARA&A, 41, 57

Lada, C. J., Muench, A. A., Rathborne, J., Alves, J. F., & Lombardi, M. 2008, ApJ, 672, 410

Larson, R. B. 1981, MNRAS, 194, 809

Larson, R. B. 1995, MNRAS, 272, 213

Larson, R. B. 2005, MNRAS, 359, 211

Lubow, S. H. & Pringle, J. E. 1993, MNRAS, 263, 701

Maíz Apellániz, J. & Úbeda, L. 2005, ApJ, 629, 873

Maschberger, T., Bonnell, I. A., Clarke, C. J., & Moraux, E. 2014, MNRAS, 439, 234

Maschberger, T. & Clarke, C. J. 2011, MNRAS, 416, 541

Maschberger, T., Clarke, C. J., Bonnell, I. A., & Kroupa, P. 2010, MNRAS, 404, 1061

Maschberger, T. & Kroupa, P. 2009, MNRAS, 395, 931

Masunaga, H. & Inutsuka, S.-I. 2000, ApJ, 531, 350

McMillan, S. L. W., Vesperini, E., & Portegies Zwart, S. F. 2007, ApJ, 655, L45

Men'shchikov, A., André, P., Didelon, P., et al. 2010, A&A, 518, L103

Moeckel, N. & Bonnell, I. A. 2009, MNRAS, 396, 1864

Motte, F., Andre, P., & Neri, R. 1998, A&A, 336, 150

Offner, S. S. R., Hansen, C. E., & Krumholz, M. R. 2009, ApJ, 704, L124

Olczak, C., Spurzem, R., & Henning, T. 2011, A&A, 532, 119

Padoan, P. & Nordlund, Å. 2002, ApJ, 576, 870

Padoan, P., Boldyrev, S., Langer, W., & Nordlund, Å. 2003, ApJ, 583, 308

Padoan, P., Cambrésy, L., Juvela, M., et al. 2006, ApJ, 649, 807

Parker, R. J., Bouvier, J., Goodwin, S. P., et al. 2011, MNRAS, 412, 2489

Peretto, N., André, P., Könyves, V., et al. 2012, A&A, 541, 63

Pineda, J. E., Rosolowsky, E. W., & Goodman, A. A. 2009, ApJ, 699, L134

Rosolowsky, E. W., Pineda, J. E., Kauffmann, J., & Goodman, A. A. 2008, ApJ, 679, 1338

Schmeja, S., Kumar, M. S. N., & Ferreira, B. 2008, MNRAS, 389, 1209

Schneider, N., Csengeri, T., Hennemann, M., et al. 2012, A&A, 540, L11

Simon, M. 1997, ApJ, 482, L81

Smith, R. J., Clark, P. C., & Bonnell, I. A. 2008, MNRAS, 391, 1091

Smith, R. J., Clark, P. C., & Bonnell, I. A. 2009, MNRAS, 396, 830

Whitworth, A. P., Bhattal, A. S., Chapman, S. J., Disney, M. J., & Turner, J. A. 1994, A&A, 290, 421

Whitworth, A. P., Boffin, H. M. J., & Francis, N. 1998, MNRAS, 299, 554

Williams, J. P., de Geus, E. J., & Blitz, L. 1994, ApJ, 428, 693

Zinnecker, H. 1982, Annals of the New York Academy of Sciences, 395, 226

Chapter 4
The Role of Feedback and Magnetic Fields

Cathie J. Clarke

In the previous chapter we considered the simplest class of simulations, i.e. 'vanilla' calculations modelling only thermal pressure, turbulence and self-gravity. During this chapter we will progress to consideration of many additional processes including various forms of feedback from massive star formation and the effects of magnetic fields. Before turning to such complex simulations we first consider the sensitivity of 'vanilla' calculations to the parameters employed.

4.1 Varying the Parameters

We will start by reviewing those simulations that are most closely allied to those discussed already (i.e. simulations with a barotropic equation of state, with no feedback and no magnetic fields).

One input parameter that can readily be varied is the power spectrum and amplitude of the initial velocity field (we still restrict ourselves for now to the case of one-off injection of kinetic energy at the outset of the simulations, i.e. so-called 'decaying turbulence'). It has already been noted that the Mach number of the turbulence has no effect on the resulting IMF (Bonnell et al. 2006); Bate (2009a) determined that the IMF is likewise insensitive to the turbulent power spectrum. This is probably because—given the rapid dissipation of undriven turbulent motions on a cloud crossing time—the initial structures produced from the input velocity field play no significant role in determining the IMF; as discussed in Chap. 3, the main effects that instead appear to control the IMF are the mean Jeans mass in the simulation combined with the effects of continued accretion.

If the amplitude of the input velocity field is varied relative to the virial velocity of the cloud, then this change in the degree of gravitational boundedness of the

C.J. Clarke (✉)
Institute for Astronomy, University of Cambridge, Cambridge, UK
e-mail: cclarke@ast.cam.ac.uk

© Springer-Verlag Berlin Heidelberg 2015
C.P.M. Bell et al. (eds.), *Dynamics of Young Star Clusters and Associations*,
Saas-Fee Advanced Course 42, DOI 10.1007/978-3-662-47290-3_4

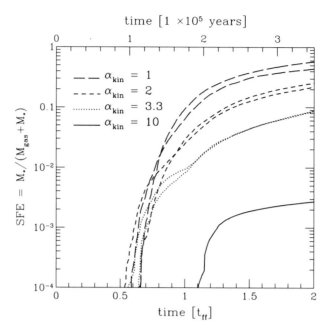

Fig. 4.1 The fraction of the initial cloud mass that has turned into stars as a function of time in the simulations of Clark et al. (2008). The cloud internal kinetic energy, which is proportional to α_{vir} increases from top to bottom (note that $\alpha_{vir} = 1$ corresponds to a state of virial equilibrium). The fraction of the cloud that turns into stars is clearly much reduced in the case of even mildly unbound clouds. Note that two different random realisations of the initial turbulent velocity field were performed for $\alpha_{vir} = 1$, 2 and 3.3. Figure from Clark et al. (2008)

initial conditions has a marked effect on the *rate* of star formation (Clark et al. 2008). Figure 4.1 demonstrates that if one triples the ratio of kinetic to potential energy compared with the situation of virial equilibrium then the number of stars formed after a couple of free-fall times by a factor of more than 5 compared with the virialised case. A further tripling of the kinetic energy decreases the number of stars formed over that period by a further factor 30. As discussed previously (see Chap. 1, Fig. 1.2) the ratio of kinetic energy to potential energy in observed clouds shows a large scatter around the situation of marginal gravitational boundedness (Solomon et al. 1987; Heyer et al. 2009), although with an ill-quantified contribution from observational uncertainties.

The fact that the star formation rate depends on a quantity with a large observed scatter is interesting, and may provide a solution to the 'star formation efficiency problem'. In this case the reason why the fraction of gas converted into stars per free-fall time is so low could stem from the fact that many clouds are globally unbound. Such clouds would then only form a few stars in locally bound regions.

We have already remarked on the difference in the clustering properties between the bound and unbound regions of the Bonnell et al. (2008) simulation. What is less clear currently is whether the degree of gravitational boundedness has a significant effect on the resulting IMF: Clark et al. (2008) and Maschberger et al. (2010) reached opposite conclusions as to whether less bound regions over- or under-produce the most massive stars.

We now turn to the body of simulations that instead drive turbulence continuously during the simulation. Here Schmeja and Klessen (2006) found that neither the power spectrum nor Mach number of the turbulence has a significant effect on the resulting clustering statistics. Moreover, Klessen and Burkert (2001) found that these quantities had no effect on the resulting IMF apart from the case where turbulence is driven at very small scales (i.e. comparable with the Jeans length) in which case it is found to disrupt the formation of the lowest mass stars. However this may well not be an issue in practice given the likelihood that the 'turbulence' in molecular clouds is actually driven on large scales and may be of Galactic origin; see Chap. 1. The insensitivity of the resulting mass spectrum to the turbulent power spectrum, even in cases where turbulent motions are continuously driven, is notable since it shows that—in the simulations at least—the gas morphology and kinematics have little effect on the pattern of stellar mass acquisition.

The amplitude and power spectrum of driven turbulence *does* however have a significant effect on the star formation efficiency: Vázquez-Semadeni et al. (2003) found that the mass fraction turned into stars generally obtains a plateau value after a few cloud free-fall times, with the value of the plateau depending most strongly on the amplitude of the turbulence and driving scale of the turbulence (see Fig. 4.2). In fact, for fixed cloud size, the plateau level depends mainly on the *sonic scale*

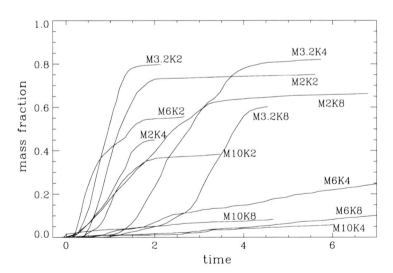

Fig. 4.2 The mass fraction of clouds converted into stars as a function of time (in units of $t_{ff}/1.5$, where t_{ff} is the free-fall time) in the simulations with continuously-driven turbulence of Vázquez-Semadeni et al. (2003). The labelling refers to the Mach number (M) and driving wavenumber (K) of the simulation, with large values of M and K corresponding to high amplitude driving and a small driving scale respectively. Figure from Vázquez-Semadeni et al. (2003)

(λ_s): if turbulent energy is injected on some driving scale and then cascades down to successively smaller scales, there exists a scale λ_s, at which turbulent support becomes negligible compared with that of thermal pressure and at this point Jeans unstable condensation can collapse. Smaller values of λ_s imply that a smaller fraction of the cloud mass is contained in fluctuations at this scale and is associated with a lower star formation efficiency. λ_s is jointly controlled by the amplitude and driving scale of the turbulence: a low λ_s is associated with a small driving scale and a high amplitude of turbulence. Since we have argued in Chap. 1 that the dynamic range of the observed size-line width relation implies that the driving scale is large (i.e. of order the cloud scale), this suggests that star formation efficiency is mainly controlled by the Mach number of the turbulence. These simulations of driven turbulence therefore bear a strong similarity to those resulting from one-off injection of turbulent energy (Clark et al. 2008) since in both cases it is the level of kinetic energy input (and hence degree of gravitational boundedness) that determines the fraction of the cloud that is converted into stars.

4.2 Putting in More Physics

4.2.1 Thermal Feedback

Bate (2009b) conducted a moderate-scale simulation in which the luminosity of accretion onto sink particles was fed back into the gas thermodynamics using flux-limited diffusion. Note that since the dominant opacity source is provided by the dust, this algorithm implicitly assumes that the dust and gas are thermally coupled (which is likely to be a good approximation above a density of $10^{-19}\,\mathrm{g\,cm^{-3}}$). The sink particle radius is 0.5 au in this simulation and thus it is only the energy liberated by accretion down to this radius that is fed back into the simulation: neither the further energy liberated by accretion down to the stellar surface nor the energy produced by the star is included and thus the magnitude of thermal feedback is, if anything, rather under-estimated by the simulation.

Nevertheless, it is found that feedback *does* have a significant effect on the resulting star formation (though *not*, interestingly, the star formation efficiency since the total mass in stars at a given time is very similar to that in barotropic calculations). The remarkable change however applies to the form of the IMF since the number of low-mass objects is drastically reduced when feedback is included. For example, the number of brown dwarfs formed per star falls to around 0.2 (comparable with the observed ratio in young clusters; Andersen et al. 2008) as compared with the value of 1.5 found in previous barotropic simulations. This is because, in the barotropic calculations (where protostellar discs are modelled as essentially isothermal gas), there is an important secondary star formation channel due to disc fragmentation: the stars formed in this way are generally of low mass. However, it is known from well-resolved studies of self-gravitating discs (e.g. Gammie 2001; Rice et al. 2003) that disc fragmentation is largely controlled by the ratio of the cooling timescale to the dynamical time (β) and is favoured by low β values where the pdV work released by collapsing condensations can be radiated away efficiently. (Note that this statement about the relative probability of fragmentation remains true despite the recent claims that discs could also fragment at long cooling time, albeit with a much smaller probability: Meru and Bate 2011, 2012; Paardekooper 2012; Hopkins and Christiansen 2013.) Isothermal conditions imply extremely efficient cooling and are therefore prone to violent fragmentation whenever discs become self-gravitating. This is however an unrealistic description of protostellar discs (Rafikov 2005, 2009; Clarke 2009) where a short cooling time and consequent violent fragmentation is only expected on scales of order 100 au or more (Rafikov 2005, 2009; Stamatellos et al. 2007; Clarke 2009).

Apart from the suppression of low-mass star formation, another notable consequence of thermal feedback is that it imprints a characteristic mass upon the star formation process. We have already discussed how—without modification to the barotropic equation of state—previous calculations had the undesirable property that the characteristic stellar mass was sensitive to mean cloud parameters. This is avoided when thermal feedback is included and Bate (2009b) produced a simple argument why feedback should stabilise the characteristic mass around 0.5 $M_\odot$.

Firstly this argument notes that the gas temperature is expected to decline with increasing distance (r) from a star. In the case where attenuation of stellar radiation over this distance can be neglected and where the gas is in thermal equilibrium with large grains heated by the stellar luminosity (L_*), the gas temperature scales as $r^{-1/2}$. This means (at fixed ρ) that the Jeans length (r_J; see Chap. 1, Eq. 1.2) declines with r as $r^{-1/4}$. Eventually the point is reached at which the Jeans length is $\sim r$ and at this point it is possible to form a new star. Thus the characteristic mass scale is the Jeans mass at a temperature where the local value of r_J is $\sim r$. Naturally, the expression for the resulting Jeans mass depends on density and stellar luminosity, but the dependence is weak, i.e. it scales as $\rho^{-1/5} L_*^{3/10}$. If one goes further and equates the stellar luminosity with the accretion luminosity resulting from collapse of material within r_J on a free-fall timescale, then higher density gas falls in faster, implying a higher value of L_*. This weakens the dependence of the Jeans mass on density still further [the final scaling in this case is Jeans mass scaling as $(M_*/R_*)^{3/7} \rho^{-1/14}$].

Bate (2009b) backed up these arguments by running simulations with feedback and a variety of initial conditions. In all cases, the stability of the resulting IMFs demonstrates that feedback breaks the sensitive dependence on initial cloud parameters that is seen in purely barotropic simulations.

Krumholz et al. (2011) have however drawn attention to a potential problem that arises when simulations of high-density gas including thermal feedback are run over longer timescales. This study contrasted the evolution of the sink IMF in the control case of isothermal gas with that of a simulation including radiative feedback (these are AMR calculations which also employ flux-limited diffusion but differ from Bate 2009b in that they include additional luminosity terms corresponding to the intrinsic luminosity of the star and to accretion luminosity that would be liberated from within the sink radius). For a suitable choice of temperature for the isothermal control case it is possible to match the sink IMF produced at a given point in the evolution. With further evolution, however, the mass scale at which the IMF peaks continues to increase in the simulations with thermal feedback. This is in contrast to the isothermal case where the location of the IMF peak does not evolve. In the isothermal case low-mass stars continue to form and so—although individual stars continue to accrete and increase in mass and the *maximum* stellar mass also increases as the simulation proceeds—the location of the IMF peak remains invariant. In the simulation with radiative feedback, however, the fresh formation of low-mass stars is suppressed once the regions of hot gas surrounding protostars start to overlap. Accretion onto existing stars however continues unabated (the temperatures produced by low-mass stars in the simulation are sufficient to inhibit fragmentation but are nowhere near those necessary to inhibit accretion). Consequently the mass of the peak in the IMF marches monotonically upwards in time. Apparently then, thermal feedback has replaced one factor that mitigates against the production of a universal IMF (i.e. excessive dependence on initial conditions) with another factor that is equally unfortunate (i.e. a sensitive dependence on time)!

Part of the solution may be the modelling of a cloud that is more realistically structured. Krumholz et al. (2012) undertook further simulations which did not start (as in those of Krumholz et al. 2011 and in all the simulations discussed hitherto) by driving turbulent motions in an initially uniform cloud. Such a situation is unrealistic in the sense that fragmentation starts to occur in density fields which are inconsistent with the velocity field; an alternative approach, adopted by Krumholz et al. (2012), is to 'settle' the turbulence in the velocity field for a crossing time or so *before* switching on self-gravity. This ensures self-consistency between the statistics of the density and velocity field at the onset of fragmentation. Clearly both approaches are to a degree unrealistic. Further improvement requires star formation simulations whose 'initial' conditions are more strongly informed by larger scale simulations which model the formation of molecular clouds from the galactic ISM (see previous discussions of simulations of Tasker and Tan 2009 and Dobbs et al. 2011, which however lack the resolution to model star formation within the clouds).

The change in the initialisation of the cloud structure adopted by Krumholz et al. (2012) produces clouds that are more structured and less centrally condensed. In this case, star formation occurs in more isolated groupings and this alleviates the 'over-heating problem' identified above: although thermal feedback may quench new star formation locally, there are other places in the clouds where low-mass stars can start to form and thus the peak of the IMF is stabilised. (Note that the over-heating problem was not apparent in the simulations of Bate 2009b because these simulations started with lower initial densities.)

Krumholz et al. (2012) found that even though this change in initial conditions produced a stabilised IMF, it was then top-heavy with respect to observations. They argued that this result can be mitigated by the effect of feedback due to outflows (see Sect. 4.2.2). However, it needs to be said that none of the non-isothermal simulations in Krumholz et al. (2012) are a good match to the observed IMF: inspection of the IMFs in differential form (see Fig. 4.3) shows that the simulations with feedback

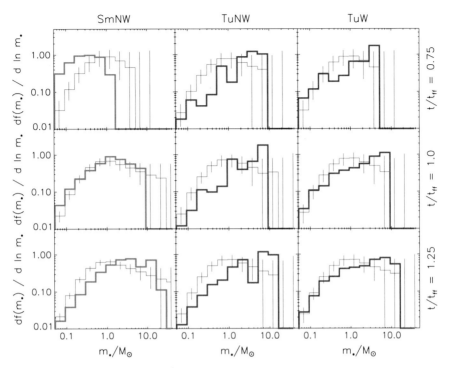

Fig. 4.3 An illustration of the 'over-heating' problem in simulations of high-density clouds with thermal feedback. The *green histograms* represent the case where the simulation starts with a uniform cloud subject to an input turbulent velocity field; the *blue histograms* are the results of simulations in which the turbulence is first 'settled' (see text) prior to the switch on of self-gravity. The simulations providing the *red histograms* additionally involve feedback from outflows. The settled turbulence stabilises the IMF peak but produces IMFs that rise with mass up to $10\,M_\odot$. Figure from Krumholz et al. (2012)

produce an IMF that is *rising* with increasing mass up to $\sim 10\,M_\odot$ (though this is formally consistent with the observed IMF on account of the large Poissionian errorbars in the high-mass bins).

We therefore have the troubling situation that the addition of more physics is apparently driving the IMF away from the 'successful' IMF form produced by the simplest barotropic simulations (note that the word 'successful' is in inverted commas because the strong dependence on initial conditions is empirically incompatible with the observed invariance of the IMF). The simulations of Bate (2009b) appeared to solve the problem because they employed a low enough density that they avoided entering the over-heating regime; the simulations of Krumholz et al. (2011, 2012) showed that the factors that determine whether or not the system enters the over-heating regime are rather complex and that one can then find that the IMF is either time-dependent or rising towards high masses. Although the number of stars formed in the computationally demanding simulations of Krumholz et al. (2011, 2012) may be too small to rule out a consistency with the observed IMF, these simulations do not appear to be able to produce a Salpeter-like tail at high masses.

At the time of writing, these studies are very recent. It is clear that the physical effects that go into these simulations are in the direction of verisimilitude (i.e. both outflows and thermal feedback are genuine effects that should be incorporated). The problems described above are not a signal to abandon such studies but rather a spur for all groups to pursue them vigorously in future.

4.2.2 Outflows

We now turn to other studies that have attempted to include the effect of outflows in star formation simulations. Dale and Bonnell (2008) added outflows to their barotropic SPH simulations of a $1000\,M_\odot$ molecular cloud: by varying the collimation angle of the outflows these could be physically associated with either jet-like outflows (see Frank et al. 2014 for a recent review) or else energetic stellar winds from massive stars. In the simulations, outflows (with constant terminal velocity and mechanical luminosity scaling as the fourth power of stellar mass) were only switched on for stars exceeding $10\,M_\odot$ in mass.

This study found that outflows only modified the IMF within the mass range for which outflows were included (i.e. $>10\,M_\odot$): outflows somewhat steepen the high-mass IMF because they inhibit accretion onto the highest mass stars. The invariance of the IMF at lower masses indicates that the feedback did little to modify the global cloud dynamics, and this is also evidenced by the minor effect of feedback upon the star formation efficiency.

It is perhaps surprising—based on momentum conservation arguments—that the effect of outflows on the cloud structure is so minor. The outflow sources produce a momentum flux which is more than enough to unbind the entire cloud and yet the mass fraction that is unbound by the winds is small (around 5 % of the cloud mass per free-fall time). This can be largely ascribed to the highly inhomogeneous 'sky' seen by each wind source: the location of the most massive stars at the intersection of massive filaments means that the distribution of surrounding material is highly anisotropic. The winds preferentially escape via low-density channels, where their speeds may significantly exceed the escape velocity of the cloud. However they entrain little dense gas in these channels and so the dynamical state of dense star-forming gas is relatively little disturbed. We will however have to revise this conclusion when we shortly consider simulations that combine outflows with magnetic fields.

4.2.3 Magnetic Fields

To date there have been relatively few calculations of star formation (as opposed to magneto-turbulent cloud structure formation) involving magnetic fields. The additional computational expense means that such simulations produce a relatively small number of stars and this makes it hard to draw statistically robust conclusions about the effect of magnetic fields at this stage.

The simulations of Price and Bate (2009) employ a range of field strengths in the super-critical regime (i.e. in the range of mass-to-flux ratios where the field is insufficient to halt collapse: note that these simulations employ ideal MHD and therefore preserve the mass-to-flux ratio). Thus magnetic fields do not prevent collapse but instead slow it down compared with field-free simulations; magnetic fields also produce smoother gas morphologies associated with magnetised shocks. Magnetic fields thus contribute to reducing the star formation efficiency: for a mass-to-flux ratio that is three times critical, the point of 5 % efficiency is attained about 0.5 free-fall times after the onset of star formation, whereas this point is achieved in less than half this time in the unmagnetised case (see Fig. 4.4). The conclusions that can be drawn from these MHD experiments are however limited by their relatively short duration and the star formation efficiency is still rising linearly with time at the end of the simulation. This is in contrast to simulations of unbound clouds (Clark et al. 2008) or turbulently driven clouds (Vázquez-Semadeni et al. 2003) which—being run for longer—have attained a *saturated* star-to-gas fraction at around a free-fall time after the onset of star formation. The computationally expensive nature of MHD calculations also prevents a meaningful study of the effect of magnetic fields upon the IMF. Nevertheless it is clear that simulations with magnetic fields *can* produce stars that span a reasonable range of masses (including also binary stars).

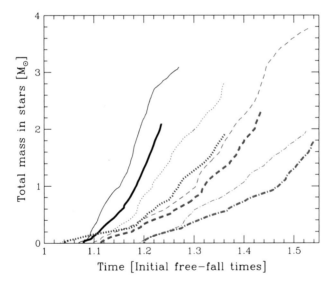

Fig. 4.4 The fraction of the initial cloud mass that has turned into stars as a function of time in the magnetised simulations of Price and Bate (2009). The mass-to-flux ratio decreases from the unmagnetised case (*left*) to a simulation where the mass-to-flux ratio is three times critical (*right*). Note that in contrast to Figs. 2.2 and 2.3 (Chap. 2), the star formation efficiency is still rising steeply at the end of the simulation. Figure from Price and Bate (2009)

Wang et al. (2010) conducted larger scale simulations including both magnetic fields and outflows and illustrated an interesting synergy between the two: they found, in contrast to the unmagnetised simulations of Dale and Bonnell (2008), that the inclusion of outflows *does* have a significant effect on the star formation rate. The simulation parameters are sufficiently different that it is hard to unambiguously determine the source of this difference. One possible explanation is that magnetic fields provide dynamical coupling, in a turbulent medium, between material in the outflow path and the rest of the cloud. This means that outflows cannot simply punch their way out through low-density channels while leaving the remaining cloud relatively unimpaired. The presence of magnetic fields at a realistic level may thus be an important ingredient in coupling the energy of outflows to the bulk of the cloud.

4.2.4 Ionising Radiation

We conclude this survey of additional physical ingredients by considering the effect of ionising radiation from the most massive stars in the simulation. Here Dale et al. (2005) found that this had a remarkably small effect on global cloud dynamics and star formation history. The reason is closely allied to our previous discussion of the (lack of) effect of outflows. In the case of ionisation feedback, the density dependence of the recombination rate means that ionising radiation is able to heat gas only in low-density channels where it accelerates modest quantities of gas to speeds of order $10 \, \text{km s}^{-1}$, which exceeds the cloud escape velocity. Within adjoining dense filaments, the impact of ionising radiation is however negligible. Consequently, although the kinetic energy absorbed by the gas exceeds the binding energy of the cloud, its selective transfer to low-density, high-velocity flows ensures that the cloud remains bound. As a result, the star formation history of the cloud is almost unchanged by ionisation feedback.

Although this early work suggested that ionisation feedback is likely to be ineffective, it is worth noting that the outcome is sensitive to the escape velocity of the cloud (Dale et al. 2012). Ionisation feedback is negligible in the case of clouds with escape velocities of $5 \, \text{km s}^{-1}$ or above but its role becomes much more pronounced in clouds with slightly lower escape velocities ($\sim 3 \, \text{km s}^{-1}$). Figure 4.5 contrasts the effect of ionising feedback on two clouds with similar free-fall times but where the escape velocity in the more massive ($10^6 \, M_\odot$) cluster (Run X) is $\sim 10 \, \text{km s}^{-1}$ as compared with $2 \, \text{km s}^{-1}$ in the low-mass ($10^4 \, M_\odot$) cluster (Run I). The gas morphology is evidently much smoother in Run I and the fraction of the cloud that is in stars when the first supernova explodes is much lower (8 % cf. > 20 %). Observed clouds have escape velocities in the range of $1–10 \, \text{km s}^{-1}$ with a mild tendency towards higher escape velocities in more massive clouds (see Fig. 4.6). The suite of simulations conducted by Dale et al. (2012) suggests that the cloud mass range $\sim 10^4 – 10^5 \, M_\odot$ is a rough boundary above which ionisation feedback is ineffective. For lower mass clouds, the development of ionised regions within the cloud does not have any immediate impact on the star formation *rate* in the dense gas but does reduce the cloud mass that is available for future star formation. We have already

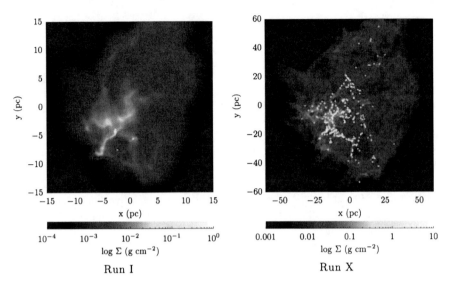

Fig. 4.5 Comparison of the gas and stellar distributions produced in two simulations including ionisation feedback. The two simulations share similar free-fall times but the masses and escape velocities of the clouds in the two runs are $10^4\,M_\odot$, $2\,km\,s^{-1}$ and $10^6\,M_\odot$, $10\,km\,s^{-1}$ (*left* and *right panels* respectively). The fraction of gas that becomes unbound in the course of the simulation is 60 and 20 % respectively. Figure adapted from Dale et al. (2012)

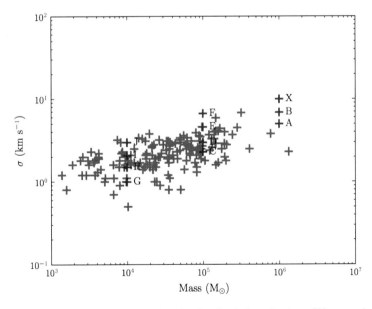

Fig. 4.6 The escape velocities of observed molecular clouds from the data of Heyer et al. (2009). Runs I and X from Dale et al. (2012, see text) can be located in this diagram from their parameters detailed in the caption to Fig. 4.5. Note that although many of the clouds in this sample have escape velocities less than $3\,km\,s^{-1}$ (and are thus susceptible to efficient ionisation radiation feedback), the bulk of star formation in the Milky Way occurs in clouds more massive than $10^5\,M_\odot$ many of which have higher escape velocities. Figure from Dale et al. (2012)

seen that the fraction of a cloud's mass that goes into stars is sensitive to how much of the cloud is bound initially: these radiation hydrodynamical simulations suggest that even initially bound clouds—at the lower end of the cloud mass spectrum—may in any case become unbound, with an associated reduction in the total mass fraction going into stars.

So far our discussion has tacitly assumed that the effects of ionisation feedback are destructive (i.e. inhibiting star formation). It is however often argued that ionisation (as well as other forms of feedback such as supernovae) can trigger star formation. It is found that such an outcome is particularly associated with situations where (a) the ionising source is external to a cloud and (b) where the cloud is unbound anyway. In such a situation, Dale et al. (2007) found that the heating of the side of the cloud adjacent to the O star impedes the cloud's expansion in that direction: escaping gas is returned to interact with the dense gas in the cloud core and the net effect is that star formation is enhanced with respect to a control simulation without the O star.

Rather disappointingly, however, there are no readily available observational diagnostics for identifying such triggered stars: they form in the same (dense) region of the cloud as in the control simulation and share similar kinematics. This is in contrast to the expectations of star formation triggered by the expansion of ionised gas within smooth and static density fields (Whitworth et al. 1994) where the stellar kinematics should indicate a clear signature of ordered expansion from the ionising source. Here however the turbulent cloud already has a (realistic) velocity dispersion of a few km s^{-1}. The free expansion speed of ionised gas is about 10 km s^{-1} so that once it has swept up significant cold gas the induced velocities are less than or similar to the cloud's original velocity dispersion. Consequently clear kinematic signatures of ionisation triggered star formation are erased in turbulent clouds.

4.3 Discussion

This survey of the effect of 'additional' physical processes (particularly feedback and magnetic fields) paints a complex picture and it is probably premature to draw conclusions at this stage. We can expect that the next few years will see a proliferation of experiments similar to those described here and that these will help to clarify the statistical significance of the trends tentatively identified in existing simulations. Of all the issues discussed above, perhaps the least well-explored is the effect of these processes on the upper IMF (simply because it is challenging to run more realistic simulations that produce large numbers of stars in the sparsely populated upper tail of the IMF; see for example the pioneering simulations of Peters et al. 2010, 2011).

Perhaps there are two areas where there is some degree of consensus. The first is that the lower end of the IMF is very sensitive to the thermal properties of the gas and that thermal feedback *from low-mass stars* is an important ingredient in determining the relative numbers of stars and brown dwarfs.

Secondly, there seems to be a variety of ways in which the star formation efficiency (fraction of gas going into stars per free-fall time) can be brought down to observationally reasonable levels (a few %). These include the effects of (i) starting with clouds that are globally unbound and/or driving motions that are larger than the escape velocity, (ii) including the effect of photoionisation feedback, and (iii) including magnetic fields with or without additional outflow feedback. Of these the third is probably the least well-explored situation. The first two effects are certainly sufficient to explain the low star formation efficiencies of nearby GMCs but they share the same problem when it comes to explaining the global star formation efficiency of the Milky Way. We have already noted that the mass function of GMCs has a slope proportional to $M^{-1.6}$ (see Chap. 1, Sect. 1.1) and that consequently the total mass budget of GMCs (and associated star formation) is dominated by the largest GMCs (i.e. those around $10^6 \, M_\odot$). On the other hand, Fig. 1.2 (Chap. 1) and Fig. 4.6 (this chapter) indicate that these largest clouds are most likely to be gravitationally bound and also have the highest escape velocities. The former makes (i) problematical in these largest clouds while the latter renders (ii) relatively ineffective (given the evidence from simulations that ionisation feedback is inefficient in clouds with escape velocities in excess of $5 \, \mathrm{km \, s^{-1}}$). It remains to be seen whether magnetic fields (even the relatively weak—supercritical—field values that are favoured by recent Zeeman measurements in GMCs, Crutcher et al. 2010; Crutcher 2012) can provide the required deceleration of star formation in the most massive clouds.

4.4 Summary

In this chapter we have built on our initial survey of cluster formation simulations by first discussing how—in the simplest simulations involving only gravity and thermal pressure assigned via a barotropic equation of state—the properties of the stars formed depend on the parameters that enter these simulations. We then reviewed simulations that include additional physics: thermal feedback, magnetic fields, outflows and feedback from ionising radiation.

It is hard to draw definitive conclusions from the simulations currently available, each of which contains only a subset of the effects listed above. One emerging consensus is that thermal feedback from young stars plays an important role in shaping the lower end of the IMF. Another conclusion is that there are multiple ways of reducing the star formation efficiency (see Chap. 1, Sect. 1.3) to acceptably low values. However it is unclear whether a leading mechanism (ionisation feedback) will be effective in the most massive clouds (where most of the star formation in the Galaxy is occurring). The mechanism for suppressing the star formation efficiency in these largest clouds thus remains an open problem.

References

Andersen, M., Meyer, M. R., Greissl, J., & Aversa, A. 2008, ApJ, 683, L183

Bate, M. R. 2009a, MNRAS, 397, 232

Bate, M. R. 2009b, MNRAS, 392, 1363

Bonnell, I. A., Clark, P., & Bate, M. R. 2008, MNRAS, 389, 1556

Bonnell, I. A., Clarke, C. J., & Bate, M. R. 2006, MNRAS, 368, 1296

Clark, P. C., Bonnell, I. A., & Klessen, R. S. 2008, MNRAS, 386, 3

Clarke, C. J. 2009, MNRAS, 396, 1066

Crutcher, R. M. 2012, ARA&A, 50, 29

Crutcher, R. M., Wandelt, B., Heiles, C., Falgarone, E., & Troland, T. H. 2010, ApJ, 725, 466

Dale, J. E. & Bonnell, I. A. 2008, MNRAS, 391, 2

Dale, J. E., Bonnell, I. A., Clarke, C. J., & Bate, M. R. 2005, MNRAS, 358, 291

Dale, J. E., Ercolano, B., & Bonnell, I. A. 2012, MNRAS, 424, 377

Dale, J. E., Ercolano, B., & Clarke, C. J. 2007, MNRAS, 382, 1759

Dobbs, C. L., Burkert, A., & Pringle, J. E. 2011, MNRAS, 417, 1318

Frank, A., Ray, T. P., Cabrit, S., et al. 2014, Protostars and Planets VI, ed. H. Beuther, R. S. Klessen, C. P. Dullemond & T. Henning, University of Arizona Press, 451

Gammie, C. F. 2001, ApJ, 553, 174

Heyer, M. H., Krawczyk, C., Duval, J., & Jackson, J. M. 2009, ApJ, 699, 1092

Hopkins, P. F. & Christiansen, J. L. 2013, ApJ, 776, 48

Klessen, R. S. & Burkert, A. 2001, ApJ, 549, 386

Krumholz, M. R., Klein, R. I., & McKee, C. F. 2011, ApJ, 740, 74

Krumholz, M. R., Klein, R. I., & McKee, C. F. 2012, ApJ, 754, 71

Maschberger, T., Clarke, C. J., Bonnell, I. A., & Kroupa, P. 2010, MNRAS, 404, 1061

Meru, F. & Bate, M. R. 2011, MNRAS, 410, 559

Meru, F. & Bate, M. R. 2012, MNRAS, 427, 2022

Paardekooper, S.-J. 2012, MNRAS, 421, 3286

Peters, T., Banerjee, R., Klessen, R. S., & Mac Low, M.-M. 2011, ApJ, 729, 72.

Peters, T., Klessen, R. S., Mac Low, M.-M., & Banerjee, R. 2010, ApJ, 725, 134.

Price, D. J. & Bate, M. R. 2009, MNRAS, 398, 33

Rafikov, R. R. 2005, ApJ, 621, L69

Rafikov, R. R. 2009, ApJ, 704, 281

Rice, W. K. M., Armitage, P. J., Bate, M. R., & Bonnell, I. A. 2003, MNRAS, 339, 1025

Schmeja, S. & Klessen, R. S. 2006, A&A, 449, 151

Solomon, P. M., Rivolo, A. R., Barrett, J., & Yahil, A. 1987, ApJ, 319, 730

Stamatellos, D., Whitworth, A. P., & Ward-Thompson, D. 2007, MNRAS, 379, 1390

Tasker, E. J. & Tan, J. C. 2009, ApJ, 700, 358

Vázquez-Semadeni, E., Ballesteros-Paredes, J., & Klessen, R. S. 2003, ApJ, 585, L131

Wang, P., Li, Z.-Y., Abel, T., & Nakamura, F. 2010, ApJ, 709, 27

Whitworth, A. P., Bhattal, A. S., Chapman, S. J., Disney, M. J., & Turner, J. A. 1994, A&A, 290, 421

Chapter 5
The Formation of Multiple Systems in Clusters

Cathie J. Clarke

Binary stars provide a rich array of observational diagnostics which can in principle be used to test and calibrate star formation simulations. Moreover, as we shall see later, binaries can be dynamically important within star clusters. In addition, the fact that binaries in certain separation ranges are either dynamically created or destroyed during star cluster evolution means that the properties of field binaries may be used to place constraints on the types of clusters in which they might have formed. We will defer a discussion of these latter points until Chap. 6, Sect. 6.3 and Chap. 7, Sect. 7.1 and start by assessing the extent to which star formation simulations can reproduce observed binary statistics.

Naturally, this can only be answered by examining simulations of appropriate resolution: in particular, the sink radius and gravitational softening length must be significantly smaller than the separations of the binaries studied (see Chap. 2). This means that we now turn from the large-scale, rather poorly-resolved simulations that we have used to discuss cluster assembly (where sink radii can be $\sim$200 au) to smaller scale simulations: the simulation of Bate (2009a) models a cloud of only 500 $M_\odot$ but employs sink radii of only 0.5 au and, with a total sample of 1250 stars and brown dwarfs produced, can well address the statistics of binaries with separations down to au scales.

Having said this, such high resolution and large sample size comes at the expense of severe simplification of the physics: currently one can only make statistically meaningful statements about binary systems using simulations that employ the simplest assumptions (i.e. barotropic equation of state, freely decaying turbulence, no feedback or magnetic fields). We will however comment on any qualitative insights into the effect on binary properties that may be obtained from more complex simulations. Although we here concentrate on the large-scale simulations of Bate (2009a),

C.J. Clarke (✉)
Institute for Astronomy, University of Cambridge, Cambridge, UK
e-mail: cclarke@ast.cam.ac.uk

© Springer-Verlag Berlin Heidelberg 2015
C.P.M. Bell et al. (eds.), *Dynamics of Young Star Clusters and Associations*,
Saas-Fee Advanced Course 42, DOI 10.1007/978-3-662-47290-3_5

we also direct the reader to a range of hydrodynamical studies of binary star formation within small-scale cores which fragment into a small number of objects: see Delgado-Donate et al. (2003), Delgado-Donate et al. (2004), Goodwin et al. (2004a), Goodwin et al. (2004b), Goodwin et al. (2006), Offner et al. (2008), Machida (2008a), Arreaga-García et al. (2010), and Walch et al. (2010).

Before proceeding to describing the extent to which the calculations succeed in reproducing binary statistics, we should mention a few caveats which make some of the predicted properties more reliable than others. As discussed fully in Bate (2009a), the finite sink particle radius obviously compromises binary statistics at separations of a few sink radii (i.e. of order an au or closer). Secondly, as described in our previous discussion of the IMF (see Chap. 3, Sect. 3.4), the isothermal equation of state over-produces low-mass stars and brown dwarfs on account of excessive disc fragmentation: this may affect the resulting mass ratio distribution since some fraction of these spurious objects end up bound to higher mass stars. Another issue is disc evolution, which—as we discussed in Chap. 2 (Sect. 2.2)—is likely to be controlled by numerical viscosity, particularly at low disc masses where poor resolution is a particular issue. Artificial viscosity accelerates angular momentum redistribution within discs and hence causes gas to accrete too quickly onto the parent star: the consequent drop in disc mass then makes the disc even more under-resolved and hence further accelerates the process. Since this effect becomes most severe when discs contain a small fraction of the (binary) star mass, it may make little difference to some binary parameters (e.g. the binary mass ratio) if the remnant disc drains too quickly. However, discs are a highly efficient sink of binary orbital angular momentum (Artymowicz and Lubow 1994, 1996) even when they contain only a small fraction of the binary's mass, since a small quantity of gas transports angular momentum to large orbital radii. Consequently the details of disc draining are probably important for a binary's final orbital elements (semi-major axis, eccentricity). Since binary orbital parameters are usually displayed in the plane of *logarithmic* semi major axis versus (linear) eccentricity, an order unity change in each quantity would have a much more pronounced effect on the eccentricity distribution. Predictions of the eccentricity distribution are thus more questionable than those for the semi-major axis distribution.

These considerations lead us to a rough hierarchy of reliability for the observational parameters produced by binary formation simulations: i.e. (in order of descending robustness) binary fractions, mass ratios, separations and eccentricities. It is for this reason that our following discussion will focus most on the earlier parts of this list and will give scant attention to the predicted eccentricity distribution.

Armed with these caveats, we can now proceed to discussing the resulting binary statistics from the simplest 'vanilla' calculations and turn in the following section to the influence of 'additional physics' on binary star formation.

5.1 The Formation of Multiple Stars in 'Vanilla' Simulations

5.1.1 Binary Star Statistics

Starting with the issue of binary fraction as a function of primary mass (see also Reid Chap. 16, Sect. 16.4) there is here remarkably good agreement between simulations and observations, not only in the simulation of Bate (2009a) but more generally in calculations of this kind: in all cases the binary fraction increases rather strongly with increasing primary mass (see Fig. 5.1). As discussed in early analytic papers (McDonald and Clarke 1993), a *dynamical* bias towards a higher binary fraction for more massive primaries is a natural expectation of any scenario in which star formation commences in small-N non-hierarchical groupings. Indeed in the case of purely N-body dynamical interactions, the resulting reconfiguration of the system into a stable binary plus ejected stars normally leads to the most massive two members being located in the binary, a situation for which the statistical consequences can be readily inferred. Such a prescription actually predicts a relationship between primary mass and binary fraction that is even steeper than that observed. McDonald and Clarke (1995) showed that the relationship is somewhat flattened if one adds in prescriptive dissipative encounters (such as would result from the presence of circumstellar discs). This result can be understood in as much as dissipative interactions can harden low-mass binaries that would otherwise have been unbound in a purely N-body scenario. It is of course a far cry from such idealised few-body integrations (with prescribed star-disc drag terms) to what is happening in the full hydrodynamical

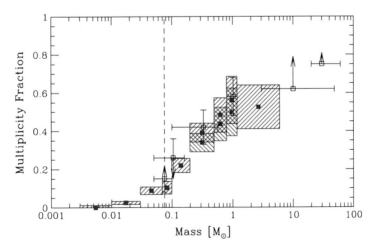

Fig. 5.1 The multiplicity fraction as a function of primary mass from the simulation of Bate (2009a) with *blue filled squares* and hatched regions representing Poissonian errors. Note that the *red filled squares* and hatched regions represent the multiplicity fraction excluding brown dwarf companions (masses $<0.075\,M_\odot$). The *black open squares* represent observed values and their uncertainties (see Bate (2009a) for the sources of observational data). Figure from Bate (2009a)

simulations, since the latter have many additional effects such as continued gas accretion, ongoing fragmentation, etc. Nevertheless it would seem likely that the dominant physical effects are the same, namely that binary pairing is controlled by few-body interactions within a gas-rich environment.

The predicted separation distribution is likewise a good match to observations in that simulations produce binaries with a wide range of separations. This is limited by finite resolution at close separations and at large separations by a mixture of the finite angular momentum of the progenitor gas together with the tendency of weakly bound pairs to be destroyed by dynamical interactions: for these reasons, the bulk of pairs form of the separation range of $1-10^4$ au for solar-mass primaries. Primaries with very-low-mass (VLM) primaries (i.e. those with masses $<0.1\,M_\odot$) are somewhat tighter, with few pairs wider than 1000 au. Again, this trend is broadly consistent with observations (see the database of VLM multiples: http://vlmbinaries. org/) and can be understood in terms of the greater fragility of wide low-mass pairs in the dynamical environment of small N clusters. Experiments with different values of the sink radius confirm that the resulting binary separation distribution is largely independent of r_{sink} apart from separations less than a few times r_{sink}.

The mass ratio distribution produced by star formation simulations is somewhat concentrated towards high mass ratio (i.e. systems with roughly equal-mass components) although a few systems with low mass ratio, $q\ (=M_2/M_1)$ are also produced. The distribution of mass ratios is found to be very insensitive to the conditions at the point of initial fragmentation and is largely determined by the subsequent accretion history onto both components. This explains the large population of binaries with $q > 0.5$ produced by the simulations (Bate 2000; Bate and Bonnell 1997): the material that collapses later onto a protobinary has higher specific angular momentum than the binary and thus tends to be preferentially accreted by the secondary (which is located further from the system centre of mass). Consequently accretion leads to an *increase* in the binary mass ratio.

It has more recently been suggested that this result could be an artifact of SPH since Ochi et al. (2005) and Hanawa et al. (2010) found that their grid-based simulations showed the opposite behaviour: although they also found that high angular momentum material is initially accreted by the secondary it then flows through the inner Lagrange point and is then accreted by the primary. They therefore found that the accretion of high angular momentum material causes a *reduction* in the mass ratio. They suggested that this flow from secondary to primary may be artificially suppressed by numerical viscosity in the case of SPH simulations.

Further investigations by the late Eduardo Delgado-Donate (with Clarke and Bonnell) however demonstrated that in fact this flow from secondary to primary is a consequence of the rather warm conditions used in the grid-based calculation (with sound speed of 25 % of the binary's orbital velocity) and that SPH calculations also replicated a similar flow with such warm gas. At realistically low temperatures, the gas that enters the Roche lobe of the secondary is retained by the secondary and the mass ratio should indeed rise. More recent calculations (Young et al. 2015) have confirmed the numerical convergence of this result for a variety of mass ratios.

It used to be thought that there was an 'extreme mass ratio problem', i.e. that simulations under-produced low q pairs compared with observations because of the effect of continued accretion onto the binary. This discrepancy was probably over-estimated however: firstly, the most recent observational study of mass ratio distributions of solar-type binaries (Raghavan et al. 2010) contains a smaller fraction of low q pairs than did its predecessor survey (Duquennoy and Mayor 1991), a result that can be largely attributed to over-generous incompleteness corrections at low q in the earlier study. In the more recent study, the observed mass ratio distribution is rather flat down to $q \sim 0.1$ (i.e. for companion masses exceeding the hydrogen burning mass limit). Secondly, there is an indication in simulations that some low q pairs are formed at later times as a result of orbital reconfiguration within non-hierarchical multiples: Moeckel and Bate (2010) found that when the products of Bate's hydrodynamical simulation were integrated as a pure N-body system for a further 10 Myr, a modest number of new low q pairs were created from the decay of unstable triples/quadruples. This orbital reconfiguration sufficed to make the resulting mass ratio distribution rather flat. Taken together, these opposite shifts in the observational data and in theoretical predictions mean that there is no longer believed to be a serious discrepancy between simulations and observations on this issue.

There may however still be some discrepancy in the case of VLM stars conventionally defined as stars and brown dwarfs of mass $<0.1\,M_\odot$, although in the opposite direction to that discussed above. Observationally, it appears that VLM stars have mass ratios that are *strongly* peaked towards unity (see also Reid Chap. 16, Sect. 16.4): indeed a number of surveys have failed to find VLM pairs with mass ratio less than 0.5 despite having ample sensitivity to lower q pairs (Close et al. 2003; Reid et al. 2006; Siegler et al. 2005). Note that these claims are generally made in the case of visual pairs where the assignment of a mass ratio requires the assumption of a mass-luminosity relation; interestingly enough, Konopacky et al. (2010) assigned lower q values to pairs for which they had obtained astrometric constraints compared with those that would apply if one used model mass-luminosity relations. The errorbars on the astrometric measurements are however currently very large; further epochs of data on these pairs are required in order to obtain good constraints through a method that is independent of the mass-luminosity relation. Star formation simulations do *not* show this strong dependence of q distribution on primary mass that is found observationally (i.e. they show no *strong* peak towards $q = 1$ in the case of VLM pairs). This therefore represents a possible area in which observations may be pointing towards some insufficiency in the simulations.

5.1.2 Disc Orientation in Protobinaries

We now turn to the issue of the orientation of circumstellar discs in binary systems. This can be constrained observationally in the case of resolved pairs through measurement of the integrated linear polarisation of the scattered starlight of each component, since this quantity should be parallel to the equatorial plane of the disc (Jensen et al. 2004; Monin et al. 2007; Wolf et al. 2001). Note that this measurement only constrains the orientation of the disc in the plane of the sky and provides no information on its inclination along the line-of-sight: Wolf et al. (2001) have however shown from statistical arguments that if the distribution of relative position angles on the sky is peaked towards zero then discs tend to be parallel in three dimensions also.

The result of these studies is that disc polarisation tends to be close to (but not exactly) parallel in binary systems but that a few objects exhibit large position angle differences of up to 90°: see Fig. 5.2. A more direct illustration of misalignment is provided by images of HK Tau (Stapelfeldt et al. 1998) in which both stars possess a disc but only one of them is edge-on (and not parallel to the position angle of the binary); T Tauri provides another similar example of an imaged misaligned system (Ratzka et al. 2009; Skemer et al. 2008). As a general rule, systems with smaller binary separations tend to be more aligned. This has been demonstrated both in the pre-main-sequence case (see also Mathieu et al. 1997) and in the case of main-sequence stars (where stellar rotation axes have been compared with that of the binary orbit; Hale 1994). A striking counter-example to this tendency is the case of DI Herculis (Albrecht et al. 2009), an early-type eclipsing binary where Rossiter-McLaughlin measurements have shown that the spins of both binary components are strongly misaligned with each other (and with the binary orbit). Since we

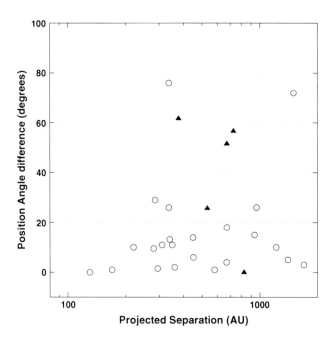

Fig. 5.2 The difference in position angle of the polarisation between binary components as a function of separation for binaries (*open circles*) and triples (*filled triangles*). Figure from Monin et al. (2007).

expect close disc-bearing systems to be rapidly aligned (see below) the origin of DI Her is puzzling, although a likely explanation is that its misalignment was acquired after disc dispersal via a tidally moderated Kozai cycle excited by a third body in the system (see Fabrycky and Tremaine 2007).

Star formation simulations yield greater misalignment between the disc planes within binaries than is seen observationally: Fig. 5.3 plots the angle between the rotation axes of each disc within binaries as a function of binary separation from Bate (2009a), note that this is a three-dimensional misalignment angle ($<180°$) and cannot be directly compared with Fig. 5.2 which is a position angle difference in the plane of the sky ($<90°$). The simulation results fill the plane and are consistent with essentially random inclinations (apart from an avoidance of systems that are very close to complete anti-alignment). Uncorrelated disc spin directions are a consequence of the turbulent conditions and the strong dynamical effects that operate within the simulations (i.e. the angular momentum vector of the accreting material is highly spatially variable; moreover tidal torques exerted by other stars within few-body groupings can readily perturb the orientations of circumstellar discs).

It is however unclear that this really represents a discrepancy with observations since within an isolated binary pair, tidal torques should produce rough alignment between the planes of discs and the binary orbital plane within about 20 orbital periods (Bate 2000; Facchini et al. 2013; Foucart and Lai 2013). This process is not well modelled in the (poorly-resolved) star formation simulations that we are

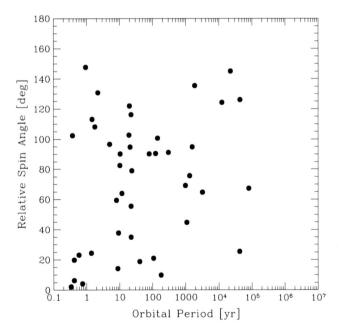

Fig. 5.3 The angle between the rotation axes of the sink particles within the binaries formed in the simulations of Bate (2009a) as a function of orbital period. Figure from Bate (2009a)

discussing here; moreover observations are based on disc-bearing systems with an average age (~few Myr) which exceeds the age at the end of these simulations by about a factor ten. Since observed systems are not perfectly aligned, then they must therefore have originated from systems that were more misaligned at birth. The simulation results are probably therefore broadly compatible with observations, although the constraints are not strong.

5.1.3 Predictions for Higher-Order Multiples

So far we have focused on the results for binary star systems, but—given the importance of few-body groupings in the evolution of the simulations—it is unsurprising that they yield a significant number of higher-order multiples. At the end of the simulation the incidence of triples (for solar-mass primaries) is around 15 %, with a similarly high fraction of higher-order multiples (quadruples and greater). These numbers are significantly larger than the observed numbers (8 and 3 %) respectively (see Raghavan et al. 2010) although these figures may be somewhat incomplete (Reipurth et al. 2014). Again, it is important to bear in mind that there would be some long timescale orbital reconfiguration within the multiples produced by the simulations and that this would lead to a secular decrease in the number of higher-order systems (see Delgado-Donate et al. 2003). This effect is evident in the study of Moeckel and Bate (2010) which involved the N-body integration of the stellar systems produced in the hydrodynamical simulations of Bate (2009a); after 10 Myr of evolution, however, the results are still not quantitatively consistent with the Raghavan et al. (2010) statistics (in particular because the main evolutionary effect is a decay of quadruples into triples, while leaving the total number of triples plus quadruples roughly constant over that timescale). Either further orbital evolution is required or else the systems produced by the simulations are over-abundant in triples/quadruples.

Having already discussed the degree of disc alignment within binary systems, we now turn to the relative alignment of orbital planes within higher-order multiples. Here the simulations and observations tell a consistent story: the simulated pairs show a mild alignment and also a mild (anti-)correlation between the mean alignment and the period ratio of the inner and outer pairs (i.e. misaligned systems are more common where the scales of the two systems are very different; see Sterzik and Tokovinin 2002). As in the case of the mutual alignment of discs within binaries, the simulated multiples also avoid the regime of strongly misaligned planes (i.e. offsets exceeding 140°).

In summary then, these simplest 'vanilla' simulations do a remarkably good job at replicating a broad sweep of binary and multiple star statistics. They are by no means perfect, but that is hardly to be expected, given the numerical issues discussed above, not to mention the omission of a number of physical ingredients (see below). In the main, discrepancies with data are quantitative rather than qualitative and their significance is compromised by less than perfect observational statistics. A possible area where there may be a *qualitative* discrepancy is in the case of the dependence of the mass ratio distribution on primary mass: there is no good evidence in the simulations for the strong tendency towards equal mass pairs observed among VLM stars.

5.2 The Effect of 'Additional Physics' on Multiple Star Formation

Turning now to the effect of additional physical processes, the problem here is one of statistics: simulations that include, for example, radiative or mechanical feedback or magnetic fields are inevitably more expensive and this has delayed the sort of detailed comparison with observations described above. For example, the simulations of Bate (2009b) that include radiative feedback form only 13 multiple systems; the similar simulation of Krumholz et al. (2012) is more populous but the minimum grid scale of 23 au prevents the resolution of hard binaries in this case. On the scale of small (few-body) cores, the most comprehensive studies of binarity in the regime of radiative feedback are those of Offner et al. (2009, 2010): these found that feedback reduced the number of multiples formed compared with barotropic simulations and that the main formation mode switched from disc fragmentation to turbulent fragmentation. More recently, Bate (2012) has conducted a large-scale simulation that is the radiative analogue of the Bate (2009a) study described above and has found that the resulting binary statistics are scarcely distinguishable from the earlier barotropic study.

On the other hand, the inclusions of magnetic fields has even raised the question of whether multiple star formation is possible in realistically magnetised cores. Hennebelle and Teyssier (2008) simulated the collapse of uniform spherical cores with uniform initial magnetic field parallel to the rotation axis. This study showed that— even in the limit of weak fields (i.e. supercritical conditions; see Chap. 1, Sect. 1.3)— there are important dynamical effects associated with the growth of toroidal fields. Even a field initially as low as 5 % of the 'critical' value was found to be sufficient to prevent fragmentation into a binary; for a field in excess of 20 % of critical, even disc formation was suppressed by the strong magnetic braking exerted by toroidal fields.

This is a clear problem since the existence of both binaries and discs is well-attested observationally; likewise the existence of magnetic fields that are substantially stronger than those employed in the simulations (see Crutcher 1999, 2012 and discussion in Chap. 1, Sect. 1.3) is indisputable. The simulations of Price and Bate (2008), by contrast, do form binaries in simulations that start off with fields that are $\sim 1/3$ of critical. Contributing factors may include differences in the field topology of collapsing cores since the fields are in this case frozen in to the large-scale turbulent motions of the parent cloud. A further difference is that, in contrast to the rather smooth initial conditions considered by Hennebelle and Teyssier (2008), the turbulent simulations of Price and Bate (2008) develop larger amplitude density fluctuations which can favour binary formation. Subsequently, several groups have examined in more detail under what conditions magnetic braking suppresses disc formation: it would seem that magnetic braking was particularly effective in the simulations of Hennebelle and Teyssier (2008) owing to their initial conditions (i.e. field aligned with the rotation axis and no turbulent motions). Joos et al. (2012) demonstrated that disc formation is not inhibited if the field is misaligned with respect to the rotation axis of the core; moreover Seifried et al. (2012) instead emphasised that turbulent motions in the environment external to the nascent disc suppress the development of toroidal fields and associated magnetic braking. Both of these effects are likely to be operative in the 'full cloud' simulations of Price and Bate (2008) and so it is perhaps unsurprising that discs do form in these simulations and that these discs can go on to form binary companions. For further explorations of the role of magnetic fields in binary star formation see Machida et al. (2008), Kudoh and Basu (2008, 2011), Boss (2009), and Commerçon et al. (2010). In conclusion, although it is premature to draw any *statistical* conclusions about the properties of binaries formed in MHD simulations, it would at least appear *likely* that clouds with a realistic degree of magnetic support should be able to form a realistic population of binary stars.

5.3 Summary

It is sometimes said that the statistics of multiple stars represent the most exacting dataset with which the results of numerical simulations can be compared and indeed the rich diversity of properties (degree of multiplicity, multiple star fraction, mass ratio distribution, period distribution, etc.) means that there is much more data to be matched than if one instead merely concentrated on matching the IMF. Having said this, it is perhaps surprising that relatively crude simulations (i.e. the 'vanilla' simulations that use a simple parametrised barotropic equation of state and which omit magnetic fields and feedback) nevertheless do an excellent job at reproducing multiple star data. In this chapter we have reviewed the match between the output of cluster formation simulations and the observed properties of binaries and higher order multiples, mainly emphasising the results of the most populous high-resolution simulation conducted to date (i.e. that of Bate 2009a). We also discussed simulations that incorporate additional physical processes and showed that these also form binaries, despite some counter-indications from the first idealised simulations including magnetic fields. We however emphasise that these more complex calculations do not currently produce a large enough sample of stars to enable a proper statistical comparison with observational data.

References

Albrecht, S., Reffert, S., Snellen, I. A. G., & Winn, J. N. 2009, Nature, 461, 373
Arreaga-García, G., Klapp-Escribano, J., & Gómez-Ramírez, F. 2010, A&A, 509, 96
Artymowicz, P. & Lubow, S. H. 1994, ApJ, 421, 651
Artymowicz, P. & Lubow, S. H. 1996, ApJ, 467, L77

Bate, M. R. 2000, MNRAS, 314, 33
Bate, M. R. 2009a, MNRAS, 392, 590
Bate, M. R. 2009b, MNRAS, 392, 1363
Bate, M. R. 2012, MNRAS, 419, 3115
Bate, M. R. & Bonnell, I. A. 1997, MNRAS, 285, 33
Boss, A. P. 2009, ApJ, 697, 1940
Close, L. M., Siegler, N., Freed, M., & Biller, B. 2003, ApJ, 587, 407
Commerçon, B., Hennebelle, P., Audit, E., Chabrier, G., & Teyssier, R. 2010, A&A, 510, L3
Crutcher, R. M. 1999, ApJ, 520, 706
Crutcher, R. M. 2012, ARA&A, 50, 29
Delgado-Donate, E. J., Clarke, C. J., & Bate, M. R. 2003, MNRAS, 342, 926
Delgado-Donate, E. J., Clarke, C. J., Bate, M. R., & Hodgkin, S. T. 2004, MNRAS, 351, 617
Duquennoy, A. & Mayor, M. 1991, A&A, 248, 485
Fabrycky, D. & Tremaine, S. 2007, ApJ, 669, 1298
Facchini, S., Lodato, G., & Price, D. J. 2013, MNRAS, 433, 2142
Foucart, F. & Lai, D. 2013, ApJ, 764, 106
Goodwin, S. P., Whitworth, A. P., & Ward-Thompson, D. 2004a, A&A, 414, 633
Goodwin, S. P., Whitworth, A. P., & Ward-Thompson, D. 2004b, A&A, 423, 169
Goodwin, S. P., Whitworth, A. P., & Ward-Thompson, D. 2006, A&A, 452, 487
Hale, A. 1994, AJ, 107, 306
Hanawa, T., Ochi, Y., & Ando, K. 2010, ApJ, 708, 485
Hennebelle, P. & Teyssier, R. 2008, A&A, 477, 25
Jensen, E. L. N., Mathieu, R. D., Donar, A. X., & Dullighan, A. 2004, ApJ, 600, 789
Joos, M., Hennebelle, P., & Ciardi, A. 2012, A&A, 543, 128
Konopacky, Q. M., Ghez, A. M., Barman, T. S., et al. 2010, ApJ, 711, 1087
Krumholz, M. R., Klein, R. I., & McKee, C. F. 2012, ApJ, 754, 71
Kudoh, T. & Basu, S. 2008, ApJ, 679, L97
Kudoh, T. & Basu, S. 2011, ApJ, 728, 123
Machida, M. N. 2008, ApJ, 682, L1
Machida, M. N., Tomisaka, K., Matsumoto, T., & Inutsuka, S.-I. 2008, ApJ, 677, 327
Mathieu, R. D., Stassun, K., Basri, G., et al. 1997, AJ, 113, 1841
McDonald, J. M. & Clarke, C. J. 1993, MNRAS, 262, 800
McDonald, J. M. & Clarke, C. J. 1995, MNRAS, 275, 671
Moeckel, N. & Bate, M. R. 2010, MNRAS, 404, 721
Monin, J.-L., Clarke, C. J., Prato, L., & McCabe, C. 2007, Protostars and Planets V, ed. B. Reipurth,
    D. Jewitt & K. Keil, University of Arizona Press, 395
Ochi, Y., Sugimoto, K., & Hanawa, T. 2005, ApJ, 623, 922
Offner, S. S. R., Klein, R. I., & McKee, C. F. 2008, ApJ, 686, 1174
Offner, S. S. R., Hansen, C. E., & Krumholz, M. R. 2009, ApJ, 704, L124
Offner, S. S. R., Kratter, K. M., Matzner, C. D., Krumholz, M. R., & Klein, R. I. 2010, ApJ, 725,
    1485
Price, D. J. & Bate, M. R. 2008, MNRAS, 385, 1820
Raghavan, D., McAlister, H. A., Henry, T. J., et al. 2010, ApJS, 190, 1
Ratzka, T., Schegerer, A. A., Leinert, C., et al. 2009, A&A, 502, 623
Reid, I. N., Lewitus, E., Allen, P. R., Cruz, K. L., & Burgasser, A. J. 2006, AJ, 132, 891
Reipurth, B., Clarke, C. J., Boss, A. P., et al. 2014, Protostars and Planets VI, ed. H. Beuther, R. S.
    Klessen, C. P. Dullemond & T. Henning, University of Arizona Press, 267
Seifried, D., Banerjee, R., Pudritz, R. E., & Klessen, R. S. 2012, MNRAS, 423, L40
Siegler, N., Close, L. M., Cruz, K. L., Martín, E. L., & Reid, I. N. 2005, ApJ, 621, 1023
Skemer, A. J., Close, L. M., Hinz, P. M., et al. 2008, ApJ, 676, 1082
Stapelfeldt, K. R., Krist, J. E., Ménard, F., et al. 1998, ApJ, 502, L65
Sterzik, M. F. & Tokovinin, A. A. 2002, A&A, 384, 1030
Walch, S., Naab, T., Whitworth, A., Burkert, A., & Gritschneder, M. 2010, MNRAS, 402, 2253

Wolf, S., Stecklum, B., & Henning, T. 2001, in IAU Symposium, Vol. 200, The Formation of Binary Stars, ed. H. Zinnecker & R. D. Mathieu, Cambridge University Press, 295

Young, M. D., Baird, J. T., & Clarke, C. J. 2015, MNRAS, 447, 2907

Chapter 6
The Role of N-body Dynamics in Early Cluster Evolution

Cathie J. Clarke

So far, we have discussed simulations that aim for the greatest realism through the modelling of hydrodynamical (and magneto-hydrodynamical) processes. These are limited in scope by their computational expense which restricts the time frame of simulations: even the most expensive simulations do not pursue cluster evolution much beyond ~ 0.5 Myr and the systems still contain substantial gas at this stage. Moreover the gravitational dynamics of the 'stars' (i.e. sinks) produced in such simulations is treated with low accuracy, with the gravitational smoothing being usually set by the gas particle smoothing length (in SPH) or by the minimum mesh scale (in AMR). Consequently there are dynamical effects associated with strongly focussed gravitational encounters that are simply missed in such simulations.

In order to follow cluster evolution for longer (e.g. over the ~ 10 Myr age range associated with 'young' pre-main-sequence stellar clusters) a number of studies have pursued the problem as an essentially stellar dynamical problem, incorporating gas, if at all, in terms of a prescribed gravitational potential. This is clearly a compromise, but has proved fruitful in pointing to some of the dynamical issues involved in young cluster evolution. It is these that we now discuss in turn.

6.1 Mass Segregation

As discussed in Chap. 3, a mass segregated system is one in which the spatial distribution of stars is a function of stellar mass. In normal usage it usually implies a situation where more massive stars are more concentrated and low-mass stars more dispersed; 'inverse mass segregation' has been applied to describe the opposite case

C.J. Clarke (✉)
Institute for Astronomy, University of Cambridge, Cambridge, UK
e-mail: cclarke@ast.cam.ac.uk

© Springer-Verlag Berlin Heidelberg 2015
C.P.M. Bell et al. (eds.), *Dynamics of Young Star Clusters and Associations*,
Saas-Fee Advanced Course 42, DOI 10.1007/978-3-662-47290-3_6

(Parker et al. 2011). Many star-forming regions show evidence for mass segregation at young ages: for example the Orion Nebula Cluster (ONC; see below) as well as the sub-clusters within Taurus (Kirk and Myers 2011).

Mass segregation may be primordial or may result from the effects of two-body relaxation (see also Mathieu Chap. 10, Sect. 10.4). The cumulative effect of the gravitational deflections induced by passing stars is that the system is driven towards a state of 'energy equipartition' (i.e. a state where the mean kinetic energy per star is independent of mass). In a self-gravitating system, the trend towards a lower velocity dispersion among massive stars has implications for their spatial distribution as slower stars tend to sink in the cluster potential. Thus two-body relaxation represents a well-studied route towards concentrating massive stars in the centres of clusters. Note that the timescale for this segregation is related to the relaxation timescale for the most massive stars: this timescale scales inversely with stellar mass and thus high-mass stars may segregate long before relaxation effects are apparent in the general stellar population.

Nevertheless, it is not clear that this timescale is short enough to explain the state of mass segregation seen in some very young clusters: in the case of the ONC, Bonnell and Davies (1998) concluded that their simulations were unable to concentrate the OB stars in the cluster core if they were initially randomly placed within the cluster. However, this conclusion may not apply if the ONC originated from cold and clumpy initial conditions as in this case cluster evolution is not only driven by two-body relaxation effects but also by so-called 'violent relaxation' (see Mathieu Chap. 10, Sect. 10.3). This latter denotes the situation when a gravitating system is started in a state that is strongly out of virial equilibrium. For example, a cluster that is highly sub-virial (i.e. dynamically cold) starts its evolution by undergoing a radial collapse in which the magnitudes of both potential and kinetic energy increase (subject, of course, to conservation of total energy). A completely cold, completely smooth spherical cluster would simply collapse to a singularity at the origin. A number of effects however act to generate a finite tangential velocity dispersion so that individual stellar orbits turn around at finite radius: the system 'bounces' and re-expands. Aarseth et al. (1988) showed that the growth of tangential velocities is mainly controlled by a collisionless fragmentation mode which is seeded by the statistical fluctuations in the initial stellar distribution; this process implies that a smooth, cold cluster collapses by a factor $\sim N^{-1/3}$ before it bounces. The bounce involves a large-scale redistribution of stellar orbits within phase space and is *not* driven by two-body relaxation but by the response of stellar orbits to the large-scale modes that develop during the collapse. The cluster then rebounds into a state of rough virial equilibrium.

What are the implications of such an evolutionary sequence for the development of mass segregation? Violent relaxation is (in ideal form) a process that is blind to stellar mass, since stars behave like test particles that respond to large-scale variations in the potential. In practice, however, violent relaxation does not always achieve a state of perfect phase mixing, particularly in the case of clumpy systems (van Albada 1982). This means that systems can retain a memory of structure and mass-positional correlations in the initial conditions. In particular a number of studies have now shown that very rapid mass segregation can result in the case of clumpy initial conditions (Allison et al. 2009, 2010; Fellhauer et al. 2009; McMillan and Vesperini 2007): the process of clump merger that occurs at the bounce tends to preferentially deposit massive stars into the core of the resulting cluster, an effect that is also seen in hydrodynamical simulations (Maschberger and Clarke 2011; Maschberger et al. 2010).

6.2 The Destruction of Binaries in Clusters

It is well-known, following Heggie (1975) and Hills (1975), that the long-term survival of binaries in clusters depends on the ratio of their internal orbital velocities to the velocity dispersion of the parent cluster. Binaries for which this ratio is greater than unity are termed 'hard'; they are tightly bound and robust against disruption by interactions with cluster members (in fact, the separation at which binaries are unlikely to be disrupted is at around 25 % of the hard-soft borderline value; see Parker and Goodwin 2012). Soft binaries are however vulnerable to disruption on a timescale that depends on the binary separation and cluster parameters. The evolution of binaries in clusters is summed up in Heggie's Law: 'hard binaries get harder, soft binaries get softer'. In old dense systems, such as globular clusters, permanent soft binaries (as opposed to temporary, weakly bound pairs) are absent while hard binaries are progressively hardened (i.e. driven into an increasingly tightly bound configuration). This hardening transfers energy into the cluster field population and drives expansion of clusters' outer regions. For an overview of the dynamics of binaries in clusters see Heggie and Hut (2003).

Kroupa et al. (2001), see also Kroupa et al. (1999, 2003); Parker et al. (2009); Marks and Kroupa (2011, 2012), conducted a series of N-body experiments designed to test the destruction of binaries within the context of *young* clusters, often using the ONC as a template of a young dense star cluster where dynamical processing is likely to be important. As anticipated, the cluster environment leads to a net decrease in binary fraction, with the largest depletion being experienced in the case of the widest pairs (and also in lower mass binary systems). Kroupa et al. (2001) argued that by comparing the statistics of pre-main-sequence binaries in different environments it is possible to place constraints on the degree of dynamical processing that the binaries have already undergone; this can then be used to reconstruct the previous dynamical history of the cluster. For example, the lower binary fraction in the ONC compared with the more diffuse environment of Taurus-Auriga (Köhler et al. 2006; Reipurth et al. 2007) can be interpreted as evidence that the ONC started from extremely compact initial conditions (see also Becker et al. 2013 for a similar inference of a high initial density in η Cha). Recent work has also demonstrated that it is not only the initial density that controls the amount of binary processing but also the degree of substructure; Parker et al. (2011) showed a more marked decline in binary fraction in the case of fractal (as opposed to smooth, centrally concentrated) initial conditions.

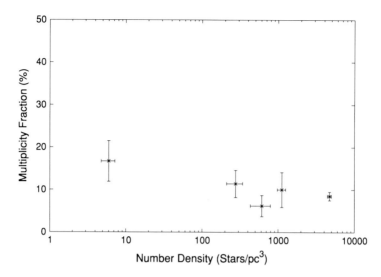

Fig. 6.1 The multiplicity fraction in the separation range 62–620 au for a range of star-forming regions: in order of ascending density (*left* to *right*) Taurus, Chamaeleon I, Ophiuchus, IC 348, and the ONC. Figure from King et al. (2012)

In principle, the detailed predictions of dynamical processing models as a function of separation can be tested against observations of binary populations in star-forming regions of various densities (see Mathieu Chap. 14, Sect. 14.2). This exercise is however non-trivial because a fair comparison needs to be restricted to separations where chance alignments can be excluded and where measurements in all the regions compared are sensitive to companions down to a given flux ratio. To date this has only allowed a comparison within the separation range 62–620 au. and the results have been surprising (see Fig. 6.1): within this range there is very little difference in the binary fraction between star-forming regions in which the stellar density varies by three orders of magnitude (King et al. 2012). Moreover the separation distributions within this range are indistinguishable in all the regions studied and are remarkably similar to the field, differing if at all, only through a mild excess in *closer* pairs (King et al. 2012). These results are apparently at odds with the expectations of dynamical processing where it is instead expected that wider pairs should be progressively depleted. Clearly this issue needs to be further investigated with studies that probe a wider separations range in all the regions compared.

There has also been some interest in exploring how dynamical processing affects very-low-mass (VLM) binaries (i.e. those with system masses $<0.1\,M_\odot$). Since VLM pairs are more easily disrupted than more massive binaries, it may be possible to explain the difference in semi-major axis distribution between VLM and higher-mass systems (i.e. the smaller separations in the VLM case) as being, at least in part, a result of dynamical processing (see Parker and Goodwin 2011). However, Parker and Reggiani (2013) found that the disruption process is rather insensitive to mass ratio (because the energy of destructive encounters is almost always significantly greater than the binding energy of the binary concerned); therefore the preference for more equal-mass companions among VLM binaries *cannot* be ascribed to dynamical processing alone. Note that it is sometimes claimed that there is a discontinuous change in binary properties in the vicinity of the hydrogen burning mass limit (Thies and Kroupa 2007) and it is certainly true that the separation distribution of VLM pairs is very different from that among M-dwarfs (Fischer and Marcy 1992). However, it needs to be borne in mind that M-dwarfs comprise a broad dynamic range of stellar masses. Bergfors et al. (2010) and Janson et al. (2012) have investigated whether there is any evidence for a trend of binary properties with spectral sub-type within the M spectral class but with indeterminate results.

We now turn to the possible *creation* of binaries in clusters, for which the usual creation mechanism is 'three-body capture' (i.e. the interaction between three mutually unbound stars which produces a bound pair as a result of energy transfer to the third body). Such energy exchange requires that the objects undergo a gravitationally focused interaction (i.e. for relative speed v and stellar mass m, all three stars need to approach within a distance $\sim Gm/v^2$). Unsurprisingly, therefore, such three-body captures tend to occur in the dense inner regions of clusters. Once formed (with a separation that is close to the hard-soft borderline), such pairs are successively hardened in accordance with Heggie's law and in the process act as a 'heat source' in the cluster core.

A more surprising type of binary formation in clusters was recently noted by Kouwenhoven et al. (2010) and Moeckel and Bate (2010) in their simulations of clusters which are in a state of expansion (either due to purely N-body relaxation effects or else as a result of being gravitationally unbound as a consequence of gas-loss). In both cases, it was noted that a population of very-soft pairs (with separations of order 10^4 au or more) were formed in the *outer* parts of the cluster (see Fig. 6.2), in apparent contradiction of Heggie's law. These pairs are initially formed from chance juxtapositions of stars with low relative velocity; in the case of non-expanding clusters, such pairs are rapidly disrupted, being very soft. However, because disruption takes a finite time, it turns out that—in the outer parts of expanding clusters—the density declines too fast for such pairs to be disrupted. Moeckel and Clarke (2011) showed that this mechanism can be expected to produce of order one pair per decade of separation per cluster. If this is the main production route of ultra-wide binaries then their incidence (at the $\sim 1\%$ level) can then be used to constrain the typical membership number (N) of the clusters whose disruption dominates the field population. This would then require that a 'typical' natal cluster numbers ~ 100 stars. This number fits in well with estimates based on observations of local star-forming regions (Lada and Lada 2003) which indicate that 'typical' embedded clusters indeed number hundreds of stars.

It is worth noting that this (cluster-based) mechanism for creating wide binaries is not strongly dependent on the masses of the stars nor on whether they are themselves tight binaries. In principle, therefore, this mechanism can be distinguished from models (Reipurth and Mikkola 2012) which involve the orbital reconfiguration of triple systems and in which one of the wide binary components must be a close binary. The study of Law et al. (2010) provided some support for the cluster dissolution

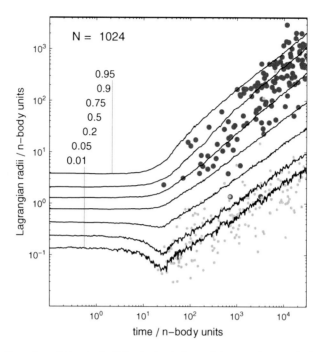

Fig. 6.2 The formation of binaries in N-body simulations of clusters whose evolution is driven by two-body relaxation: each binary that exists at the end of the simulation is plotted at its moment of creation and according to its distance from the cluster centre at that point. The *lines* represent Lagrange radii (i.e. radii enclosing a fixed fraction of the total mass of the cluster). The *grey* points represent binaries formed through three-body capture in the cluster core while the *red* points are the 'permanent' wide binaries formed in the outer regions of the cluster as described in the text. Note that the binaries plotted result from 48 random realisations of the same initial conditions. Figure from Moeckel and Clarke (2011)

mechanism in that it showed that the primaries of very-wide pairs are themselves no more likely to be binaries than in the case of isolated field stars. On the other hand, both mechanisms would apparently struggle to reproduce the observed (flat) distribution of mass ratios (q) in ultra-wide pairs (Tokovinin and Smekhov 2002): in the triple reconfiguration mechanism, there is a tendency for the outlying star to be of lower mass (Delgado-Donate et al. 2004) whereas the cluster dissolution mechanism also predicts a distribution that is rising towards low q, in rough accord with the random pairing hypothesis. The origin of very-wide pairs with almost equal mass components is thus not readily explained at present.

6.3 Stellar Dynamics Plus Gas: Stellar Collisions

Gas modifies stellar dynamics in a variety of ways. We start by examining the effect of gas addition—i.e. accretion onto individual stars in a cluster. In the case that the gas originates from outside the stellar cluster, has zero momentum and is accreted *slowly* (i.e. on a greater than dynamical timescale) then it can be shown that the cluster responds *adiabatically* and shrinks such that its radius R scales with mass M according to $R \propto M^{-3}$. This implies that the stellar density rises very steeply with M, i.e. $\rho \propto M^{10}$. Bonnell et al. (1998) suggested that this could in principle lead to such high densities that stars would collide.

Moeckel and Clarke (2011) conducted N-body simulations in which mass was added prescriptively to the stars at a constant rate, while a background potential (ostensibly representing the effect of distributed gas) was correspondingly reduced (see also Baumgardt and Klessen 2011; Bonnell and Bate 2002; Davis et al. 2010). After 1 Myr of evolution, the gas was instantaneously removed, to mimic the onset of stellar feedback (see below). The cluster evolution has three characteristic phases. Prior to gas expulsion, the cluster contracts homologously in response to adiabatic accretion; following gas expulsion, the loss of the gas potential increases the role of two-body relaxation effects and the cluster undergoes *core collapse* shortly thereafter. Core collapse is a process that is well-studied in the context of purely stellar dynamical cluster modelling (Gürkan et al. 2004) and is a consequence of the outward transport of energy from the inner regions of the cluster by two-body relaxation. In these simulations, where gas accretion has brought the stars into a very compact configuration, the two-body relaxation timescale is short at the point of gas expulsion and so the cluster goes into core collapse very soon afterwards. The density attained in the cluster core at this point is limited by the effect of three-body capture binaries (see above), which provide an energy source for re-inflation of the cluster core. Thereafter the entire cluster undergoes a slow self-similar expansion which is driven by energy extracted from binaries in the cluster core.

Although the simulations show that the high densities attained at core collapse can initiate a chain of successive stellar collisions, the quantitative results are strongly dependent on the simplifying assumptions involved in modelling the outcome of collisions. Nevertheless, the simulations do yield some useful qualitative insights. Firstly, one of the main channels for stellar collisions is found to be from the hardening of massive binaries in the cluster core. Secondly, once initiated, the collisional process in the core tends to run away in the sense that the same stars are involved in a number of successive collisions. This means that the imprint of stellar collisions on the IMF is likely to involve the creation of a single product of multiple collisions which ends up much more massive than the other stars in the cluster. In principle such a mechanism could lead to the production of an intermediate-mass black hole in the cluster core; it does *not* however provide a viable mechanism for creating a range of stellar masses with which to populate the upper IMF. It therefore seems unlikely that stellar collisions are a primary route for massive star formation, as originally suggested by Bonnell et al. (1998).

There are a few additional points to note about the possible role of stellar collisions in *young* clusters (note that on long timescales, there is a finite probability of collisions at much lower densities and it is well established that occasional collisions in globular clusters are manifest as 'blue stragglers', see Bailyn 1995; Lanzoni et al. 2007; Perets and Fabrycky 2009). The attainment of the necessary high densities in the cluster core within the first 1 Myr of a cluster's life requires the system to be first shrunk by gas accretion and then to undergo stellar dynamical core collapse. The depth of core collapse however increases with the number of stars in the cluster (N): at small N, three-body capture binaries—which reverse the collapse—are formed at lower densities. Such arguments suggest that—if stellar collisions are important anywhere—it is likely to be in *populous* clusters ($N > 10^4$; Clarke and Bonnell 2008; Davis et al. 2010). Moeckel and Clarke (2011) showed that collisions are very unlikely in clusters on the scale of the ONC ($N \sim 10^3$) but may arguably play a limited role in clusters such as the Arches. However, even in the Arches, the age and current stellar density are such that no evolutionary paths are consistent with such collisions having already occurred in the cluster's past (although they may do so in the *future*).

6.4 Stellar Dynamics with Gas Removal: Infant Mortality

There is a simple dynamical argument that can be used to assess the effect of instantaneous mass-loss from a cluster (see also Mathieu Chap. 10, Sect. 10.7). Consider a cluster (containing a mixture of gas and stars) that is originally in virial equilibrium, i.e. where the mean kinetic energy per star (T) and the mean potential energy per star (W) are related by $2T + W = 0$. Now consider the loss of gas from the cluster which lowers the mass of the cluster to a factor ϵ times its original value. This reduces the mean potential energy per star by a factor ϵ also, so that the total mean energy per star is now $T + \epsilon W = (\epsilon - 0.5)W$ where the latter equality follows from the initially virialised state of the cluster. Since $W < 0$ it follows that the mean energy per star is positive (i.e. the cluster is unbound) if $\epsilon < 0.5$, i.e. if more than 50 % of the cluster mass is lost (Hills 1980). (Note that this derivation assumes instantaneous mass removal, or at least, gas removal on a timescale that is much less than the cluster dynamical time. In the opposite limit the mass-loss is adiabatic and the cluster expands but remains bound because there is time for it to revirialise as the cluster responds to mass loss.)

The above simple arguments have been broadly confirmed by a range of numerical simulations which model instantaneous gas expulsion starting from a state of initial virial equilibrium with gas and stars well mixed (Boily and Kroupa 2003a, b; Lada et al. 1984). Although the analytic argument must hold for *average* quantities, it does not reflect the fact that in a simulation the stars respond differently to mass-loss depending on their initial location in the cluster. Thus *some* stars are lost from the outer regions of the cluster even when $\epsilon > 0.5$ while *some* stars remain in a bound core even for $\epsilon < 0.5$ (see Adams 2000; Baumgardt and Kroupa 2007; Boily and Kroupa 2003a, b; Goodwin 1997). This is something of a detail, however, since the fraction of stars retained falls very steeply with declining ϵ, reaching $\sim 10\%$

for $\epsilon = 0.3$ and dropping to zero for $\epsilon = 0.2$ (see Fig. 6.3). As expected, higher bound fractions are obtained at given ϵ if the gas is expelled slowly or if the stellar population is initially sub-virial (Geyer and Burkert 2001). It however requires some fine tuning for gas-loss to occur when the stars are significantly sub-virial: a sub-virial stellar population should virialise on a dynamical timescale and it is hard to argue that feedback—which requires the output of energy and momentum from star formation—should be effective on less than a dynamical time.

The above picture suggests that clusters are likely to be unbound by gas-loss unless star formation is locally efficient, with more than half the gas mass going into stars. Star formation efficiencies of 50 % or more are far higher than the values inferred on the scale of entire molecular clouds (Evans et al. 2009); see discussion in Chap. 1, Sect. 1.3). On the other hand, it is often argued that such a high threshold in star formation efficiency is consistent with the fact that many clusters do *not* survive. This is the argument for 'cluster infant mortality': most stars are 'clustered' within star-forming regions and yet by an age of 10 Myr only about 10 % of stars are in clusters. Apparently, the remaining 90 % of clusters dissolve within the first 10

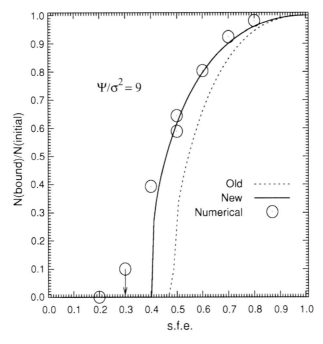

Fig. 6.3 The fraction of stars remaining bound as a function of the initial star to total mass ratio assuming well-mixed gas and stars and instantaneous gas expulsion. The *open circles* represent the result of *N*-body simulations while the *solid* and *dashed lines* are the results of applying semi-analytic algorithms: for details see Boily and Kroupa (2003a). Figure from Boily and Kroupa (2003b)

Myr of their lives and it has become standard to invoke low star formation efficiencies combined with gas-loss in order to explain this phenomenon (Lada and Lada 2003).

Various pieces of observational evidence have been claimed in support of this scenario. Direct confirmation of cluster dissolution would derive from measuring high ratios of kinetic energy to potential energy (i.e. low values of the ratio of the cluster mass to the dynamical mass required for virial equilibrium). Goodwin and Bastian (2006) drew attention to just such an effect in the case of a number of clusters at an age of ~10 Myr. However, the subsequent analysis of Gieles et al. (2010) has shown that the high velocity dispersion measured in these clusters is likely to be due to the contamination of the radial velocity signal by red supergiant binaries which contribute strongly at such ages. Other recent analyses of populous young clusters suggest that they are remarkably close to virial equilibrium (Cottaar et al. 2012; Kouwenhoven and Grijs 2008), raising the obvious question of why we are not observing—in addition to such bound examples—a much more numerous population of dissolving (unbound) clusters. On the other hand, there are features in the light profiles of some clusters that are well matched to those expected in dissolving clusters: Bastian and Goodwin (2006) showed that simulations of unbound clusters develop 'shoulders' in the cluster light profile which are consistent with those seen in several extragalactic clusters.

Whatever the observational situation regarding the boundedness of clusters, the requirement of $\epsilon > 0.5$ is often regarded as something of an obstacle to cluster survival. However, it is worth recalling that a key assumption in deriving this criterion is that *stars and gas are well-mixed*. If one relaxes this assumption and segregates the stars at small radii with respect to the gas then the effect of gas-loss is much less severe: gas exterior to the stars does not contribute to the gravitational force experienced by the stars and they are therefore unaffected by its removal. Several idealised simulations have achieved such segregation of the stars with respect to the gas by allowing the stellar distribution to shrink—either by accretion (Moeckel and Clarke 2011) or by cold collapse (Smith et al. 2011)—while artificially not allowing the gas potential to follow. Unsurprisingly, subsequent removal of the extended gas potential does not unbind the cluster.

What is more interesting is that a similar effect appears to be operating in hydro-dynamic simulations which do not hold the gas potential fixed but which follow the interplay of stellar dynamics and gas dynamics. We have already noted the finding (Girichidis et al. 2012; Kruijssen et al. 2012; Moeckel and Bate 2010, see Chap. 3, Sect. 3.3) that the clusters within hydrodynamical simulations are locally gas-poor. It is however currently unclear whether this property is shared by real clusters: it is in fact remarkably hard to distinguish the situation where clusters are embedded in three dimensions from that in which they are merely embedded in projection.

In the case of clusters created in hydrodynamical simulations it is of course possible to explore the future evolution of the system following instantaneous expulsion of the residual gas. We have already discussed how such an exercise (performed in the case of the star cluster formed in Bate 2009 at a point when the gas removed constituted ~60 % of the mass of the system) led to the dynamical creation of ultra-wide binaries (Moeckel and Bate 2010) in the expanding halo of unbound stars.

Moeckel et al. (2012) conducted a similar experiment using the simulation of Bonnell et al. (2008) as a starting point and thus investigated the evolution of an ensemble of smallish-N clusters (typically numbering hundreds of stars) subject to gas-loss. Since (as pointed out by Kruijssen et al. 2012) these clusters are known to be internally gas-poor, it might be expected that removal of the gas between the clusters (which comprised around 85 % of the system mass just prior to gas expulsion) would leave the clusters individually intact. In fact, these clusters expand dramatically over 10 Myr; their expansion is however driven not by gas expulsion but by two-body relaxation, since the latter is relatively rapid in such modest-N systems. Indeed Moeckel et al. (2012) demonstrated that such clusters undergo core collapse soon after gas removal, this process being accelerated by the fact that the systems created in the hydrodynamical simulations are already mass segregated. They then enter a phase of *self-similar* expansion, driven by the extraction of energy from hard binaries in the cluster core. (Here the term 'self-similar' implies that the clusters expand homologously, i.e. such that the density profile at any time represents a scaled version of its form at previous times; in this phase the two-body relaxation timescale at the half-mass radius is always of order the system age.)

This self-similar expansion implies that the expansion of clusters due to two-body relaxation effects is rather insensitive to initial conditions. A compact cluster undergoes core collapse and starts its self-similar expansion earlier than a more diffuse system but both converge on the same self-similar evolutionary path. For parameters typical of young star clusters this evolutionary convergence occurs at ~1 Myr; thereafter all clusters of given N have the same scale at a given age (see Fig. 6.4). Gieles et al. (2012) argued that the *observed* surface density distribution of stars in star forming clouds (Bressert et al. 2010) provides evidence that clusters are indeed in such a state of relaxation driven expansion: the mean surface density ($\sim 20\,\mathrm{pc}^{-2}$) is exactly what is expected in the case of self-similar expansion of systems numbering a few hundred stars at a few Myr.

Where do these recent simulations leave the issue of cluster 'infant mortality'? The paucity of gas on the scale of the stars within the simulations means that gas-loss is relatively unimportant dynamically. However this does *not* mean that clusters remain tightly bound at their initial sizes since the clusters expand due to two-body relaxation. Since two-body relaxation is an N dependent phenomenon, the small-N groupings that typify observed star-forming regions expand rather fast until they reach the point that they either merge with other clusters or are else tidally disrupted by the background potential. Thus for these systems, 'infant mortality' (i.e. disruption within the first 10 Myr or so) seems to be readily achievable by two-body effects and does *not* rely on the gas-loss which has long been held responsible. This is an important distinction because whereas the efficiency of gas-loss might depend on environment or epoch (for example, metallicity might control the efficacy of stellar feedback), N-body processes are instead independent of such factors. However, unlike gas-loss, relaxational effects *are* N dependent: although small-N clusters are rapidly dispersed, the same does not hold for their larger-N counterparts.

One might therefore try to understand the observational situation as simply showing that low-N clusters (e.g. those with $< 10^3$ members) do dissolve within ~ 10 Myr

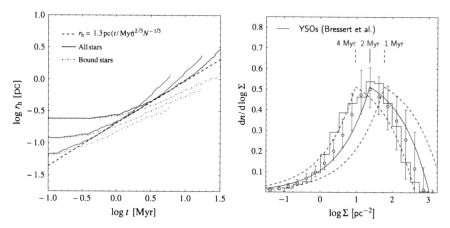

Fig. 6.4 *Left panel*: The expansion of star clusters containing $N = 256$ stars due to two-body relaxation as a function of initial radius. The *solid lines* shows the evolution of the half-mass radius for all stars in the simulation, while the *dotted lines* refer only to the stars that are gravitationally bound. The plot illustrates that the evolution of clusters of widely differing initial radii converge after about 1 Myr. *Right panel*: A Monte Carlo demonstration of the distribution of stellar surface densities that would be produced by clusters (with N in the range 50–500) evolving as shown in the *left panel*. The distribution is broadened by the range of surface densities within each cluster and the peak of the predicted distribution moves to lower values of the surface density as the population ages. The histogram shows the observed distribution from Bressert et al. (2010). Figure adapted from Gieles et al. (2012)

whereas more populous clusters do not. Given the predominance of low-N systems in censuses of nearby star forming regions (Lada and Lada 2003 this would account for the dissolution of the majority of clusters; similar evidence for cluster infant mortality is found in dwarf starburst galaxies where again the observed clusters are typically of rather low mass (Tremonti et al. 2001). On the other hand, such an N dependence for cluster survival time would also be compatible with the observation that a number of higher-N systems (Cottaar et al. 2012; Kouwenhoven and Grijs 2008) have been shown to be *bound* at ages of 10s of Myr.

To test this hypothesis further, one has to look to other galaxies which have a larger census of larger-N clusters at a range of ages: is the data compatible with a scenario in which the majority of clusters with high N ($\sim 10^4$ or above) are robust on timescales of 10 Myr and longer? This is a contentious issue (Bastian et al. 2005; Chandar et al. 2010a,b; Gieles and Lamers 2007; Maschberger and Kroupa 2011). At its simplest, the required test just involves counting the clusters in a given mass range in the logarithmic age range 10^6–10^7 years and checking whether this exceeds 10 % of the corresponding clusters with logarithmic age range 10^7–10^8 years (here the value of 10 % just reflects the shorter age range in the former bin and hence—at a constant cluster formation rate—a correspondingly smaller number of clusters).

This is however a non-trivial exercise for several reasons. Firstly, one needs a robust measure of the cluster mass (i.e. good models for the fading of clusters of constant mass due to pure stellar evolutionary effects). Such effects (in a magnitude limited sample) mean that it may be necessary to apply corrections to the numbers of clusters at higher ages due to sample incompleteness. Secondly, this test is predicated on the assumption of constant cluster formation rate; unfortunately, the galaxies that yield large numbers of clusters are also often those that exhibit starburst activity and therefore this assumption is particularly questionable. The reader is directed to the references given above in order to sample a diversity of opinions on this issue.

6.5 Summary

In this chapter we have reviewed what can be learned by calculations which focus on clusters as *N*-body systems; such studies are particularly useful for exploring processes that occur over a number of dynamical timescales and which are driven by the cumulative effect of many small-angle gravitational deflections (so-called two-body relaxation effects). These effects are important, for example, in driving cluster core collapse and subsequent re-expansion and also shape the properties of surviving binary populations through dynamical destruction of weakly bound pairs. On the other hand, recent work has shown how stable but very-wide binaries can form in the environment of an expanding cluster. We also discussed the issue of 'cluster infant mortality' through reviewing how the ultimate survival of star clusters depends on the distribution of their natal gas and how this is dispersed. We highlighted recent work which shows that protoclusters are internally gas-poor *in simulations*; if real clusters share this property then it implies that gas-loss may play a smaller role in cluster dispersal than believed hitherto.

References

Aarseth, S. J., Lin, D. N. C., & Papaloizou, J. C. B. 1988, ApJ 324, 288
Adams, F. C. 2000, ApJ 542, 964
Allison, R. J., Goodwin, S. P., Parker, R. J., et al. 2009, MNRAS, 395, 1449
Allison, R. J., Goodwin, S. P., Parker, R. J., Portegies Zwart, S. F., & de Grijs, R. 2010, MNRAS, 407, 1098
Bailyn, C. D. 1995, ARA&A, 33, 133
Bastian, N., Gieles, M., Lamers, H. J. G. L. M., Scheepmaker, R. A., & de Grijs, R. 2005, A&A, 431, 905
Bastian, N. & Goodwin, S. P. 2006, MNRAS, 369, L9
Bate, M. R. 2009, MNRAS, 392, 590
Baumgardt, H. & Klessen, R. S. 2011, MNRAS, 413, 1810
Baumgardt, H. & Kroupa, P. 2007, MNRAS, 380, 1589
Becker, C., Moraux, E., Duchêne, G., Maschberger, T., & Lawson, W. 2013, A&A, 552, 46
Bergfors, C., Brandner, W., Janson, M., et al. 2010, A&A, 520, 54
Boily, C. M. & Kroupa, P. 2003a, MNRAS, 338, 665
Boily, C. M. & Kroupa, P. 2003b, MNRAS, 338, 673
Bonnell, I. A. & Bate, M. R. 2002, MNRAS, 336, 659
Bonnell, I. A., Bate, M. R., & Zinnecker, H. 1998, MNRAS, 298, 93

Bonnell, I. A., Clark, P., & Bate, M. R. 2008, MNRAS, 389, 1556
Bonnell, I. A. & Davies, M. B. 1998, MNRAS, 295, 691
Bressert, E., Bastian, N., Gutermuth, R., et al. 2010, MNRAS, 409, L54
Chandar, R., Fall, S. M., & Whitmore, B. C. 2010a, ApJ 711, 1263
Chandar, R., Whitmore, B. C., & Fall, S. M. 2010b, ApJ 713, 1343
Clarke, C. J. & Bonnell, I. A. 2008, MNRAS, 388, 1171
Cottaar, M., Meyer, M. R., Andersen, M., & Espinoza, P. 2012, A&A, 539, 5
Davis, O., Clarke, C. J., & Freitag, M. 2010, MNRAS, 407, 381
Delgado-Donate, E. J., Clarke, C. J., Bate, M. R., & Hodgkin, S. T. 2004, MNRAS, 351, 617
Evans, II, N. J., Dunham, M. M., Jørgensen, J. K., et al. 2009, ApJS, 181, 321
Fellhauer, M., Wilkinson, M. I., & Kroupa, P. 2009, MNRAS, 397, 954
Fischer, D. A. & Marcy, G. W. 1992, ApJ 396, 178
Geyer, M. P. & Burkert, A. 2001, MNRAS, 323, 988
Gieles, M., Lamers, H. J. G. L. M., & Portegies Zwart, S. F. 2007, ApJ 668, 268
Gieles, M., Moeckel, N., & Clarke, C. J. 2012, MNRAS, 426, L11
Gieles, M., Sana, H., & Portegies Zwart, S. F. 2010, MNRAS, 402, 1750
Girichidis, P., Federrath, C., Banerjee, R., & Klessen, R. S. 2012, MNRAS, 420, 613
Goodwin, S. P. 1997, MNRAS, 284, 785
Goodwin, S. P. & Bastian, N. 2006, MNRAS, 373, 752
Gürkan, M. A., Freitag, M., & Rasio, F. A. 2004, ApJ 604, 632
Heggie, D. & Hut, P. 2003, The Gravitational Million-Body Problem: A Multidisciplinary Approach
    to Star Cluster Dynamics, ed. Heggie, D. and Hut, P., Cambridge University Press
Heggie, D. C. 1975, MNRAS, 173, 729
Hills, J. G. 1975, AJ, 80, 809
Hills, J. G. 1980, ApJ 235, 986
Janson, M., Hormuth, F., Bergfors, C., et al. 2012, ApJ 754, 44
King, R. R., Goodwin, S. P., Parker, R. J., & Patience, J. 2012a, MNRAS, 427, 2636
King, R. R., Parker, R. J., Patience, J., & Goodwin, S. P. 2012b, MNRAS, 421, 2025
Kirk, H. & Myers, P. C. 2011, ApJ 727, 64
Köhler, R., Petr-Gotzens, M. G., McCaughrean, M. J., et al. 2006, A&A, 458, 461
Kouwenhoven, M. B. N. & de Grijs, R. 2008, A&A, 480, 103
Kouwenhoven, M. B. N., Goodwin, S. P., Parker, R. J., et al. 2010, MNRAS, 404, 1835
Kroupa, P., Aarseth, S., & Hurley, J. 2001, MNRAS, 321, 699
Kroupa, P., Bouvier, J., Duchêne, G., & Moraux, E. 2003, MNRAS, 346, 354
Kroupa, P., Petr, M. G., & McCaughrean, M. J. 1999, New Astron., 4, 495
Kruijssen, J. M. D., Maschberger, T., Moeckel, N., et al. 2012, MNRAS, 419, 841
Lada, C. J. & Lada, E. A. 2003, ARA&A, 41, 57
Lada, C. J., Margulis, M., & Dearborn, D. 1984, ApJ 285, 141
Lanzoni, B., Sanna, N., Ferraro, F. R., et al. 2007, ApJ 663, 1040
Law, N. M., Dhital, S., Kraus, A., Stassun, K. G., & West, A. A. 2010, ApJ 720, 1727
Marks, M. & Kroupa, P. 2011, MNRAS, 417, 1702
Marks, M. & Kroupa, P. 2012, A&A, 543, 8
Maschberger, T. & Clarke, C. J. 2011, MNRAS, 416, 541
Maschberger, T., Clarke, C. J., Bonnell, I. A., & Kroupa, P. 2010, MNRAS, 404, 1061
Maschberger, T. & Kroupa, P. 2011, MNRAS, 411, 1495
McMillan, S. L. W., Vesperini, E., & Portegies Zwart, S. F. 2007, ApJ 655, L45
Moeckel, N. & Bate, M. R. 2010, MNRAS, 404, 721
Moeckel, N. & Clarke, C. J. 2011a, MNRAS, 410, 2799
Moeckel, N. & Clarke, C. J. 2011b, MNRAS, 415, 1179
Moeckel, N., Holland, C., Clarke, C. J., & Bonnell, I. A. 2012, MNRAS, 425, 450
Parker, R. J., Bouvier, J., Goodwin, S. P., et al. 2011a, MNRAS, 412, 2489
Parker, R. J. & Goodwin, S. P. 2011, MNRAS, 411, 891
Parker, R. J. & Goodwin, S. P. 2012, MNRAS, 424, 272

Parker, R. J., Goodwin, S. P., & Allison, R. J. 2011b, MNRAS, 418, 2565
Parker, R. J., Goodwin, S. P., Kroupa, P., & Kouwenhoven, M. B. N. 2009, MNRAS, 397, 1577
Parker, R. J. & Reggiani, M. M. 2013, MNRAS, 432, 2378
Perets, H. B. & Fabrycky, D. C. 2009, ApJ 697, 1048
Reipurth, B., Guimarães, M. M., Connelley, M. S., & Bally, J. 2007, AJ, 134, 2272
Reipurth, B. & Mikkola, S. 2012, Nature, 492, 221
Smith, R., Fellhauer, M., Goodwin, S., & Assmann, P. 2011, MNRAS, 414, 3036
Thies, I. & Kroupa, P. 2007, ApJ 671, 767
Tokovinin, A. A. & Smekhov, M. G. 2002, A&A, 382, 118
Tremonti, C. A., Calzetti, D., Leitherer, C., & Heckman, T. M. 2001, ApJ 555, 322
van Albada, T. S. 1982, MNRAS, 201, 939

Chapter 7
Concluding Issues

Cathie J. Clarke

We have now presented a survey of the results of a range of gas dynamical and stellar dynamical simulations: these model the formation of stars in clusters and trace the evolution of clusters over the first few Myr of their existence. So far we have mainly focussed our observational comparisons on the statistical properties of the stars (and multiple systems) formed within the clusters and have not attempted any detailed comparisons between simulations and individual clusters. We now turn to this issue, discussing how simulations compare with observations of the youngest gas-rich clusters. We then discuss more generically whether the properties of field stars bear the imprint of an origin in a clustered environment and then re-focus the argument by trying to assess what can be said about the birth environment of the Sun.

7.1 Modelling Individual Clusters

7.1.1 Gas-Free Studies

The most popular object for N-body studies is the Orion Nebula Cluster (ONC) since it is well-studied observationally, relatively nearby and, by the standards of clusters within 500 pc of the Sun, relatively populous (containing ~ 4000 stars within a region ~ 5 pc across). Dynamical studies that have attempted to constrain the early history and future evolution of the ONC through models that match its current properties (at an age of ~ 2 Myr) include Kroupa et al. (2001); Scally and Clarke (2001, 2002); Scally et al. (2005); Proszkow et al. (2009); Allison and Goodwin (2011).

C.J. Clarke (✉)
Institute for Astronomy, University of Cambridge, Cambridge, UK
e-mail: cclarke@ast.cam.ac.uk

© Springer-Verlag Berlin Heidelberg 2015
C.P.M. Bell et al. (eds.), *Dynamics of Young Star Clusters and Associations*,
Saas-Fee Advanced Course 42, DOI 10.1007/978-3-662-47290-3_7

These modelling attempts indicate considerable degeneracy with respect to initial conditions: because the cluster is dense (with central densities of $10^5 \, \text{pc}^{-3}$, McCaughrean and Stauffer 1994; Hillenbrand and Hartmann 1998) the associated dynamical times are short and this allows ample time for traces of initial conditions to be erased. For example, it is easy to accommodate a variety of clumpy, sub-clustered origins for the ONC despite its present day smoothness (Scally and Clarke 2002; Allison et al. 2010). It is therefore not a good testbed with which to either confirm or refute the hypothesis of hierarchical cluster assembly that is suggested by hydrodynamical simulations of cluster formation (e.g. Bonnell et al. 2008). On the other hand, it is well known from simulations of cluster merging on a larger scale (Fellhauer and Kroupa 2002) that kinematic signatures of sub-clustering are considerably more durable than traces in the spatial distribution of stars. Here however, current modelling efforts are frustrated by the lack of kinematic data (see below).

The observational situation in the ONC is that the stellar population is well characterised by the seminal studies of Hillenbrand and Hartmann (1998) as recently updated by Da Rio et al. (2012). Moreover, recent investigations (Fűrész et al. 2008; Tobin et al. 2009) have also provided a good measure of the stellar radial velocity distributions in the ONC. There are however two problems with interpreting kinematic data (see also Mathieu Chap. 13). Firstly, the only proper motion data available is that of Jones and Walker (1988). In this study, any net contraction or expansion of the cluster was subtracted from the data because of an uncertainty in the absolute plate scale between the two epochs. Secondly, it is hard to interpret the radial velocity data unambiguously. Fűrész et al. (2008) and Tobin et al. (2009) report a velocity gradient along the major axis of the cluster (the ONC is mildly flattened on the sky with an aspect ratio of 2−3:1; Hillenbrand and Hartmann 1998). Although these authors interpret their kinematic data in terms of a collapsing filament it is equally compatible with a state of expansion.

These difficulties mean that we do not currently have a good measure of the virial state of the cluster nor of whether it is expanding or contracting: this is evidently a matter that will be addressed by *Gaia* over the coming years.

Another system which has proved a fruitful object for dynamical modelling is the nearby η Cha association, which is somewhat older and considerably sparser than the ONC. At an age of 6−7 Myr it contains 18 systems within a parsec. The core of the system contains 4 stars with masses in excess of 1.5 $M_\odot$; there are apparently no stars associated with η Cha which have masses less than 0.1 $M_\odot$. This mass distribution is conspicuously top-heavy with respect to the canonical IMF (Kroupa et al. 1993) and raises the question whether such a distribution can be explained in terms of dynamical evolution: specifically, has the missing complement of brown dwarfs been ejected from the cluster by two-body relaxation? Becker et al. (2013) studied this hypothesis in detail via a suite of N-body simulations which started from a range of densities and virial states; they concluded that (assuming a normal IMF) there is *no* dynamical history that can simultaneously account for both the concentration of massive stars in the core and the observed lack of brown dwarfs. η Cha thus represents a rare case of a system in which there is good evidence for a deviant initial mass function (IMF) (i.e. one whose discrepancy cannot simply be ascribed to finite sampling effects).

7.1.2 Embedded Star-Forming Regions

We now turn to the issue of how well simulations reproduce the properties of regions that are still heavily embedded in their natal gas. Figure 7.1 compares the results of the simulation of Bonnell et al. (2008) with *Herschel* maps of Aquila by Könyves et al. (2010) and Bontemps et al. (2010). The resemblance is striking, at least superficially: both show a clustered core of stars and a further population of sources organised along filaments which (in the simulations) are in the process of infalling into the cluster core. In Aquila, the distributed population in the filaments is younger (pre-stellar); this is consistent with the simulations, where the stars in clusters are those that formed first (Maschberger et al. 2010).

One of the first embedded regions to be qualitatively compared with simulations is the core of ρ Ophiuchus. Figure 7.2 (from André et al. 2007) presents a millimetre map of the L1688 region that is colour-coded according to line-of-sight velocities of pre-stellar gas condensations derived from N_2H^+ measurements. These condensations (designated as 'MM' objects in Fig. 7.2) are organised in groupings (A–F).

Does this image bear out the predictions of hydrodynamical modelling? André et al. (2007) drew attention to the rather small global velocity dispersion of the cores in the region and used this to argue that '...the condensations do not have time to interact with one another before evolving into pre-main sequence objects'. This data

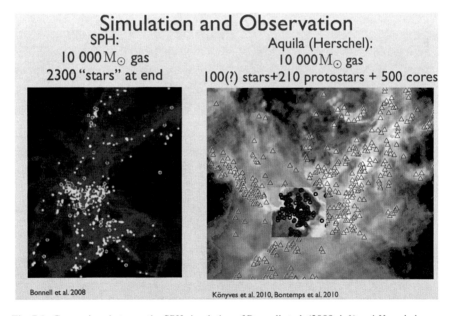

Fig. 7.1 Comparison between the SPH simulation of Bonnell et al. (2008, *left*) and *Herschel* maps of Aquila (*right*). In the *left panel* the *yellow filled circles* represent stars while the *blue filled circles* denote brown dwarfs. In the *right panel* the stars and protostars from the survey of Bontemps et al. (2010) are represented by the *red circles* in the central inset while pre-stellar cores from the survey of Könyves et al. (2010) are denoted by *blue triangles*

has thus been used to argue for a quasi-static picture of clump collapse which is apparently at odds with the dynamical picture emerging from simulations. If cores indeed lacked significant relative bulk motions and did not exhibit orbital motions in the local potential, then this would be remarkable result, raising questions about what processes could stop cores from responding to the local gravitational field.

Closer examination of the numbers however reveals a situation which, reassuringly, is broadly compatible with the simulation results. The measured one-dimensional velocity dispersion ($0.4 \, \mathrm{km \, s^{-1}}$) corresponds to a three-dimensional velocity dispersion of $0.7 \, \mathrm{km \, s^{-1}}$; this is roughly the free-fall velocity given the masses and sizes of the core groupings (labelled A–F in Fig. 7.2). Moreover the crossing timescale within such groupings is rather short (a few times 10^5 years): such cores will thus be able to traverse their natal groupings on a timescale comparable with their internal collapse times. In addition, the Ophiuchus map also provides observational support for hierarchical cluster formation as manifest in the simulations: the velocity differential ($\sim 1 \, \mathrm{km \, s^{-1}}$) between the groupings to the NW and SE is such that these may well merge on a timescale of $\sim 1 \, \mathrm{Myr}$.

The detailed comparison between simulations and the structure and kinematics of *gas* in embedded regions is still relatively in its infancy: see Offner et al. (2009) for an analysis of the relative kinematics of the gas and stars in simulations and Kirk et al. (2010) for an observational study of the relative kinematics of dense cores and distributed gas in Perseus.

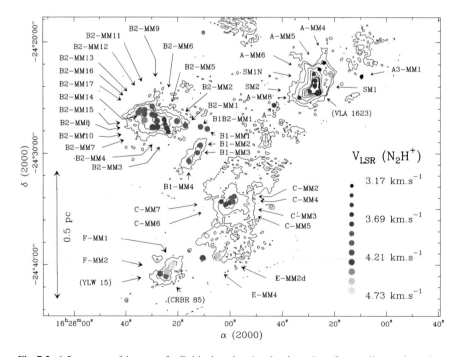

Fig. 7.2 1.2 mm map of the core of ρ Ophiuchus showing the clustering of pre-stellar condensations and their kinematic properties as traced by $N_2H^+(1–0)$ emission. Figure from André et al. (2007)

7.2 Imprint of Cluster Origin on Field Star Populations

There has been much discussion over the years as to whether there is any difference in the properties of stars that form in clusters (which may subsequently dissolve) compared with those that form in isolation. This question however has to be updated to reflect recent observational and theoretical insights. Firstly, there is considerable observational evidence that *most* stars in star-forming regions are 'clustered' in some sense (Lada and Lada 2003; Bressert et al. 2010) whatever the dynamical status of these groupings: under these circumstances it is hard to define a control sample with which the cluster population should be compared. Secondly, simulations point to cluster formation as being a hierarchical process so that stars mostly form in small-N groupings which then—depending on the environment—follow an upward progression through the cluster merger tree, being incorporated into successively larger structures (Maschberger et al. 2010; see Fig. 7.3). This implies that it is hard to define what is meant by a star 'born in a cluster'.

Instead we have to frame some more nuanced questions. These include 'Are there properties of stars (in general) which bear evidence of dynamical interactions in their early history?' as well as 'Are there properties of stars that depend on the scale of the cluster in which they at some stage find themselves situated?' Here we shall look at the latter question with regard to a possible imprint on the IMF.

It is long been noted that the maximum stellar mass within young clusters has a generally positive correlation with the cluster mass. This must at least in part be a statistical effect—i.e. if one thinks of a star formation event as drawing stars from an underlying distribution then one is more likely to select stars high up in the steep (Salpeter) tail of the distribution if one is selecting a large number of objects. The magnitude of this effect can be readily quantified (see below) in order to assess whether (for an assumed universal IMF) the statistics of maximum stellar mass versus N conform with expectations.

Weidner and Kroupa (2004, 2006) have argued that the data do *not* conform with the statistics of random drawing and argue that instead there is an additional *system-atic* dependence of maximum stellar mass on cluster mass. The sign of the claimed dependence is positive (i.e. it has the same sign as the stochastic effect described above) so the effects within individual clusters are rather subtle. Nevertheless, there are profound differences between these two hypotheses when one stacks up an inte-grated IMF (averaged over all clusters: henceforth termed the IGIMF). In the case of random drawing, the IGIMF is of course identical to the input IMF by construction. In the case of there being a systematic, cluster mass dependent upper mass limit per cluster, the effect of stacking up an ensemble of truncated power-laws is that the IGIMF can end up being steeper than the input IMF. The magnitude of this effect depends not only on the assumed relationship between maximum stellar mass and cluster mass but also on the assumed *cluster* mass function: a pronounced influence on the IGIMF requires the integrated population to be dominated by small-N clus-ters, so that (for a power-law cluster mass function) the slope needs to be steeper than -2 to have any significant effect.

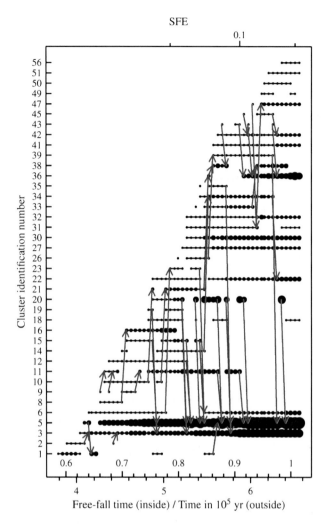

Fig. 7.3 An illustration of hierarchical cluster assembly within the simulations of Bonnell et al. (2008). Clusters identified via the minimum spanning tree are depicted with the symbol size representing the mass of the most massive star and the arrows represent cluster merging events. Figure from Maschberger et al. (2010)

The issue of the IGIMF is important because, on the scale of entire galaxies, it controls the normalisation between observed star formation diagnostics (produced by massive stars) and the overall star formation rate. It is hard to assess this relationship a priori on galactic scales because of the large number of observational uncertainties (in addition to the IGIMF) which be-devil the analysis (see Elmegreen 2006; Pflamm-Altenburg et al. 2007; Selman and Melnick 2008 for contrasting conclusions on the empirical status of the IGIMF, as well as the discussion in Reid Chap. 16, Sect. 16.6).

Fortunately, however, we can attack the problem from the other end by assessing the direct observational evidence for truncated IMFs within clusters: this can be achieved by using simple binomial statistics to work out the expected distribution of the maximum stellar mass as a function of cluster membership number N and then enquiring where the observed datapoints are located with respect to the centiles of the predicted distribution. Note that it is important to consider the data in this way instead of comparing the data with the *expectation value* (i.e. mean) of the maximum stellar mass at a given N. This is because the predicted distributions are very asymmetric: the median is much less than the mean and this implies that with sparsely sampled datasets the data values are likely to be significantly less than the mean in the majority of samplings. This does *not* mean, on its own, that the IMF is necessarily truncated.

The results of this exercise are shown in Fig. 7.4; taken at face value, they are entirely consistent with the results of random selection. However, there are a couple of notable features about the observational data. Firstly, the position of the datapoints with respect to the centiles depends on the selection criteria employed: i.e. whether the data involved a measurement of stellar maximum mass in already identified clusters (green and blue points) or instead the identification of clusters around already identified massive stars (red points). Unsurprisingly, the latter points tend to lie higher on the centiles; this emphasises the importance of unbiased target selection in constructing such a diagram. Secondly, it is worth noting that the most observationally

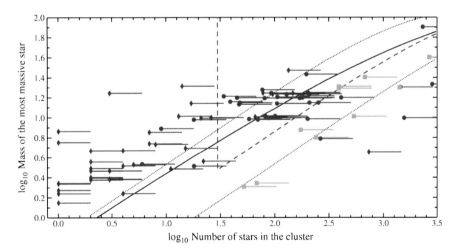

Fig. 7.4 Data on the maximum stellar mass as a function of cluster membership number (see Maschberger and Clarke 2008 for the data sources). The *solid line* is the median value based on random sampling of an untruncated IMF and the *dotted lines* the 1/6th and 5/6th quantiles of the same distribution. Note the fact that the membership numbers suffer from poorly-determined incompleteness (a notional factor of two one-sided errorbar is added to each point). Note also that the location of the points in the diagram depend on whether the data is selected by most massive star or by cluster (see text). Figure from Maschberger and Clarke (2008)

discrepant part of the diagram is that at low mass (low N) where, contrary to the IGIMF theory in the form usually proposed, the observational data is actually too *high* relative to the centiles for random drawing. We will return to the fact that massive stars are apparently to be found in surprisingly sparse clusters when we come to assess the birthplace of the solar system. It is however worth stressing that the values of N in the plot are lower limits since they have been obtained (see Testi et al. 1997, 1998) from deep near-infrared imaging of apparently isolated, but relatively distant, Herbig Ae/Be stars. These values are thus likely to suffer from ill-quantified incompleteness.

The only regime for which the observational data is arguably too low compared with the centiles is for massive clusters ($> 10^3 \, M_\odot$) where maximum stellar masses of around $30 - 40 \, M_\odot$ are a little low compared with the model (Weidner et al. 2010). In this regime, however, there is a further uncertainty: the lifetime of stars of this age is short (a few Myr). Given the uncertainties in measuring the ages of clusters at these youngest ages, it is then hard to find a sample of clusters that are sufficiently young that one can be sure that the most massive members have not already exploded as supernovae.

Finally, it might be argued that trying to answer this question using only one star per cluster (the most massive) is wasteful of statistical information and requires an unacceptably large ensemble of clusters in order to measure subtle effects. Alternatively, one can search for truncation of the IMF within an individual cluster: see Koen (2006) and Maschberger and Kroupa (2009) for statistical tests that are sensitive to the extremes of the distribution and are hence suitable for detecting evidence of truncation. Nevertheless it should be stressed that however good the statistical tools employed, the significance of the answer also relies on robust mass determinations for massive stars. These are not generally available, particularly in the absence of spectroscopic data (see Burkholder et al. 1997; Massey 2002; Weidner and Vink 2010).

7.3 Imprint of Cluster Birthplace on Discs

The 'proplyds' in the ONC present a vivid demonstration of how the properties of circumstellar discs may be modified in a rich cluster environment. 'Proplyds' are young stars with associated ionisation fronts that are significantly offset with respect to their parent stars (and also, with respect to their protoplanetary discs, which are detected in silhouette against the bright nebular background emission, O'Dell et al. 1993; Bally et al. 2000). This ionised emission is well accounted for by the interaction between ionising radiation from the O6 star ($\Theta_1 C$) in the cluster core and a neutral wind that is photoevaporated from the disc by the softer (non-ionising) ultraviolet flux of $\Theta_1 C$. Theoretical photoevaporation models (Johnstone et al. 1998) predict disc mass-loss rates that are similar to those inferred from radio free-free emission (Churchwell et al. 1987): these rates are high, being a few times $10^{-7} \, M_\odot \, yr^{-1}$ and imply that a circumstellar disc with the mass of the 'minimum mass solar nebula' (i.e. the mass of hydrogen that would—at solar abundances—have originally accompanied the solid components of the planets in the solar system) would be photoevaporated in a mere 0.1 Myr. Since this timescale is $< 10\%$ of the age of the ONC, there is little doubt that the cluster environment (specifically the presence of a strong ultraviolet source) must have a major impact on planet formation. Indeed, the short timescale associated with photoevaporation in the ONC suggests that we are witnessing a brief episode at a privileged epoch. In fact this is backed up by the observed paucity of proplyds in other regions (Stapelfeldt et al. 1997; Stecklum et al. 1998; Balog et al. 2006).

On the other hand, it is worth emphasising that the strong effect of $\Theta_1 C$ is pretty localised, with the high photoevaporion rates cited above being restricted to the inner ~ 0.3 pc of the cluster. Fatuzzo and Adams (2008) have conducted population synthesis studies in which they examine the global impact of photoevaporation by massive stars in clusters, given observationally motivated assumptions about the mass spectrum and stellar content of clusters. Their conclusion (based on the assumption that the field population is derived from the loose clusters seen in star-forming regions) is that the overall impact on discs (and hence on potential planet formation) is rather modest: only about 25 % of stars in the solar neighbourhood would have suffered a 'significant' disc mass-loss (i.e. photoevaporation down to ~ 30 au) over a 10 Myr timescale.

Another potential environmental effect in dense clusters (such as the ONC) is the stripping of discs by dynamical encounters. It is well-known that stellar fly-bys cause discs to be stripped down to a fraction of the closest encounter distance, this fraction depending on the mutual orbital inclination, mass ratios and velocities of the stars (see Clarke and Pringle 1993; Moeckel and Bally 2006; Pfalzner et al. 2006; Olczak et al. 2006). Scally and Clarke (2001) undertook N-body calculations of the evolution of the ONC, keeping track of the closest encounter distance for every star. Although a few stars in the dense central regions pass within ~ 100 au of each other (with consequently severe consequences for their planet forming discs), the bulk of stars in the ONC do *not* undergo such close encounters (see Fig. 7.5 and de Juan Ovelar et al. 2012). Encounters are more significant in the case of massive stars (Moeckel and Bally 2006; Pfalzner et al. 2006) since these are dynamically segregated to the central, densest regions; nevertheless there are probably other effects (associated with the strong winds driven by ionising radiation from massive stars; Hollenbach et al. 1994) which are also important in limiting disc lifetimes in this case.

7.4 The Birth Environment of the Sun

The properties of the solar system place a number of constraints on the environment in which its planetary system was born and has evolved: for further background, the reader is directed to the excellent review of this subject by Adams (2010).

One important constraint is the fact that—unusually among exoplanetary systems—the solar system is very dynamically cold, with its planetary orbits

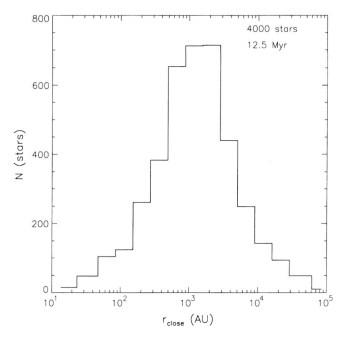

Fig. 7.5 A histogram of the closest encounter distance recorded per star during 12.5 Myr of N-body evolution of a cluster that is initiated with properties similar to the observed ONC. A small fraction of the stars in the cluster will have have encounters within 100 au during the typical lifetime of circumstellar discs. Figure from Scally and Clarke (2001)

being nearly circular and virtually co-planar. This is a property that argues for a rather isolated environment. On the other hand, meteoritic samples contain elements that are daughter products of short-lived radio nuclides (e.g. ^{60}Fe, ^{26}Al; see McKeegan and Davis 2003; Wasserburg et al. 2006; Wadhwa et al. 2007; Gounelle and Meynet 2012); the necessity of condensing these nuclides into grains within their half-lives implies that the primordial solar nebula was rather close to the site of a supernova: this argues generically for a clustered environment. We will quantify the above remarks in order to place limits on the likely range of conditions that are suitable birth environments for the Sun.

Turning first to the constraints offered by the low inclinations and eccentricities of the solar planetary system, large suites of Monte Carlo simulations (e.g. Adams and Laughlin 2001; Heggie and Rasio 1996; Malmberg and Davies 2009) have been used to investigate the types of encounters that are required in order to induce a significant (e.g. factor two) change in these quantities. The results of these calculations imply that the Sun cannot have undergone any encounters with pericentre less than about ~ 200 au. If we combine this result with analyses of close encounter distances in simulations of the ONC (Scally and Clarke 2001) we find that this condition is not particularly constraining: around 90 % of stars in the ONC would not have had such a close encounter.

On the other hand, we can turn the question around and enquire what are the features in the solar system which *can* be explained by encounters. For example, we can enquire how close a stellar fly-by is required in order for this effect to account for the observed drop-off in the density of Kuiper belt objects at 50 au (Allen et al. 2007). The answer to this question (closest approach of $\sim 200-300$ au) is uncomfortably close to the limit obtained above. This suggests that the outer limit of the Kuiper belt should not be explained in these terms because it then requires some orbital contrivance in order to achieve this without 'heating' the planetary orbits excessively. Alternatively, it has been suggested that an encounter is required in order to lift or scatter Sedna into its current orbit (Kenyon and Bromley 2004; Morbidelli and Levison 2004; Brasser et al. 2006): the required encounter distance for this to work is in the range $400-800$ au, which fits in better with the constraints on planetary orbits. Encounter distances in this range are comfortably provided by moderately-dense clusters (e.g. 25 % of stars in the ONC have suffered encounters in this range).

Turning now to the constraints provided by radionuclides, we consider the argument first put forward by Cameron and Truran (1977) which invoked a nearby supernova in order to explain the over-abundance of decay products of ^{60}Fe in meteoritic samples. The inferred high value of the ^{60}Fe to ^{56}Fe ratio within meteoritic material (compared with its value in the ISM) requires that the mass of the supernova progenitor is $\sim 25\,M_\odot$ and that the supernova explodes within about 0.2 pc of the Sun (this latter being required in order that the protosolar nebula acquires a sufficient complement of ^{60}Fe). However, the supernova cannot have exploded within about 0.1 pc of the Sun because of the consequent damage to the primordial nebula via blast wave stripping.

There are a variety of environments that can provide a supernova in the near vicinity without inflicting excessive blast wave stripping: for example, the ONC provides a suitable environment. Adams (2010) argues that the requirement of a $25\,M_\odot$ progenitor requires a rather populous birth environment, using the expectation value of the maximum stellar mass as a function of cluster N in order to place a lower limit on N of 10^3-10^4. However, this may be unnecessarily stringent, since empirical data (see Fig. 7.4) suggests that stars of $\sim 25\,M_\odot$ *may* occur in much smaller-N systems.

Putting all this together, the best evidence that the Sun was born in a cluster is the radionuclide data, since this requires that a supernova occurred within 0.2 pc of the young Sun. We have argued that although this is compatible with the Sun being formed in a populous cluster, it does not necessarily require this, since there is observational evidence for suitably massive stars in relatively small-N groupings. It does however place an obvious requirement on the stellar density, since it requires interstellar separations of order 0.2 pc; this is met over most of the ONC, for example (Hillenbrand and Hartmann 1998). It is also met in 25 % of the star-forming regions whose surface densities were compiled by Bressert et al. (2010), although this estimate relies on uncertain de-projection factors. Finally, there appear to be no observed stellar birth environments that are *too* dense to be compatible with the birthplace of the solar system. Even though *some* of the stars in the core of the ONC undergo encounters which are too close to leave the planetary system dynamically cold, there are plenty of stars—even in the central regions—that do not encounter another star within 1000 au.

Since these lectures were delivered, the claimed high inferred value of the initial ^{60}Fe to ^{56}Fe ratio in meteorites has been challenged by the recent measurements of Tang and Dauphas (2012). These authors infer a value that is compatible with that in the ISM and therefore argue against contamination of the primordial solar nebula by the products of a supernova explosion. On the other hand, the high initial ratio of ^{26}Al to ^{27}Al that have been inferred in meteoritic data still requires the proximity of a massive star (in this case a Wolf-Rayet star). In this revised picture, the agent of contamination is via winds rather than an explosive event: a massive star ($>30M_\odot$) is still required however. The requirements on the Sun's natal cluster environment is thus not much modified from those discussed above.

7.5 Summary

In this concluding chapter we first examined attempts to match the results of simulations to modelling specific young clusters and associations. We then turned to a discussion of the ways in which birth in a clustered environment may shape stellar properties. We focussed in particular on the possible relationship between cluster membership number, N, and the maximum stellar mass in a cluster, as well as the extent to which protoplanetary discs are likely to be disrupted by dynamical and feedback effects within a cluster environment. We concluded with a discussion of whether the solar system bears evidence of birth in a cluster environment. Although the paradigm of supernova contamination of the protoplanetary disc is not borne out by recent meteoritic analyses, there is still evidence for the Sun's formation in the vicinity of a massive star. This is the strongest evidence for the Sun's formation in a cluster. However the apparent occurence of suitably massive stars in rather small-N clusters means that the constraints on the properties of the Sun's natal cluster are rather weak.

References

Adams, F. C. 2010, ARA&A, 48, 47
Adams, F. C. & Laughlin, G. 2001, Icarus, 150, 151
Allen, L., Megeath, S. T., Gutermuth, R., et al. 2007, Protostars and Planets V, ed. B. Reipurth, D. Jewitt & K. Keil, University of Arizona Press, 361
Allison, R. J. & Goodwin, S. P. 2011, MNRAS, 415, 1967
Allison, R. J., Goodwin, S. P., Parker, R. J., Portegies Zwart, S. F., & de Grijs, R. 2010, MNRAS, 407, 1098
André, P., Belloche, A., Motte, F., & Peretto, N. 2007, A&A, 472, 519
Bally, J., O'Dell, C. R., & McCaughrean, M. J. 2000, AJ, 119, 2919
Balog, Z., Rieke, G. H., Su, K. Y. L., Muzerolle, J., & Young, E. T. 2006, APJ, 650, L83

Becker, C., Moraux, E., Duchêne, G., Maschberger, T., & Lawson, W. 2013, A&A, 552, 46
Bonnell, I. A., Clark, P., & Bate, M. R. 2008, MNRAS, 389, 1556
Bontemps, S., André, P., Könyves, V., et al. 2010, A&A, 518, L85
Brasser, R., Duncan, M. J., & Levison, H. F. 2006, Icarus, 184, 59
Bressert, E., Bastian, N., Gutermuth, R., et al. 2010, MNRAS, 409, L54
Burkholder, V., Massey, P., & Morrell, N. 1997, APJ, 490, 328
Cameron, A. G. W. & Truran, J. W. 1977, Icarus, 30, 447
Churchwell, E., Felli, M., Wood, D. O. S., & Massi, M. 1987, ApJ, 321, 516
Clarke, C. J. & Pringle, J. E. 1993, MNRAS, 261, 190
Da Rio, N., Robberto, M., Hillenbrand, L. A., Henning, T., & Stassun, K. G. 2012, ApJ, 748, 14
de Juan Ovelar, M., Kruijssen, J. M. D., Bressert, E., et al. 2012, A&A, 546, L1
Elmegreen, B. G. 2006, ApJ, 648, 572
Fatuzzo, M. & Adams, F. C. 2008, ApJ, 675, 1361
Fellhauer, M. & Kroupa, P. 2002, Ap&SS, 281, 355
Fűrész, G., Hartmann, L. W., Megeath, S. T., Szentgyorgyi, A. H., & Hamden, E. T. 2008, ApJ, 676, 1109
Gounelle, M. & Meynet, G. 2012, A&A, 545, 4
Heggie, D. C. & Rasio, F. A. 1996, MNRAS, 282, 1064
Hillenbrand, L. A. & Hartmann, L. W. 1998, ApJ, 492, 540
Hollenbach, D., Johnstone, D., Lizano, S., & Shu, F. 1994, ApJ, 428, 654
Johnstone, D., Hollenbach, D., & Bally, J. 1998, ApJ, 499, 758
Jones, B. F. & Walker, M. F. 1988, AJ, 95, 1755
Kenyon, S. J. & Bromley, B. C. 2004, Nature, 432, 598
Kirk, H., Pineda, J. E., Johnstone, D., & Goodman, A. 2010, ApJ, 723, 457
Koen, C. 2006, MNRAS, 365, 590
Könyves, V., André, P., Men'shchikov, A., et al. 2010, A&A, 518, L106
Kroupa, P., Aarseth, S., & Hurley, J. 2001, MNRAS, 321, 699
Kroupa, P., Tout, C. A., & Gilmore, G. 1993, MNRAS, 262, 545
Lada, C. J. & Lada, E. A. 2003, ARA&A, 41, 57
Malmberg, D. & Davies, M. B. 2009, MNRAS, 394, L26
Maschberger, T. & Clarke, C. J. 2008, MNRAS, 391, 711
Maschberger, T., Clarke, C. J., Bonnell, I. A., & Kroupa, P. 2010, MNRAS, 404, 1061
Maschberger, T. & Kroupa, P. 2009, MNRAS, 395, 931
Massey, P. 2002, ApJS, 141, 81
McCaughrean, M. J. & Stauffer, J. R. 1994, AJ, 108, 1382
McKeegan, K. D. & Davis, A. M. 2003, Treatise on Geochemistry, 1, 431
Moeckel, N. & Bally, J. 2006, ApJ, 653, 437
Morbidelli, A. & Levison, H. F. 2004, AJ, 128, 2564
O'Dell, C. R., Wen, Z., & Hu, X. 1993, ApJ, 410, 696
Offner, S. S. R., Hansen, C. E., & Krumholz, M. R. 2009, ApJ, 704, L124
Olczak, C., Pfalzner, S., & Spurzem, R. 2006, ApJ, 642, 1140
Pfalzner, S., Olczak, C., & Eckart, A. 2006, A&A, 454, 811
Pflamm-Altenburg, J., Weidner, C., & Kroupa, P. 2007, ApJ, 671, 1550
Proszkow, E.-M., Adams, F. C., Hartmann, L. W., & Tobin, J. J. 2009, ApJ, 697, 1020
Scally, A. & Clarke, C. 2001, MNRAS, 325, 449
Scally, A. & Clarke, C. 2002, MNRAS, 334, 156
Scally, A., Clarke, C., & McCaughrean, M. J. 2005, MNRAS, 358, 742
Selman, F. J. & Melnick, J. 2008, ApJ, 689, 816
Stapelfeldt, K., Sahai, R., Werner, M., & Trauger, J. 1997, in Astronomical Society of the Pacific Conference Series, Vol. 119, Planets Beyond the Solar System and the Next Generation of Space Missions, ed. D. Soderblom, 131
Stecklum, B., Henning, T., Feldt, M., et al. 1998, AJ, 115, 767
Tang, H. & Dauphas, N. 2012, Earth and Planetary Science Letters, 359, 248

Testi, L., Palla, F., & Natta, A. 1998, A&AS, 133, 81

Testi, L., Palla, F., Prusti, T., Natta, A., & Maltagliati, S. 1997, A&A, 320, 159

Tobin, J. J., Hartmann, L., Furesz, G., Mateo, M., & Megeath, S. T. 2009, ApJ, 697, 1103

Wadhwa, M., Amelin, Y., Davis, A. M., et al. 2007, Protostars and Planets V, ed. B. Reipurth, D. Jewitt & K. Keil, University of Arizona Press, 835

Wasserburg, G. J., Busso, M., Gallino, R., & Nollett, K. M. 2006, Nuclear Physics A, 777, 5

Weidner, C. & Kroupa, P. 2004, MNRAS, 348, 187

Weidner, C. & Kroupa, P. 2006, MNRAS, 365, 1333

Weidner, C., Kroupa, P., & Bonnell, I. A. D. 2010, MNRAS, 401, 275

Weidner, C. & Vink, J. S. 2010, A&A, 524, 98

Part II
Kinematics of Star Clusters and Associations

Chapter 8
Introduction to Open Clusters

Robert D. Mathieu

8.1 Introduction

Every astronomical journey should begin with beautiful images, lest we forget the romance of the Universe amidst our analytic thinking. Figure 8.1 shows the 150 Myr open cluster M35 behind which is the 1 Gyr open cluster NGC 2158. While in the stellar dynamics world M35 is considered young, in the context of this School even M35 is old and thus both M35 and NGC 2158 are classical open clusters. In the next several chapters of this book we focus on groups of stars that only recently formed compared to these classical open clusters.

8.2 Classical Open Clusters

8.2.1 Definition

In the list below I present a few properties that allow us to identify stellar systems as classical open clusters in the Milky Way. These properties, and especially their limits, are more conceptual for understanding than definitive for classification.

- Age $\gg$ timescale for loss of natal gas (few Myr)
- Age $\gg$ dynamical timescale ('crossing time')
- $10 < M_{\text{cluster}} \lesssim 10^4 \, M_\odot$
- Metallicity $\sim$ solar
- Location in Milky Way disc

R.D. Mathieu (✉)
Department of Astronomy, University of Wisconsin, Madison, WI, USA
e-mail: mathieu@astro.wisc.edu

© Springer-Verlag Berlin Heidelberg 2015

C.P.M. Bell et al. (eds.), *Dynamics of Young Star Clusters and Associations*,
Saas-Fee Advanced Course 42, DOI 10.1007/978-3-662-47290-3_8

Fig. 8.1 Classical open clusters. The 150 Myr M35 (located at a distance of ∼850 pc) is shown in the *upper left*, whereas the more compact 1 Gyr NGC 2158 (located at four times the distance of M35) is shown in the *lower right*. Figure courtesy of D. Willasch

Essentially by definition, classical open clusters have ages greater than the duration of the formation of all the individual stars. Thus their ages are much larger than the timescale for the loss of the natal gas. This timescale is a few Myr.[1]

The age of a classical open cluster must also be larger than its dynamical timescale, or the crossing time ($t_{cross} \sim 2R_{cluster}/v_{dispersion}$). This will become clearer when I provide an overview of collisional stellar dynamics below, but essentially this criterion ensures that the cluster is gravitationally bound.

The next definitional properties are less criteria than observed properties of classical open clusters in the Milky Way. I do not think the approximate upper limit of $10^4 M_\odot$ will surprise anyone. The lower limit of $10 M_\odot$ may be a bit more unexpected. I selected this lower limit to make the point that the difference between an open cluster and a multiple stellar system is somewhat arbitrary. However, in this context we can reflect on the fact that one system is characterised by evolving stellar orbits due to multiple dynamical encounters and the other is characterised by a hierarchical system with stable Keplerian orbits. This distinction also reflects the difference in stability between small-N open clusters and multiple systems (recognising that the evolution of the former eventually leads to the latter). Thus $10 M_\odot$ is rather low compared to commonly studied systems; $\sim 100 M_\odot$ is more typically given as the lower limit for open clusters. This reflects our ability to distinguish them in the field as well as their longevity.

The heavy element abundances of members (metallicities, typically cited as [Fe/H], in logarithmic units relative to the abundances of those elements in the Sun) of open clusters tend to be within a factor of 3 of the solar value. Again, this is not so much a defining property as an observed property, but as described in the Chapters

---

[1]Estimating stellar ages (and thus the clusters in which they reside) is a complex topic and the reader is referred to Soderblom (2010) for a comprehensive review.

of Reid, in the Milky Way it allows us to distinguish most (young) open clusters from most (old) globular clusters. And in the same spirit the spatial distribution of open clusters in the Milky Way also distinguishes them from globular clusters as discussed below.

There are roughly 2,000 open clusters identified, with a modern useful catalogue today being that of Dias and colleagues (Dias et al. 2012). A complete sample within 850 pc yields about 250 open clusters.

8.2.2 Global Properties

How does our census of open clusters compare to recent infrared surveys of the Milky Way that penetrate through more of the obscuring dust?

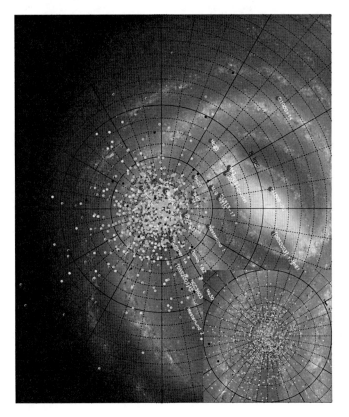

Fig. 8.2 Open clusters from the Dias et al. (2012) catalogue superimposed on a recent schematic model of the Milky Way based on *Spitzer*/GLIMPSE data. The *dots* are colour coded by age from log $\tau = 5$ (*blue*) to 9 (*red*). The inset image shows only the vicinity of the Sun. Figure courtesy of R. Benjamin

It is clear from Fig. 8.2 that our current open cluster database is very biased toward the solar region. In Fig. 8.2 we also zoom in on the solar region to look for an association of the Dias et al. catalogue with the nearby spiral arms. It is not clear that such an association is evident, which I suspect reflects more on the quality of the estimated distances as on the physics of star formation in the Milky Way.[2] *Gaia* will revolutionise Fig. 8.2. Recently, Piskunov et al. (2006) performed a detailed analysis of the open cluster spatial distribution, ultimately based on the All-Sky Catalogue of Stars. Figure 8.3 shows surface density versus distance from the Sun. Within roughly 850 pc the census is approximately complete.

In Fig. 8.4 I present another view of the Galactic distribution of open clusters, taken from Portegies Zwart et al. (2010). The blue dots are young open clusters, while the red dots are the oldest open clusters. You will notice that the older clusters tend to be found out of the disc. Likely this is an evolutionary selection effect, in the sense that clusters which orbit within the plane of the Galactic disc are continually buffeted by molecular clouds and other dynamical interactions leading them to rapidly evaporate

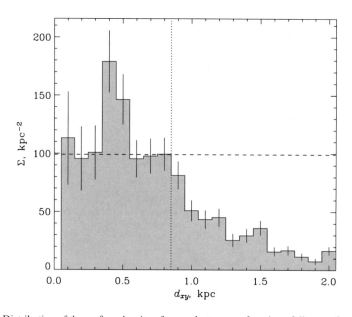

Fig. 8.3 Distribution of the surface density of open clusters as a function of distance from the Sun projected onto the Milky Way plane. The *dotted line* indicates the completeness limit, whereas the *dashed horizontal line* corresponds to the average density of open clusters. Figure from Piskunov et al. (2006)

---

[2]Ivan King once cautioned to be very careful with compilations. They are extremely valuable, but they are inherently heterogeneous in terms of both the content and the quality of the entries.

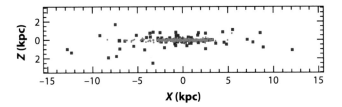

Fig. 8.4 The distribution of open clusters in the z-direction of the Milky Way. *Blue dots* represent young open clusters ($\tau < 100$ Myr), whereas the *red dots* denote older open clusters ($\tau > 3$ Gyr). Figure adapted from Portegies Zwart et al. (2010)

and disappear. Clusters whose orbits lead them to spend most of their time out of the plane of the Milky Way have calmer lives and live longer.[3]

More quantitatively, Röser et al. (2010) derive a scaleheight for all open clusters of about 50 pc cf. 300 pc for the thin disc; see Reid Chaps. 15 and 19, a surface density of about 100 kpc^{-2}, and a volume density of about 1,000 kpc^{-3} (but of course the open clusters do not occupy 1 kpc in height). Considering the actual volume filled by open clusters, one finds a total population in the Milky Way of order 100,000 open clusters.

Figure 8.5 shows the luminosity functions of open clusters, from Piskunov et al. (2008). While the mass distributions of open cluster systems are of considerably more interest to star formation researchers, it is difficult to construct them based on observations as discussed below. Luminosity functions can be constructed directly form observations with relative ease and compared between different regions of the Milky Way or different galaxies. Provided the vagaries mentioned below present consistent challenges between two cohorts, the comparison can be interesting. It is worth noting that the luminosity function of the open clusters in the Milky Way has a slope very similar to what is found for extragalactic clusters, though in the Milky Way we can extend the luminosity function to smaller clusters. Currently the turnover is thought to be real, and probably due to dynamical dissolution processes which we will come back to in Chap. 10. However, I remain somewhat skeptical: we may wish to revisit this issue after *Gaia* observations are available.

How do we measure the mass of these open clusters? It is not easy (in the absence of data from *Gaia*!). One approach is to do a complete census—simply count every single cluster member and assign a mass to each star. Determining which stars projected towards an area of interest on the sky are cluster members is determined probabilistically based on space motions, position in the colour-magnitude diagram, and other factors (e.g. elemental abundances). One has to assess completeness for the lowest-luminosity stars, and extend your census over a fixed area to the true outer radius of the cluster. Thus corrections for incompleteness are required, and your determined cluster mass will be sensitive to the mass function and spatial distrib-

---

[3] All of the clusters in Fig. 8.4 are in motion about the Milky Way. Those clusters seen far from the disc mid-plane are on orbits with larger average z-components than most of those seen in the disc. Nonetheless, they also pass through and interact with the disc, just on a less frequent basis.

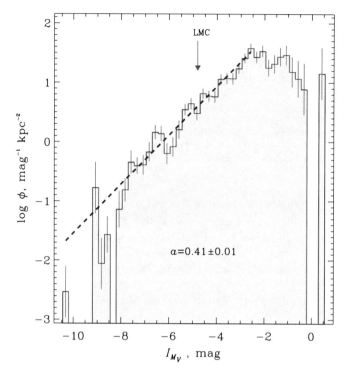

Fig. 8.5 Luminosity function of open clusters. The *dashed line* shows a linear fit for the brighter part of the histogram where a is the corresponding slope. Figure from Piskunov et al. (2008)

ution you assume (consistent with the observations) but extrapolated to parameter space not covered in the survey.

One must be especially cautious in interpreting tabulated values for cluster radii: many published values from the historical literature are meaningless. They are often the result of the visual impression derived by someone of an image, dictated by observational constraints, that does not correspond to anything quantitative or astrophysical. If you are going to work on open clusters, make sure that you are using a (trustworthy) core radius, a half-mass radius, or a tidal radius.

Another way to determine a cluster mass is dynamical, perhaps by simply using the virial theorem or by fitting more sophisticated dynamical models for clusters. As discussed below, this is difficult for open clusters, mainly because the stellar velocity dispersions are so small and difficult to measure. One also has to consider whether the sample used to fit the model is complete or representative, as a function of relative brightness and spatial distribution. Again, *Gaia* will help, although we will discuss later how the frequency of binary stars are going to affect such analyses.

Finally, Ivan King suggested an intermediate approach between a full census and dynamical modelling. Put very simply, if you determine the tidal radius and know the galaxy gravitational potential, then you can derive the cluster mass. It is an elegant idea. Unfortunately the mass depends on the tidal radius to the third power, and thus deriving adequately precise tidal radii is a challenge for deriving useful masses.

So in the end, it is not trivial to determine the masses of any clusters. If you want to do it accurately (as compared to precisely), it requires great technical skill and care. Piskunov et al. (2008) used masses estimated from tidal radii to derive the mass distribution of current open clusters (which due to evolution is not their initial mass distribution). As mentioned earlier, their range is 10 to perhaps $10^5\,M_\odot$, with the majority between 100 and $10^4\,M_\odot$ and an average mass of about 700 $M_\odot$.

Figure 8.6 from Portegies Zwart et al. (2010), shows the half-mass radius versus mass for open clusters, globular clusters, and the recently discovered young massive clusters that are the subject of their review. The half-mass radius is that physical radius within which is located half of the total mass of the cluster. Notice that the half-mass radii of open clusters, young massive clusters, and globular clusters are roughly similar. Of course, the masses of open clusters and globular clusters are different, $10-10^4\,M_\odot$ compared with $10^5-10^6\,M_\odot$ respectively. Even so, their stellar densities within half-mass radii are not as distinct as often presumed; indeed across the distributions they overlap. Yet the central densities of globular clusters, and especially post-collapse globular clusters, can be much higher than found in open clusters.

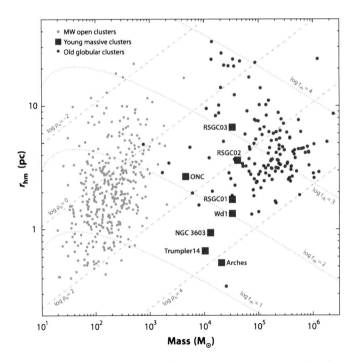

Fig. 8.6 The mass-radius diagram of Milky Way open clusters, young massive clusters, and old globular clusters. Figure from Portegies Zwart et al. (2010)

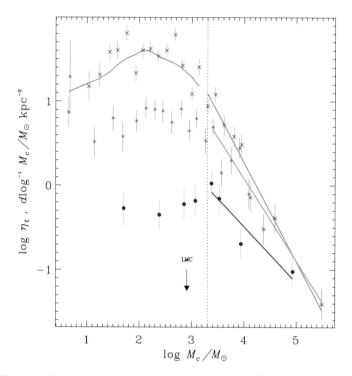

Fig. 8.7 Evolution of the mass function of Galactic open clusters. Different symbols denote samples with different upper limits of cluster ages. Blue filled circles represent clusters with ages log τ < 6.9, green stars for ages log τ < 7.9 and magenta crosses for log τ < 9.5. The *arrow* indicates the lower mass limit reached for open clusters in the LMC. Figure from Piskunov et al. (2008)

Turning to the open cluster mass function, one finds a power-law distribution that tends to flatten to lower masses (see Fig. 8.7). Again, this power law is similar to what is seen in other galaxies. Piskunov et al. (2008) show the mass function for clusters with ages less than 10 Myr as proxy for a cluster initial mass function (IMF). They find the mass function slope to decrease, as expected: as clusters age, they evaporate, losing stars through dynamical interactions and stellar evolution. The largest clusters i.e. 'disappear', and all clusters evolve into smaller ones leading to a steepening of the slope with time. Perhaps most interesting is that the estimated i.e. 'initial' power-law slope of -1.7 is very similar to that found for embedded clusters, as we will discuss later.

Piskunov et al. (2008) use their proxy for the cluster IMF to derive a formation rate for classical open clusters of 0.4 kpc^{-2} Myr^{-1}. This is a factor of 10 smaller than the formation rate from embedded clusters of $2 - 4$ kpc^{-2} Myr^{-1}, which we will discuss in Chap. 12. It is this order-of-magnitude difference that contributes to the frequent statement that roughly 10 % of the stars are made in open clusters. (In fact, Röser et al. (2010) conclude that 37 % of thin disc stars are made in open clusters.)

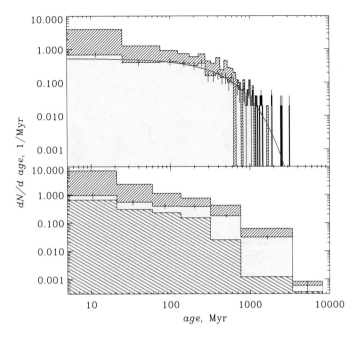

Fig. 8.8 Distributions of open cluster age. The *upper panel* shows the distributions of Piskunov et al. (2006), with the filled distribution representing their complete sample. The *solid curve* represents a fit to the age distribution curve. The *lower panel* shows the evolution in the data since the classic paper of Wielen (1971, hatched). Figure from Piskunov et al. (2006)

Finally, the age distribution of open clusters is a venerable field of study in which the classic works of Wielen (e.g. Wielen 1971) should be particularly noted. Piskunov et al. (2006) revisited the question with their modern cluster database, as shown in Fig. 8.8. There is some difference between the 1971 findings and today, with the currently derived mean lifetime being about 300 Myr. The sharp drop in the number of clusters with ages greater than a few Gyr has long been taken to be evidence for dynamical evolution (see Chap. 10).

8.2.3 Internal Properties

The global properties of clusters tend to be of great interest to those who study the Milky Way, and other galaxies, while the internal properties draw the attention of stellar dynamicists and those studying stellar evolution. Much like macro- and micro-economics, the two are distinct but intimately connected.

To introduce you to the internal properties of open clusters, I will use a cluster—NGC 188—which at the moment happens to be a target of much current research (however not especially young with an age of 7 Gyr).

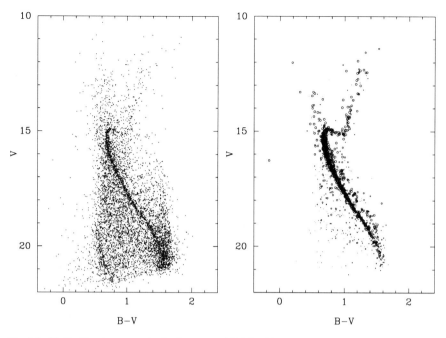

Fig. 8.9 V, $B-V$ colour-magnitude diagrams of NGC 188. *Left panel*: All stars in the field of NGC 188 within the magnitude limits. *Right panel*: 1490 probable cluster members, with proper motion members probabilities between 10 and 99 %. The size of each circle is proportional to the membership probability. Figure adapted from Platais et al. (2003)

Before attempting any astrophysical study with an open cluster, one has to first address the issue of cluster membership. If you take all of the stars in the field of NGC 188 out to a radius of ∼17 pc and make a V, $B-V$ colour-magnitude diagram, it looks like the left panel of Fig. 8.9. There is no doubt that the cluster is there, and likely there is a giant branch; although it would be hard to select which of the stars are cluster members versus non-members (dominated by field star giants).

So how does one determine which stars are cluster members? The answer depends a bit on the intended scientific study. In the chapters of I. Neill Reid it is suggested to use the intersection of the many expected properties of cluster members—kinematic, photometric, spectroscopic and more (see Chap. 16). This will certainly provide you with a very secure sample of members. On the other hand, if you require that all of these properties to indicate membership, then you are going to miss the unusual—and often the most interesting—stars, the gems among the common pebbles.

Whatever the scientific goal, one property that is clearly necessary for membership is kinematic association in three dimensions, and preferably distance association, if you have adequate precision such as *Gaia* will provide. In Fig. 8.10 I show three key figures in the membership process, taken from the proper motion study of NGC 188 of Platais et al. (2003). The left panel of Fig. 8.10 is the proper motion vector-point diagram. The cluster proper motion centroid is evident. Equally evident is that the

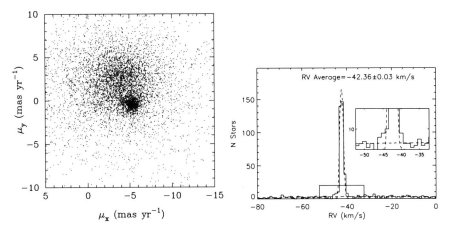

Fig. 8.10 *Left panel*: The proper motions of all stars in the field of NGC 188 within the magnitude limits. The concentration of cluster members is evident against the more dispersed field stars (from Platais et al. 2003). *Right panel*: A one-dimensional distribution of radial velocities of stars in the field of NGC 188. Gaussian fits to the field and cluster distributions, used for membership probability determinations, are shown (from Geller et al. 2008)

cluster centroid lies within the Milky Way proper motion distribution. Thus inevitably there are field stars with the same proper motions as the cluster. The right panel of Fig. 8.10 shows the proper motion distribution in one dimension, again showing the narrow cluster proper motion distribution and the broad field distribution. Typically two two-dimensional Gaussian functions are simultaneously fit to both the cluster and the field, and the membership probability for a star of any given proper motion is defined as the ratio of the value of the cluster Gaussian divided by the sum of the cluster and field Gaussians, all evaluated at that proper motion. The result is a membership probability. Clearly the higher the precision of the proper motion measurements and the larger the difference in the systemic velocities of the cluster and field stars, the better able we are to distinguish cluster and field members. Finally, the same process can be done with precise radial velocities, albeit with some minor complication from spectroscopic binaries, in order to provide three-dimensional kinematic selection (e.g. Geller et al. 2008) or valuable one-dimensional kinematic information when appropriate data for proper motion analyses are not available (e.g. Milliman et al. 2014).

The value of this work is evident in the right panel of Fig. 8.9, where the giant branch and the blue straggler population are now clearly evident. Even so, I stress that no one star can have 100 % membership probability based on kinematic data alone. If you have a thousand stars with 99.7 % membership probability, do not forget that a few of them will be field stars. And if you decide to write a paper on a fascinating star that you have found in a star-forming region, you had better be very careful to wrestle with this statistical uncertainty.

Note that once the kinematic measurement precision is better than the internal velocity dispersion of the cluster or star-forming region, additional precision is not of help, unless the goal is detailed investigation of sub-structure (e.g. spatially dependent mass segregation). Thus *Gaia* improvement in proper motion precision will be of limited help for bulk dynamics of nearby clusters, but of great help in more distant clusters. Furthermore, *Gaia* will certainly help in terms of distance determinations to clusters (either through direct parallax measurements or estimated using convergent-point methods), distance determinations for stars within (nearby) clusters, assessing mass segregation and/or bulk rotation, and in providing comprehensive data for all clusters.

Now with membership probabilities in hand, let us turn to the spatial distributions of stars in open clusters. For classical clusters we will consider only radial distributions, recognising that non-radial effects are expected from both rotation and the Galactic tidal field. For well-relaxed classical open clusters, multi-mass King models fit the stellar spatial distributions well, with one example being shown in the left panel of Fig. 8.11 for the open cluster M11. Such models provide measures of the core radii, and with adequate radial extent of the data also measures of tidal radii.

The large range of stellar masses in open clusters have always made them prime laboratories for studying mass segregation as a consequence of energy equipartition processes (see Mathieu 1984; Chap. 3). Mass segregation means the greater central concentration of more massive stars. While evident in the left panel of Fig. 8.11, cumulative distributions are more effective presentations of mass segregation, such as shown in the right panel of Fig. 8.11 where the more massive stars are evidently more centrally concentrated. This approach also immediately allows the application of the Kolmogorov-Smirnov test to determine whether two sub-samples have been drawn from the same parent population.

Of course the spatial distributions of stars are an instant in time reflection of the motions of the stars in their self-gravitating potential. One would like to measure energy equipartition and tidal truncation directly in the velocity distributions. This turns out to be very challenging because the stellar velocity dispersions, both in these clusters and in star-forming regions, are very small, of order $1\,\mathrm{km\,s^{-1}}$ or less in one dimension. The radial distribution of velocity dispersions in NGC 188 are shown in Fig. 8.12, for a reasonably massive open cluster.

It is important to remember that many stars in the Milky Way are members of multiple systems and this fact will have an impact on single-epoch observations of star cluster kinematics. Suppose you are granted time on the VLT with the FLAMES multi-object spectrograph. You place 300 fibres on stars in a young star-forming region and from these spectra you measure highly precise radial velocities, compute a velocity dispersion, analyse the physical implications, and publish the results. With high probability, your analysis of the physical implications will be wrong! Because within all of those velocities are the orbital motions of the $\sim$50 % of the stars that are binaries. So what you are measuring is the internal motions of the cluster itself convolved with the orbital motions of the binaries. Now if you make multiple observations, you will be able to identify and remove the short-period binaries (or obtain centre-of-mass velocities from orbital solutions). But the short-period binaries

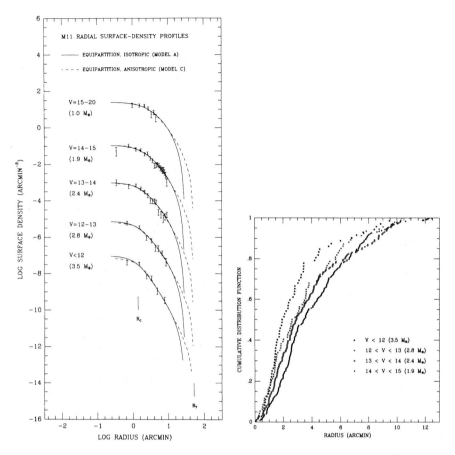

Fig. 8.11 Spatial distribution of stars in the open cluster M11. *Left panel*: Multi-mass King model fits to stellar surface densities. *Right panel*: Cumulative radial distributions of stellar positions, clearly showing the presence of mass segregation. Figure adapted from Mathieu (1984)

are not your greatest problem; because of their high orbital velocities, single velocity measurements were likely significant outliers from the observed velocity distribution of the star-forming region. Thus you would likely identify them (incorrectly, of course) as non-members. The greatest problem you face are the long-period binaries that have orbital motions of a few km s^{-1} and periods of many tens of years. They are the ones that are populating the 2–3σ tail of your velocity distribution. And even with multiple measurements on the timescale of a dissertation, you are not going to identify them as binaries! If you adopt a binary population, you can correct for their influence (e.g. Mathieu 1985; Geller et al. 2010; Cottaar et al. 2012). Of course, the issue of undetected binary companions is a general issue in stellar astronomy, and one recognised only relatively recently in the study of young stars. You ignore them at your peril.

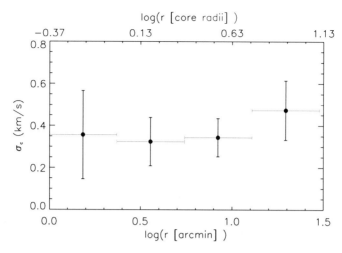

Fig. 8.12 Radial velocity dispersion as a function of radius in NGC 188. The *horizontal bars* show the region included in each measurement. Figure adapted from Geller et al. (2008)

Next let us turn to the stellar mass functions in open clusters. Figure 8.13 shows a whole set of open cluster mass functions, along with those for associations and globular clusters. Typically, cluster mass functions at the high-mass end tend to be similar to each other and the field, to within the effects of stellar evolution, and well fit by similar power-laws. The low-mass end is more difficult to determine technically and results have tended to show significant differences between clusters. In addition dynamical evolution effects make the interpretation of observed differences at the low-mass end problematic. (Note also the lack of low-mass stars among the globular clusters, likely due to preferential evaporation of low-mass members.) All this said, some of the more recent results seem to suggest that the low-mass end is also fairly stable (De Marchi et al. 2010).

8.2.4 OB Associations

Let me briefly bring OB associations into our discussion, with an homage to Adrian Blaauw. The ages of OB associations are larger, but not much so, than the timescale for the loss of the natal gas. Near almost all associations, there are still regions actively forming stars. Typically, association ages are less than 25 Myr or so, for an important physical reason. Ambartsumian argued definitively from the densities of OB associations that they are not bound (Ambartsumian 1947). If we take their three-dimensional internal motions to be 4 km s^{-1} (a bit higher than reality), in 25 Myr the stars travel 100 pc. So this upper limit on their ages is effectively set by their dissolution time. Of course, many are much younger.

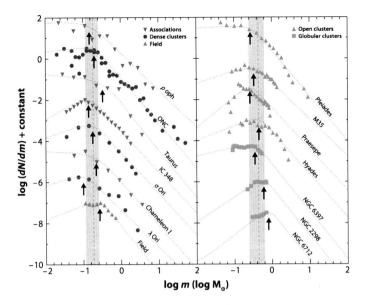

Fig. 8.13 The derived present-day mass function of a sample of open clusters spanning a large age range and old globular clusters. The *black arrows* show the characteristic mass of each fit. Figure from Bastian et al. (2010)

OB associations are associated with molecular clouds. They are located in the Milky Way disc with a scaleheight similar to the young open clusters. Their total masses tend to be similar to the open clusters, although more to the higher end. (This last is likely a statistical phenomenon related to the infrequency of OB stars in the IMF. Smaller mass associations without OB stars are known, but are more difficult to identify.)

Figure 8.14 shows a post-*Hipparcos* map of the OB associations in the solar neighbourhood. It is an honour to the work of Prof. Blaauw that the map was little changed from his earlier work (e.g. Blaauw 1991). What did improve, of course, is our knowledge of the systemic motions of the associations and the identification of members, which improved tremendously with *Hipparcos*, as shown in Fig. 8.15.

Figure 8.15 also makes the point that Upper Scorpius is much more concentrated than Upper Centaurus, which is more concentrated than Lower Centaurus. This is generally interpreted as the sequential dissolution of unbound systems, with Upper Sco being the most recently unbound and the currently embedded ρ Ophiuchus region soon to be the next in the sequence. We are actually seeing the systems unbind.

The memberships of young associations are one aspect of the dynamics of star-forming regions where *Gaia* will make a huge difference. Only because of apparent brightness, *Hipparcos* was unable to provide kinematics and membership for stars of later than spectral type A, i.e. for all the lower-mass stars. But they are assuredly there, for example as shown in Fig. 5 of Preibisch et al. (2002) where there certainly is no deficit of low-mass stars. These deep ground-based surveys which show that

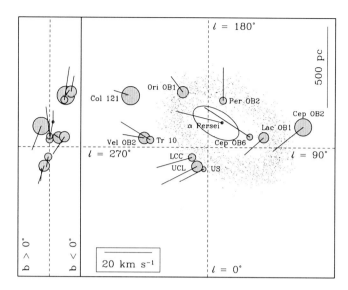

Fig. 8.14 Locations of the kinematically detected OB associations projected onto the Galactic plane. *Circles* represent the physical dimensions, the *ellipse* represents the Cas-Tau association, and the *vectors* represent the common streaming motions. Figure from de Zeeuw et al. (1999)

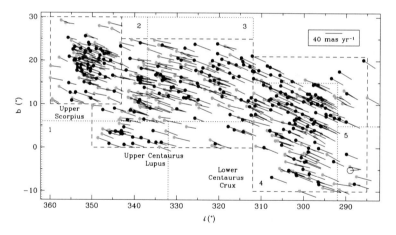

Fig. 8.15 Proper motions for 532 members of the Scorpius-Centaurus association, selected from 4156 candidate *Hipparcos* stars. The *dashed* and *dotted lines* are schematics boundaries of the three sub-associations. Figure adapted from de Zeeuw et al. (1999)

in fact associations have the entire IMF down to at least 0.1 M$_\odot$ are very hard work, but they are absolutely critical from the star formation point of view. *Gaia* should greatly expand our understanding of the global properties of low-mass star formation and mass functions.

8.3 Closing Thought

Do most stars form in OB associations? The distinction between OB associations and clusters is of historical origin. As we shall see, whether it remains an important physical distinction today in terms of the formation and evolution of star-forming regions is not clear. Open clusters are bound, OB associations are unbound. In 1930, that was a profound statement. But now, if you take a broader view of star formation and molecular clouds, and you see star formation as occurring throughout the molecular clouds with greater rates in some areas, then it becomes clear that the young stars in certain locations a few pc in size will end up being bound in clusters, and the rest of the young stars in the cloud will necessarily disperse as the gas disappears. Associations, OB or otherwise, are nothing more than the inevitable consequence of star formation (of low efficiency) going on in molecular clouds without global densities high enough to remain bound. Whether a particular grouping is an OB association depends on whether or not OB stars happened to have formed there.

So I think the real question is: are most stars formed in unbound groupings? The comparison of embedded star formation to the open cluster statistics suggests the answer is yes. However, we need to understand what fraction of stars are formed in small-scale, high-star-formation-efficiency regions in clouds and whether they dissolve before we can find them. Whether they are OB associations or T associations or clusters is, perhaps, historical jargon that is best left to history.

References

Ambartsumian, V. A. 1947, Stellar Evolution and Astrophysics, Armenian Acad. of Sci.
Bastian, N., Covey, K. R., & Meyer, M. R. 2010, ARA&A, 48, 339.
Blaauw, A. 1991, in NATO ASIC Proc. 342: The Physics of Star Formation and Early Stellar Evolution, ed. C. J. Lada & N. D. Kylafis, 125.
Cottaar, M., Meyer, M. R., & Parker, R. J. 2012, A&A, 547, 35.
De Marchi, G., Paresce, F., & Portegies Zwart, S. 2010, ApJ, 718, 105.
de Zeeuw, P. T., Hoogerwerf, R., de Bruijne, J. H. J., Brown, A. G. A., & Blaauw, A. 1999, AJ, 117, 354.
Dias, W. S., Alessi, B. S., Moitinho, A., & Lepine, J. R. D. 2012, VizieR Online Data Catalog, 1, 2022.
Geller, A. M., Mathieu, R. D., Harris, H. C., & McClure, R. D. 2008, AJ, 135, 2264.
Geller, A. M., Mathieu, R. D., Braden, E. K., et al. 2010, AJ, 139, 1383.
Mathieu, R. D. 1984, ApJ, 284, 643.

Mathieu, R. D. 1985, in IAU Symposium, Vol. 113, Dynamics of Star Clusters, ed. J. Goodman & P. Hut, Cambridge University Press, 427

Milliman, K. E., Mathieu, R. D., Geller, A. M., et al. 2014, AJ, 148, 38.

Piskunov, A. E., Kharchenko, N. V., Röser, S., Schilbach, E., & Scholz, R.-D. 2006, A&A, 445, 545.

Piskunov, A. E., Kharchenko, N. V., Schilbach, E., et al. 2008, A&A, 487, 557.

Platais, I., Kozhurina-Platais, V., Mathieu, R. D., Girard, T. M., & van Altena, W. F. 2003, AJ, 126, 2922.

Portegies Zwart, S. F., McMillan, S. L. W., & Gieles, M. 2010, ARA&A, 48, 431.

Preibisch, T., Brown, A. G. A., Bridges, T., Guenther, E., & Zinnecker, H. 2002, AJ, 124, 404.

Röser, S., Kharchenko, N. V., Piskunov, A. E., et al. 2010, Astronomische Nachrichten, 331, 519.

Soderblom, D. R. 2010, ARA&A, 48, 581.

Wielen, R. 1971, Ap&SS, 13, 300.

Chapter 9
Overview of Multiple Star Systems

Robert D. Mathieu

9.1 Introduction

In what ways are multiple star systems relevant to the dynamics of young star clusters and associations? Here we review several aspects of multiple star systems in order to provide a framework for some answers which we discuss at the end of this section. In what follows, the term 'binary stars' is often used. However, we should keep in mind that multiple star systems include triples, as well as higher-order bound systems. This complicates the statistics of multiple systems as discussed below. Keeping careful track of all the data in a systematic way is vital to compare observational results to theories of the formation and evolution of multiple systems. This overview of multiple star systems is intended to serve as an introductory reference to the field. For a more comprehensive discussion on the topic, the reader is referred to the recent review of Duchêne and Kraus (2013).

9.2 Field Solar-Type Binary Population

We begin with the field solar-type multiple population not only because of its obvious relevance to our Sun, but also because it is by far the best characterised and serves as a comparison for all other populations. Here 'solar-type' includes stars from perhaps $0.8–1.3\,M_\odot$. There is a long history of studies of binary populations in this domain, reaching all the way back to William and Caroline Herschel. Yet it was Helmut Abt, who initiated the modern era of comprehensive and systematic studies of binary populations as a particular class of stars (e.g. Abt and Levy 1976). A major step forward in terms of both the caliber of the data and the quality of the analysis was

R.D. Mathieu (✉)
Department of Astronomy, University of Wisconsin, Madison, WI, USA
e-mail: mathieu@astro.wisc.edu

© Springer-Verlag Berlin Heidelberg 2015
C.P.M. Bell et al. (eds.), *Dynamics of Young Star Clusters and Associations*,
Saas-Fee Advanced Course 42, DOI 10.1007/978-3-662-47290-3_9

the paper by Antoine Duquennoy and Michel Mayor, from Geneva Observatory (Duquennoy and Mayor 1991). This superb piece of research that has drawn more than 1,500 citations and set the stage for much modern research.

Even so, techniques and technology improve, and recently Raghavan et al. (2010) have published a major study that is now the standard reference. For their target sample, Raghavan et al. (2010) sought to have a volume-limited sample, and so took 450 stars from the *Hipparcos* catalogue out to a distance of 25 pc (parallaxes greater than 40 mas). They restricted the stellar colours to $0.5 < B-V < 1.0$ mag, or F6 to K3 dwarfs and subdwarfs. And then they applied to the sample an array of ground-based binary discovery techniques in order to find every companion that they could, including high-precision radial velocities, optical interferometry with the CHARA facility, speckle imaging, and multi-epoch imaging for common-proper-motion pairs. Note that literature observations also remained very relevant to this study. That is a beautiful thing about binary stars—accurate measurements retain their value indefinitely for subsequent orbital solutions.

Beginning with the most basic result, the frequencies of single and multiple systems, Raghavan et al. (2010) find that $56 \pm 2\%$ of the stars are observed to be single (over the range of mass ratios to which the study was sensitive, specifically $0.1 < q = M_{companion}/M_{primary} < 1.0$, and a broad range of semi-major axes that cover the vast majority of potential systems). Similar values have been found over the years, so roughly speaking a 1:1 ratio of single and multiple systems among solar-type stars is worth remembering. Raghavan et al. (2010) find the frequency of binary stars to be 33%, triples 8% and higher order systems 3%, or in total a frequency of multiple systems of 44%.[1]

The period distribution of binary pairings is shown in Fig. 9.1.[2] Note that this is a distribution in log period, ranging from orbital periods of less than one day to orbital periods greater than 1000 days. The mean period of a binary is about 10^5 days or 300 years, equivalent to a semimajor axis of order 50 au. This log-normal (Gaussian in the log of the orbital semi-major axis) distribution is very similar to that found by Duquennoy and Mayor (1991), shifted somewhat to longer periods.

There are many measures of binary frequencies provided or cited in the literature that lack mention of the period range (and mass-ratio) over which the frequency is measured. Any given technique only samples a part of the period and mass-ratio distribution. Thus, without specific and careful characterisation of period and mass ratio ranges to which the survey is sensitive, the quoted frequency is in fact of little value.

The 'e-log P diagram' shown in Fig. 9.2 has become a fundamental figure in binary studies, ever since a meeting at Bettmeralp, Switzerland, 25 years ago in honour

---

[1] There are many ways to do the accounting of multiplicity. Here we use a simple approach: a 'star system' may be single (no detected companions), or binary, or triple, or quadruple, or perhaps a higher-order system. The frequency of multiple systems is defined as the number of multiple systems divided by all such 'star systems' examined.

[2] Almost all field star multiple systems are hierarchical, as required for dynamical stability. Thus there is no ambiguity, for example, in identifying two distinct binary pairings within a triple system, or three within a quadruple system.

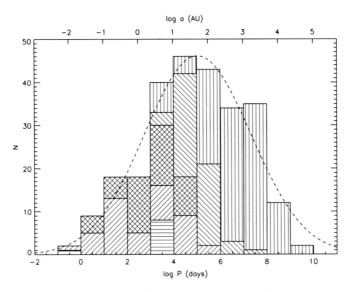

Fig. 9.1 Period distribution for binary pairings within solar-type multiple systems. Spectroscopic binaries are identified by *positively sloped lines*, visual binaries by *negatively sloped lines*, companions found by both spectroscopic and visual techniques by *cross-hatching*, common proper motion pairs by *vertical lines*, and unresolved companions via proper motion accelerations by *horizontal line* shading. The semi-major axes shown in au at the top correspond to the periods on the x-axis for a system with a mass sum of $1.5\,M_\odot$, the average value for all the pairs. The *dashed curve* shows a Gaussian fit to the distribution, with a peak at $\log P = 5.0$ and standard deviation of $\sigma_{\log P} = 2.3$. Figure from Raghavan et al. (2010)

of Roger Griffin's one-hundredth published orbit solution. The figure shows a plot of orbital eccentricity versus log period in days, and in explaining it one must consider numerous astrophysical issues. For example, most binaries with periods shorter than 12 days have circular orbits, presumed to be circularised by tides between the companions. At longer periods there is a broad flat distribution of eccentricities, which in the field reflects the primordial distribution. Notably, nature rarely forms binaries with circular orbits, and at the same time very-high eccentricity orbits also are not favoured (indicating that the distribution is not relaxed by dynamical encounters). Overall the orbital eccentricity distribution can be represented by a very-broad Gaussian with a mean eccentricity of ∼0.4.

The field distribution of mass ratios (i.e. companion mass over primary mass) in multiple systems is shown in Fig. 9.3. The mass ratio distribution is largely flat, with two notable deviations. One is that there seems to be a deficiency of small mass ratios, which Raghavan et al. (2010) argue is real; and second, there seems to be an excess of nearly equal-mass systems, often called twins. Both effects are predominantly found in the short-period binaries, the spectroscopic binaries. Historically this question of whether there is a preference for equal-mass systems has been around for a long time, and the pendulum has swung back and forth as to whether such an excess exists or

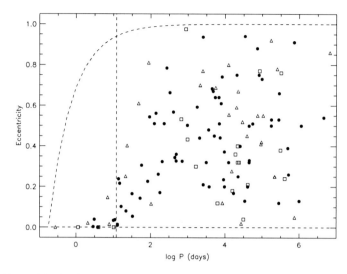

Fig. 9.2 Eccentricity-log P diagram for field solar-type binary stars. Components of binaries are plotted as *filled circles*, of triples as *open triangles*, and of quadruple systems as *open squares*. The *vertical dashed line* marks an approximate tidal circularisation period of 12 days. The *dashed curve* represents a boundary, to the left of which pairs of 1.5 $M_\odot$ total mass will pass within 1.5 $R_\odot$ at periastron and hence are vulnerable to collision. Figure from Raghavan et al. (2010)

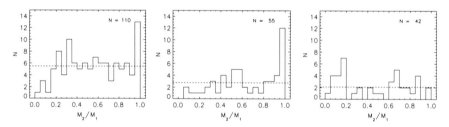

Fig. 9.3 Mass ratio distribution for binaries (*left*), pairs in higher-order multiple systems (*middle*), and composite-mass pairs in multiple systems (*right*). For example, in a triple system composed of a spectroscopic binary Aa, Ab, and a visual binary AB, the M_{Aa} to M_{Ab} ratio is included in the middle panel and the $M_{(Aa+Ab)}$ to M_B ratio is included in the right panel. The *dashed lines* mark the average value for each plot. Figure from Raghavan et al. (2010)

not. For example, Duquennoy and Mayor (1991) did not find such an excess of equal-mass systems and claim the companion mass ratio distribution is consistent with the field star IMF. Answers to these questions are important in order to test theories. For example, if there is a preference to make equal-mass binary stars, that would be a very significant datum for either binary formation or possibly early dynamical evolution in star-forming regions.

Figure 9.4 shows a result from Reggiani and Meyer (2011, see also Reggiani and Meyer 2013). They have taken the IMF of Bochanski et al. (2010) and explored

Fig. 9.4 Companion mass ratio distributions (CMRD) compared with random draws from the field IMF of Bochanski et al. (2010). From *top* to *bottom* the comparison between the observed CMRD and the field IMF is shown for M dwarfs, G stars in the field, and for a sample of intermediate-mass stars in Sco OB2, respectively. The hatched histogram represents the observed CMRD for the respective dataset of binary systems. Superimposed with a dashed line is the CMRD generated for the same number of objects through random pairing from the field IMF. Figure from Reggiani and Meyer (2011)

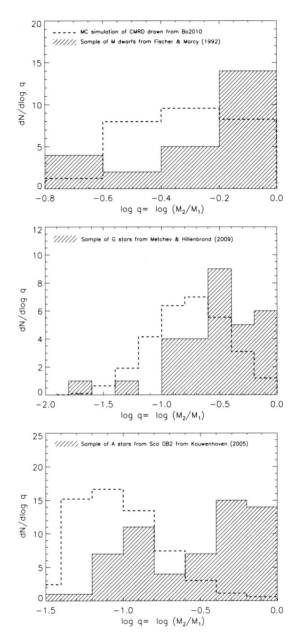

with Monte Carlo methods whether the mass ratio distribution can be explained by random drawings from such an IMF. They find for the M stars only a 1 % chance of this origin, a 0.001 % chance for the G stars, and for the A stars it was even more unlikely. So their essential conclusion is that the secondary mass ratio distribution is flat, regardless of the primary star mass, a key clue to the process of binary formation; they do not simply result from a post facto pairing of stars.

Now we summarise what is known about multiplicity for the field solar-type stars: (1) The frequency of multiple systems is 44%; (2) the period distribution is log-normal in shape, with a mean period of 300 years; (3) the orbital eccentricity distribution is a broad Gaussian (and indeed is consistent with a uniform distribution), but circular binaries above the tidal circularisation limit, and very-high eccentricities are notably absent; and (4) the companion mass ratio distribution is roughly uniform, possibly with an excess population of twins in mass.

9.3 Field OB Binary Population

Studies of the OB star binary population have a long history, including an unpublished manuscript by Adriaan Blaauw that was perhaps one of the most widely circulated unpublished papers. However this long history has not led to a clear convergence on a set of well-defined binary population properties, for OB stars pose many challenges to the observer of companions. These include their spectral properties (few absorption lines, rapid rotation, emission lines); intrinsically small sample sizes; and a large dynamic range in primary-secondary luminosity ratio. Let us restrict this discussion to stars with spectral types earlier than B3 as in the recent update of Mason et al. (2009) and overview of Sana and Evans (2011). These stars have masses greater than $8\,M_\odot$ or so.

Mason et al. (2009) divide the OB sample into those that are in clusters and associations; those that are in the field; and then the runaway OB stars, which is another fascinating group in itself. The most striking result is simply the very high binary frequency of 75% among OB stars in clusters and associations. Indeed, Mason et al. (1998) suggested that the frequency may be as high at 100% when biases and incompleteness are considered. Interestingly, the binary frequencies among field and runaway OB stars are less, 59 and 43% respectively. These may reflect on the dynamics of young star clusters and associations. For example, there is a long literature discussing the origin of OB runaways via dynamical ejection or the release of binary companions of supernovae, which we will not discuss here (see introduction of Gvaramadze et al. 2012). Particularly notable, Mason et al. (2009) find a frequency of 57% spectroscopic binaries in clusters and associations.[3] Similarly, Fig. 9.5 shows binary frequencies among O stars in open clusters with orbital periods up to 10^3 days. The average frequency is 44%, and the sensitivity to companion mass ratios has not been extensively explored yet. The majority—but not all according to Sana and Evans (2011)—of OB stars in clusters and associations have such close companions.

---

[3] This result harkens back to Blaauw's manuscript in which the vast majority of OB stars in associations were claimed to be spectroscopic binaries. There is cause to have caution regarding such results. If one happens to be optimistic about one's external precision, particularly if it is determined from internal precisions, one's spectroscopic binary detection rate will be inflated. Nonetheless, this result of high binary frequency among OB stars has survived the years.

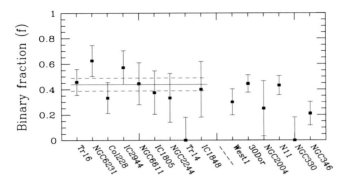

Fig. 9.5 Spectroscopic binary frequency among O stars in nearby (*left*) and distant/extragalactic (*right*) clusters. The *solid line* and *dashed lines* indicate the average frequency and 1σ dispersion computed from the nearby cluster sample. Figure from Sana and Evans (2011)

Furthermore, among the Galactic clusters Sana and Evans (2011) argue that the hypothesis that the OB short-period binary frequency is the same in all clusters cannot be rejected. As a reminder the binary frequency for periods $<10^3$ days among field solar-type stars is 11 % (see Fig. 9.1). This difference in binary frequency as a function of stellar mass is another key observational fact with which binary formation, and dynamical evolution theories, must address (see e.g. García and Mermilliod 2001 and Gvaramadze et al. 2012).

Moving on to orbital parameters, we begin to tread on less certain ground. Only 40–50 % of the candidate OB spectroscopic binaries overall have orbital solutions, although the percentage is much higher among the subset of binaries in nearby O star-rich clusters. Figure 9.6 shows a cumulative distribution of orbital periods. Notice that the 60 % point in the distribution is at 10 days, indicating a remarkably high frequency of very close companions. One can not help but be concerned about biases against deriving orbital solutions for—or detecting—longer period binaries with smaller orbital amplitudes, given the achievable radial velocity precisions for OB stars. On the other hand, in some clusters orbits have been obtained for most of the identified binaries, and indeed for most of the O stars (García and Mermilliod 2001). The eccentricity distribution of O star spectroscopic binaries is notable in that 25 % of the orbits are circular (see Fig. 9.7; also Fig. 4 in Sana and Evans 2011). While this might be expected given the high frequency of short-period binaries, the explanation is not that straightforward. The e-log P diagrams of early-type stars are strikingly different from such diagrams for solar-type stars (e.g. Matthews and Mathieu 1993; Sana et al. 2008). First, binaries with periods longer than even the field solar-type tidal circularisation period are found to be circular. If these circular orbits are the result of tidal circularisation this is unexpected, both because of the much younger ages of OB stars and because early-type stars do not have convective envelopes for effective tidal dissipation. Second, there are many binaries with short

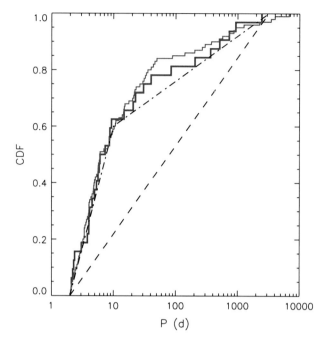

Fig. 9.6 Cumulative distribution function of O star spectroscopic binary periods. The *solid magenta* and *blue lines* represent the cumulative distribution functions of the Galactic O star sample and O star-rich clusters, respectively. Remarkably, the 60th percentile occurs at a period of 10 days. The *dashed line* is uniform in log period. See Sana and Evans (2011) for the *ad hoc* alternative model also shown (*dot-dashed*). Figure from Sana and Evans (2011)

periods and high eccentricities, which is equally puzzling: the orbital evolution of OBA-type binaries presents interesting challenges, and may be important probes of both stellar interiors and evolution.

The mass ratio distribution of O spectroscopic binaries appears to be uniform (see Fig. 4 in Sana and Evans 2011; also Mason et al. 1998). There is no indication of an excess of twins, nor does it represent random pairings from an IMF. The latter derives from an observed over-abundance of O–OB pairings. Reggiani and Meyer (2011) also find that the companion mass ratio distribution of binaries with A-type primaries is not drawn from random pairings from the IMF with very high confidence.

To summarise, current evidence indicates that OB stars, especially in clusters and associations, have a much higher binary frequency than do solar-type stars; they have a very high frequency of short-period binaries; have a greater diversity in orbital eccentricity with period; and have a mass ratio distribution thought to be uniform (consistent with lower-mass primaries).

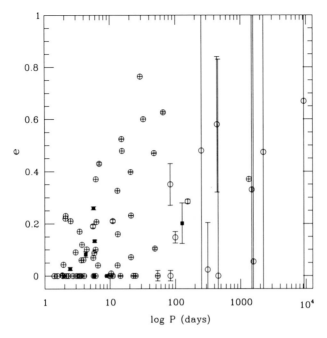

Fig. 9.7 Eccentricity-log P distribution for the O-type spectroscopic binaries in the 9th Spectroscopic Binary Catalogue. Figure from Sana et al. (2008)

9.4 Open Cluster Solar-Type Binary Population

Young open clusters are found in star-forming regions, but as yet their binary populations are not as well-characterised as those in classical open clusters. Here we discuss properties of the open cluster M35 with an age of 150 Myr to represent the primordial open cluster solar-type binary population. At an age of 150 Myr, N-body simulations of rich clusters like M35 show little evolution in the initial hard binary population, while the soft binary population is rapidly destroyed by dynamical encounters (A. Geller, private communication). We will discuss the concept of hard and soft binaries, and their dynamical evolution, in the next chapter. For our purposes here it will suffice to say that the hard binaries (those most difficult to alter dynamically) are those with periods shorter than $\sim 10^5$ days, and so the discussion here will focus only on the short-period binaries. The period distribution of solar-type binaries in M35 is shown in Fig. 9.8.[4] Within the period domain out to 10^4 days, the distribution is fully consistent with the period distribution for field solar-type binaries (see Fig. 9.1). The orbital eccentricity distribution for the binaries of M35 is also shown in Fig. 9.8 with periods greater than the tidal circularisation cut-off

---

[4]These data and results for M35 are those of the WIYN Open Cluster Study and the radial velocity measurements are published in Geller et al. (2010). The papers with the binary orbits and population distributions are from (Leiner et al. 2015).

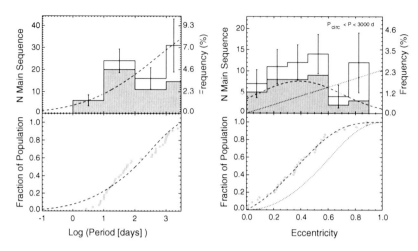

Fig. 9.8 Histogram and cumulative presentations of the period (*left*) and orbital eccentricity distributions (*right*) for binary stars in the open cluster M35. In both upper panels the hatched regions are the observed distributions and the open regions are incompleteness-corrected. The eccentricity distribution is only shown for periods greater than the tidal circularisation period. *Dashed lines* represent the distributions of field solar-type binary stars. The *dotted line* in the *right panel* shows the dynamically relaxed eccentricity distribution

period. The eccentricity distribution is broad and consistent with a Gaussian, as it is for the field binary population.

The e-log P diagram for M35 is essentially indistinguishable from the field (see Fig. 9.2), with the one notable exception that the tidal circularisation period for M35 solar-type binaries is 10 days compared to 12 days for the field solar-type binaries (Meibom and Mathieu 2005). These differences fit into a larger picture of tidal circularisation in cluster binaries, as shown in Fig. 9.9. The increase in tidal circularisation period with age for ages above 1 Gyr is a direct measure of tidal circularisation rate. Equally interesting is the lack of increase in tidal circularisation period for the younger clusters. Possibly the tidal circularisation period is set during the pre-main-sequence stage of evolution when the stars are large and deeply convective, and subsequent main-sequence tidal circularisation does not have an observable effect until the passage of $\sim$1 Gyr. Perhaps most significant is that current tidal circularisation theory does not self-consistently explain these observations, and as such the use of tidal theory in determining the outcomes of close binary encounters in young small-N systems must be considered with care.

Finally, the secondary mass distribution for solar-type binaries in M35 is shown in Fig. 9.10. It is not well fit by the IMF, and is consistent with a uniform distribution like that found for solar-type stars in the field (see Fig. 9.3). However, in M35 there is no suggestion of an excess of twins as found by Raghavan et al. (2010) in the field. To summarise, recognising that we can only make the comparison in the period domain of less than 10^4 days, the global properties of the M35 solar-type binary population

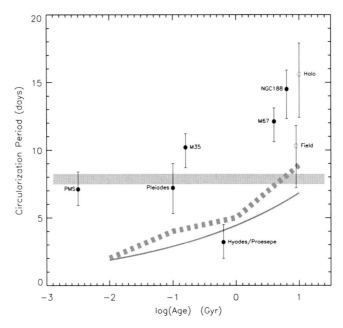

Fig. 9.9 Distribution of circularisation periods with age for solar-type binary populations in clusters and the field. The *solid curve* shows the predicted cut-off period as a function of time based on main-sequence tidal circularisation using the revised equilibrium tide theory. The *broad dashed band* represents the predicted cut-off periods calculated in the framework of the dynamical tide model including resonance locking. The *horizontal grey band* represents tidal circularisation being significant only during the pre-main-sequence phase. Figure from Meibom and Mathieu (2005), where a more detailed discussion of the data and models can also be found

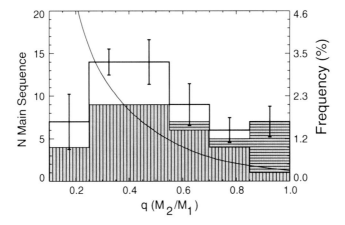

Fig. 9.10 The mass ratio distribution for solar-type binaries in the young cluster M35. The hatched region is the observed distribution and the open region is incompleteness-corrected. *Horizontal hatching* represent double-lined binaries while the *vertical hatching* represents statistical derivation from single-lined binary mass functions

are essentially indistinguishable from the field solar-type binary population. The differences in the details, however, are the subject of fascinating stellar astrophysics.

9.5 Closing Thoughts

What then is the significance of these binary populations to the dynamics of young star clusters and associations? Drawing from the closing discussion and contributions from the students of the 42nd Saas–Fee Advanced School:

Constraint on star formation processes: As shown in Cathie Clarke's section, there are now very impressive hydrodynamical simulations of the formation and early evolution of stars and star-forming regions. However, if these simulations ultimately do not make these observed binary populations, then there are clearly fundamental issues, and perhaps fundamental processes, which have been omitted.

Dynamical impact within young small-N systems: The encounter time in small-N systems is very short, indeed comparable to the orbital periods in their global gravitational potential. N-body simulations have long shown that such systems rapidly form binaries dynamically as they evolve, often one at a time. The insertion of a high frequency of primordial binaries into such systems assuredly will lead to their rapid dissolution, with some stars ejected at high velocities.

Tracers of origin: If we find out that the pre-main-sequence binary populations differ between different sites of star formation, for example between associations, embedded clusters and rich clusters, then that star formation mode which produces a field-like binary population may be identified as the dominant producer of field stars. In other words, the binaries may be a tracer of how and where most field stars form.

And as a final point:

Unknown participants in your data: One ignores the fact that many 'stars' are binaries at one's own peril. If you determine the mass of a 'star' from its luminosity and colour, but you are actually analysing the light of two stars, your derived mass and resulting analyses of the star may be very wrong. If you publish a paper on X-ray flares discovered from a young O star, you risk finding later that the flares originate in a late-type companion. If you compare your models of mass growth from disc accretion with observations of T Tauri stars, be aware that few of their accretion discs are not modified by companions. But perhaps this point is best made more simply. Until roughly the mid-1980's, most theories of star formation made one star. Often these theories were compared to T Tau itself. T Tau is at least triple.

References

Abt, H. A. & Levy, S. G. 1976, ApJS, 30, 273.

Bochanski, J. J., Hawley, S. L., Covey, K. R., et al. 2010, AJ, 139, 2679.

Duchêne, G. & Kraus, A. 2013, ARA&A, 51, 269.

Duquennoy, A. & Mayor, M. 1991, A&A, 248, 485.

García, B. & Mermilliod, J. C. 2001, A&A, 368, 122.

Geller, A. M., Mathieu, R. D., Braden, E. K., et al. 2010, AJ, 139, 1383.

Gvaramadze, V. V., Weidner, C., Kroupa, P., & Pflamm-Altenburg, J. 2012, MNRAS, 424, 3037.

Leiner, E. M., Mathieu, R. D., Gosnell, N. M., & Geller, A. M. 2015, AJ, 150, 10.

Mason, B. D., Gies, D. R., Hartkopf, W. I., et al. 1998, AJ, 115, 821.

Mason, B. D., Hartkopf, W. I., Gies, D. R., Henry, T. J., & Helsel, J. W. 2009, AJ, 137, 3358.

Matthews, L. D. & Mathieu, R. D. 1993, in Astronomical Society of the Pacific Conference Series, Vol. 35, Massive Stars: Their Lives in the Interstellar Medium, ed. J. P. Cassinelli & E. B. Churchwell, 223.

Meibom, S. & Mathieu, R. D. 2005, APJ, 620, 970.

Raghavan, D., McAlister, H. A., Henry, T. J., et al. 2010, ApJS, 190, 1.

Reggiani, M. M. & Meyer, M. R. 2011, ApJ, 738, 60.

Reggiani, M. & Meyer, M. R. 2013, A&A, 553, A124.

Sana, H. & Evans, C. J. 2011, in IAU Symposium, Vol. 272, Active OB Stars: Structure, Evolution, Mass Loss, and Critical Limits, ed. C. Neiner, G. Wade, G. Meynet, & G. Peters, Cambridge University Press, 474.

Sana, H., Gosset, E., Nazé, Y., Rauw, G., & Linder, N. 2008, MNRAS, 386, 447.

Chapter 10
Overview of Collisional Stellar Dynamics

Robert D. Mathieu

10.1 Introduction

Here I discuss collisional stellar dynamics in the absence of gas; elsewhere in this volume we will also consider the complications added by considering gas and hydro-dynamics (see Clarke Chap. 3). Before delving into the dynamics, let me make an essential overarching point. When considering an image of a beautiful cluster, such as M35 (Chap. 8), it is critical to recognize that this is not a representation of a static situation. Select any star—its current location is not where the star was not so long ago nor is it where it's going to be not so long from now. If you wish to understand and study the dynamics of star-forming regions, it is crucial to see them as dynamic, changing places, not only globally but also at the level of individual stars. Every object in the system was somewhere else before, to within the timescales that we will discuss next. This basic concept that the young stars are moving is fundamental, and one ignores it at peril.

As a simple framework for us to consider, I propose to separate the evolution of star-forming regions and star clusters into three time intervals. The first interval is while the stars are embedded in their natal molecular gas, when the gas mass and the stellar mass are comparable: the gas contributes meaningfully to the global gravitational potential for a timescale of a few Myr. Of course, in reality the system evolution depends on stellar dynamics, gas dynamics, magnetic fields, radiative transfer and many other variables (see Clarke Chaps. 2 and 4). The next interval, which we do not discuss in much detail here, is the period when the system is no longer embedded but is still losing significant mass due to the evolution of the massive stars. The evolution of the system at that point is very sensitive to the initial mass function, and occurs on a timescale set by the lifetimes of massive stars, of order 10–100 Myr. The third period is what I refer to as pure stellar dynamics, i.e.

R.D. Mathieu (✉)
Department of Astronomy, University of Wisconsin, Madison, WI, USA
e-mail: mathieu@astro.wisc.edu

© Springer-Verlag Berlin Heidelberg 2015   137
C.P.M. Bell et al. (eds.), *Dynamics of Young Star Clusters and Associations*,
Saas-Fee Advanced Course 42, DOI 10.1007/978-3-662-47290-3_10

the dynamics of point sources. Of course, stars are not point sources, and one of the fascinating topics of recent cluster dynamical work concerns stellar collisions. But here I will introduce point-source stellar dynamics, since these processes will fold back into our discussions of the dynamical evolution of star-forming regions.

Very little that I present here is new; my goal is to clearly provide a working knowledge of the key concepts. There are many fine books that will allow you to delve deeper into stellar dynamics, for example Spitzer (1988) and Binney and Tremaine (2008).

10.2 Timescales

It is important to learn to think in terms of relevant timescales. They determine the relative significance of varied processes, and thereby provide important insights on how different systems are going to evolve. For dynamical systems the key timescales are the dynamical timescale, the two-body relaxation timescale, the evolution timescale, and the age of the system.[1]

The dynamical time scale t_d is essentially a crossing time. Thus one straightforward approach to its evaluation is taking twice the radius r of the system divided by the typical velocity dispersion about the systemic velocity of the system σ_v. For a system in virial equilibrium, this becomes:

$$t_d \equiv \frac{2r}{\sigma_v} = \left(\frac{8r^3}{Gnm} \right)^{1/2} \simeq \frac{1}{\sqrt{G\rho}}, \tag{10.1}$$

where n is the number density within r, m is the average particle mass, and ρ is the mass density within r. The latter expression is an extremely useful tool for estimation. Whilst we are on the subject of useful tools, it is worth remembering that if one uses units of $1\,M_\odot$, $1\,pc$ and $1\,Myr$, then the gravitational constant G is simply $1/233$. Finally, note that a velocity of $1\,km\,s^{-1}$ translates to a distance of approximately $1\,pc$ over a timescale of $1\,Myr$. Among other things, in the study of star-forming regions this allows you to consider, in an approximately quantitative way, my first point: that each star was and will be someplace else on dynamical timescales.

Now, returning to the physical meaning of the dynamical time scale, essentially t_d is the time in which a particle responds to the global gravitational potential. For a system in dynamical equilibrium, such as an older star cluster, it is essentially the orbital time. The dynamical timescale is sometimes called a mixing time, reflecting that it is also the timescale on which a system starts to lose non-equilibrium internal structures. And if a system is not bound, it is an estimate of the dissolution time. Roughly speaking, t_d is the timescale on which a system that is no longer bound is going to expand, dissolve, or disappear. For all of these reasons, the dynamical

---

[1]For completeness we also introduce the free-fall time, the timescale for a system to collapse under its own gravitational potential. For a uniform density sphere of any radius without any internal support, $t_{ff} \sim 1/\sqrt{\rho}$.

time scale is perhaps the most important timescale for the evolution of star-forming regions.

The two-body relaxation time goes back to Chandrasekhar and Spitzer, and there have been many derivations and expressions of it over the years, such as:

$$t_r = \frac{1}{25} \sqrt{\frac{N r^3}{Gm}} \frac{1}{\log_{10}(\frac{N}{2})}. \tag{10.2}$$

Note that here N is the number of stars in the system, not a number density. So, critically, in virial equilibrium the relaxation time increases with the number of stars. Physically, the relaxation time is the average time to transfer significant energy of orbital motion between two stars. Formally it is often defined as the time over which the cumulative gravitational perturbations due to all particles in a system change the energy of a body by an amount roughly comparable to its orbital energy. This is actually a divergent integral; the effects of all distant stars are more important than nearby stars.[2] Practically speaking, most dynamical systems have natural physical limits on the integral. Two-body relaxation establishes a near-Maxwellian[3] velocity distribution, which I emphasise here because in multi-mass systems a Maxwellian implies particle velocity distributions dependent on the masses of the particles. Two-body relaxation is an energy equipartition process, and in equilibrium yields mass segregation in dynamical systems. In terms of the overall evolution of a dynamical system, two-body relaxation is the timescale for energy flow. Thus the relaxation time is the equivalent of a thermal timescale, such as dictates the time for energy to flow from the core to the surface in a star. Essentially the same energy flow occurs within star clusters, except it is the transfer of orbital energy of stars.

The evolution timescale is the time for secular evolution of the cluster properties such as mass and structure. Physically this is the time for a global change in the energy structure of a cluster. I will show shortly that the energy flows due to two-body relaxation lead to systemic energy changes, for example requiring that cores have to collapse and particles have to escape. Roughly speaking, a star cluster loses 1 % of its stars every relaxation time, so that in of order 100 relaxation times, a cluster disappears except for the last remaining tight binary at the centre. Thus an evolution timescale of $100 t_r$ can be defined by the evaporation time for the cluster.

The final timescale is the age of the system, which is not defined by dynamics but is critical for dynamical analyses. Indeed, it is the addition of age to the mix that makes the topic of this volume so interesting: we focus here on systems where the dynamical times, or the crossing times, and the ages are comparable, a few Myr for both. So we can not presume that these systems are well-mixed. In fact, we can presume that they are not well-mixed in most cases! Further many well-studied nearby star-forming regions contain a modest number of stars and richer events are often composed of

---

[2]Note that this statement is not true in the case of binary stars, where the encounters are tidal in nature and go as r^3, where r is roughly the impact parameter. We will return to this later.

[3]'Near-Maxwellian' in part because a cluster has an escape velocity that limits the maximum extent of velocity distributions.

smaller sub-groups with significant sub-structure. For $N \lesssim 100$, the dynamical and relaxation times are comparable in near-virial systems. In other words, a few-body system relaxes on the same timescale that its constituents cross the system. And so it is quite possible that we are going to be discussing systems where the crossing times, the relaxation times, and the ages are all of the same timescale. From a theoretical and observational point of view, this comparability of the timescales makes the dynamics of young stellar systems both demanding and interesting.

10.3 Violent Relaxation

A challenge for early stellar dynamical theory was that the two-body relaxation times for galaxies and the most massive globular clusters are very long, indeed longer than the age of the Universe. And yet the structures of these systems appeared dynamically relaxed, with at least Gaussian velocity distributions. And so Donald Lynden-Bell and several of his colleagues in the early 1960s developed the concept of violent relaxation, which is going to be very important to our discussion here. The essential idea of violent relaxation is that the energy of a particle as it orbits a cluster is only conserved if the global potential is fixed. If the potential is varying, then the energy of the particle will also vary, quite independent of any two-body interactions. And so the essence of violent relaxation is that, if a system is not in equilibrium, then dE/dt over a particle orbit is not equal to zero because the system structure—and thus the gravitational potential—is going through large macroscopic changes. In this situation, the specific energy change of a star over its orbit, in other words per mass, is comparable to the change in the global potential. The violent relaxation time is defined as the time where change in the specific energy of this star is comparable to its specific energy. (Very similar to the definition for two-body relaxation.) But the change in the potential is really the same as the change in the mass distribution, and for systems far out of equilibrium changes in mass distributions happen on dynamical timescales, or crossing times. So roughly speaking we can expect violent relaxation of a system to occur on a dynamical timescale (or a free-fall time). It can happen very fast, much faster than a two-body relaxation time.

Importantly, velocity distributions and spatial distributions are independent of mass after violent relaxation, just as the orbital properties of Jupiter in the Sun's potential do not depend on the mass of Jupiter to high significance. Thus violent relaxation sets up Gaussian-like velocity distributions, not Maxwellian velocity distributions, and the consequent core-halo structures are independent of mass. Finally, velocity anisotropy, for example from formation conditions or initial collapse phases of a system, tend to be preserved.

Violent relaxation approaches but never achieves equilibrium, because as a system approaches equilibrium the changes in the gravitational potential become ever smaller and the violent relaxation timescale stretches out. So the system never actually reaches complete equilibrium. Finally, the nature of violent relaxation in the context of large (gaseous) sub-structures and merging clumps has been an active subject in this field for the last few years (see Clarke Chap. 6). Again, star-forming regions bring fascinating subtleties and new understanding to classical stellar dynamics.

10.4 Energy Equipartition and Mass Segregation

Two-body relaxation due to gravitational interactions between particles is an energy-equipartition process producing a near-Maxwellian velocity distribution. Thus the kinetic energy distributions of particles vary inversely with their masses, and equivalently the velocity distributions of particles vary as the inverse square root of the masses, or:

$$f(E_{\text{star}}) \propto e^{E/\sigma^2} \Rightarrow f(E[v_i]) \propto e^{-\frac{1 m_i v_i^2}{2\sigma^2}} \Rightarrow \frac{<v_i^2>}{<v_j^2>} = \frac{m_j}{m_i} \tag{10.3}$$

Thus, for example, the equipartition timescale for massive particles is very short (which is essentially a result of dynamical friction being very efficient). If there are $10\,M_\odot$ stars in a system comprised predominantly of solar-mass stars, then the equipartition time for the $10\,M_\odot$ stars is a factor of 10 shorter than the nominal relaxation time. The structural consequence of energy equipartition is mass segregation, or the greater central concentration of higher-mass stars. Mass segregation is often seen in well-relaxed star clusters, such as the old open star cluster M67 (see Chap. 8) but only down to some stellar mass.

This question of energy equipartition, and thus mass segregation, timescales is more complex in very young clusters. There is no doubt that the Trapezium Cluster shows mass segregation, most notably through the presence of the four massive Trapezium stars at its centre. Accurate mass-dependent equipartition timescales and histories are critical to understanding whether or not the Trapezium stars formed *in situ* at the centre, or whether they fell to the centre very quickly through dynamical processes. To date, dynamical studies of that question have failed to reach a consensus.

10.5 Evolution of Dynamical Systems: Some Fundamental Physics

This section is dedicated to Lyman Spitzer, who provided to me as a young student this beautiful demonstration of fundamental physical thinking.

The probability of a star being in a particular state of energy E_i is, as in statistical mechanics:

$$P_i = C g_i e^{-k E_i}, \tag{10.4}$$

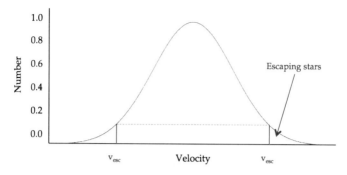

Fig. 10.1 Schematic representation of a thermal velocity distribution, the location of the escape velocity in a virialised system, and the fraction of stars (0.74 %) that escape per relaxation time

where g_i is the amount of phase space available, k is the Boltzmann constant and C is a normalisation factor for the distribution. So how do we maximise that probability?

And this is where I get to say that gravity is the coolest force in the Universe, because it has this wonderful property of having negative heat capacity, with even more wonderful implications. For a gravitational system, there are two ways that one can maximise the above probability. First, one can increase the binding energy. In other words, you can make E_i very, very negative (with point masses one can in principle make them infinitely negative). Second, one can maximise the phase space that is available. But there is no upper limit on the spatial coordinates; the particle distribution can extend to infinity in principle. And so how do you maximise the probability in a gravitational system? You can not. There is no maximum entropy of a gravitational system given finite mass and energy. And so bound gravitational systems necessarily evolve continuously; there is no equilibrium state. Furthermore, in its attempt to maximise entropy and minimise energy the system does two things at the same time. It collapses the core to zero size but finite mass, while taking the rest of the cluster mass to infinity. In other words, any gravitational system is going to collapse at its core and spew outward the rest of the material. It does not matter if it is a star. It does not matter if it is a cluster. Ultimately, it does not matter if it is a galaxy. Inevitably, the evolution of gravitational systems is characterised by collapsing cores and expanding halos.

10.6 Evolution of Dynamical Systems: Two-Body Processes

The mechanism of energy exchange by which these inevitable gravitational events happen within stellar systems is two-body relaxation. We know the outcome will be escaping stars and collapsing cores. Let us consider escaping stars—and cluster evaporation—first.

Fig. 10.2 Schematic energy level diagram for a star cluster in virial equilibrium. *Black* energy levels represent an initial state, *red* energy levels represent a subsequent state after the loss of energy (e.g. due to stellar evaporation)

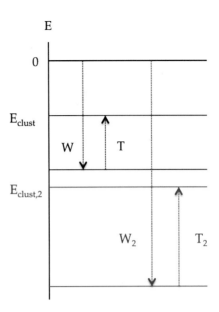

The key concept of this is shown in Fig. 10.1, which goes back to Ambartsumian and Spitzer (independently). Two-body relaxation seeks to establish a Maxwellian velocity distribution. But a gravitational system has an escape velocity.[4] As a consequence, the tail of the velocity distribution will be lost (roughly on a dynamical timescale). Two-body processes continuously re-populate that tail on a relaxation timescale, and the process continues until almost all stars have escaped. That is cluster evaporation.

A rough estimate of the evaporation timescale is straightforward to calculate. The distribution beyond the escape velocity turns out to be about 1 % of the total population. Every relaxation time the cluster loses that much mass, so roughly speaking the evolution evaporation time for the cluster is about 100 times the two-body relaxation time.

Critically for the cluster, the escaping stars carry away energy and mass. Necessarily, this loss of energy leads to collapse of the cluster. (Which by now should come as no surprise.) Consider an energy level diagram for a cluster in virial equilibrium (see Fig. 10.2). The energy of the bound cluster (E_{clust}) is of course negative; the associated ratio of the kinetic (T) and potential energy (W) reflects the virial theorem. Now if escaping stars carry away energy, lowering the cluster energy ($E_{clust,2}$), then the cluster reconfigures itself on a dynamical timescale to re-establish the virial balance. As a result two things happen. One, the potential energy (W_2) becomes deeper. In other words, the cluster (more specifically, the core) contracts. And second, the kinetic energy (T_2) has actually increased. In other words, as you

---

[4]This discussion is intentionally simplified. The escape velocity differs throughout a cluster, as will the local Maxwellian velocity distribution when the cluster is not isothermal.

remove energy from a gravitational system, it heats up! And that of course is negative heat capacity.

This is not a comfortable situation for a cluster. As energy is removed from that cluster, the core heats up, which leads to yet more energy flow outward. This is the runaway that we predicted from our intuitive physical analysis. Now we know the driving process—two-body relaxation—and the evolutionary timescale—of order $100t_r$.

There are a multitude of excellent papers examining these processes and timescales in more detail. One particularly worth noting again goes back again to Donald Lynden-Bell, who showed that this core collapse can be an unstable process. Consider a star in the halo of a cluster—the escape velocity from the halo is smaller than the core, and if the star is to be bound it cannot have a large velocity. On the other hand, stars in the core can have much higher velocity dispersions. So that means that the halo has to be colder than the core, and energy has to flow from the core to the halo. Thus the inner region contracts and heats up. The halo, which is not self-gravitating because it responds to the core potential, receives that energy and also heats up. But if the inner region heats up more than the outer region, you have a thermal runaway. For idealised conditions, this instability depends on the central concentration of the core. If it is a highly centrally concentrated core, the gravothermal catastrophe (one of the finest phrases in astrophysics!) occurs, with core collapse on the order of 15 relaxation times. In this situation, the core collapse happens much faster because of the energy draw that the halo places on the core.

There is one more instability, known as the mass-segregation instability, which may again be relevant to the Trapezium. Consider a system which has only a few stars that are much more massive than the typical star, such that these massive stars are not dictating the gravitational potential. These stars will migrate to the centre of the potential very quickly as a result of energy equipartition. Once there, they form their own self-gravitating dynamical system in the core. Schematically, the cluster potential is broad with a localised dip at its deepest point. As the rest of the cluster keeps removing energy from that small self-gravitating core of massive stars, the core has to collapse rapidly to provide the energy. And so mass segregation can actually accelerate the gravothermal catastrophe, because all of the energy that the halo is demanding from the core is being drawn from just a few massive stars at the centre.

So how do star clusters solve this fundamental gravitational problem, in all its forms? Well, we have seen this problem before in stellar evolution. And how do stars solve this problem? Well, they solve the problem at least temporarily by providing a fuel source. Instead of providing the energy demanded by the envelopes of the stars from the gravitational energy of the stellar core, nature provides an energy source—nuclear fusion. Clusters do the same thing in principle, but they do it with binaries. We will return to this later.

10.7 Evolution of Dynamical Systems: Cluster Dissolution

The final topic that I will touch on is cluster dissolution. We know that clusters have limited lifetimes. For example, in Chap. 8, I showed that the number of clusters has dropped significantly by 800 Myr or so, much sooner than the age of the Galactic disc. This almost certainly is the result of internal processes such as evaporation, likely accelerated by the Galactic tidal field and tidal impact encounters with molecular clouds.

However, for our purposes we are more concerned with the loss of the binding energy of the gas in which the clusters form than about evaporation (except for the smallest-N systems). The loss of the natal gas is certainly going to happen, and the response will be on a dynamical timescale.

If you consider a simple virialised system and you remove a fraction ε of the mass rapidly compared to the dynamical time, then it is very easy to compare the final and initial radii r of the cluster:

$$\frac{r_f}{r_i} = \frac{1 - \varepsilon}{1 - 2\varepsilon} = \frac{\eta}{2\eta - 1}. \tag{10.5}$$

Often the fraction of gas lost, ε, is rewritten in terms the star formation efficiency, η, the fraction of gas that gets converted into stars. As is well-known from introductory physics, if more than half of the total mass is lost, the system becomes unbound. But this is only true if the mass loss happens rapidly. If it happens slowly with respect to the dynamical timescale, then the system will adiabatically expand:

$$\frac{r_f}{r_i} = \frac{1}{1 - \varepsilon} = \frac{1}{\eta} = \frac{m_i}{m_f}. \tag{10.6}$$

In principle all of the mass could be lost and the system could expand to infinity without becoming unbound. If the mass-loss is slow compared to a dynamical time, then the ratio of the final mass to the initial mass is the inverse of the star formation efficiency. Thus, unless the star formation efficiency is zero, in other words unless all of the gas is converted to stars, which is not a terribly interesting scenario dynamically, the system remains bound, but it does expand and it gets very large. Of course in reality during expansion the system runs into physical issues with its environment within the molecular cloud. My main point here is that large mass-loss does not necessarily imply that a system is going to fly apart on a dynamical timescale.

This analytic analysis was carried out by several of us in the early 1980s, from which we concluded that the existence of bound clusters implied at least some locations of high star formation efficiency. Charlie Lada and his colleagues ran N-body models and added an important nuance to the discussion, that cores are more tightly bound than are clusters globally. Thus even high fractions of rapid mass-loss may leave behind bound cores that ultimately become lower mass star clusters.

References

Binney, J. & Tremaine, S. 2008, Galactic Dynamics: Second Edition, ed. J. Binney & S. Tremaine, Princeton University Press

Spitzer, Jr., L. 1988, Dynamical Evolution of Globular Clusters, ed. L. Spitzer Jr., Princeton University Press

Chapter 11
λ Ori: A Case Study in Star Formation

Robert D. Mathieu

11.1 Introduction

In previous chapters we have discussed the products of star formation, from binary stars to field stars to open clusters to the Milky Way itself. Now we turn to an overview of the star-forming regions themselves. I will begin with an intensive look into one star-forming region that has long been a favourite of mine, and then later in Chap. 12, I will give a more general overview of star-forming regions. The region for this in-depth look is the eponymous λ Orionis association, named for the brightest star in the head of Orion. It is not actually an extremely bright star (which often leads to jokes about Orion's intelligence). However the last laugh is Orion's, for λ Ori lies at the centre of one of the most spectacular shells in the IRAS 100 μm maps. λ Ori also drives a particularly beautiful Sharpless HII region. Finally, the λ Orionis region is important because the evidence suggests that a recent supernova has cleared much of the region of its natal gas. Thus λ Ori is a star-forming region that yields a final census of the outcome, and thus a near-final story of how the region evolved. Furthermore, the region is currently dissolving into the Galactic field, so that we are seeing the interface between star-forming regions and the disc of the Milky Way.

11.2 Overview

Figure 11.1 places the λ Orionis region in the setting of the nearby Galaxy, as revealed in IRAS 100 μm dust emission. Such emission generally traces the interstellar medium with temperatures >30 K. As this is significantly higher than the ambient 3 K cosmic microwave background, as well as dense molecular clouds heated only by cosmic rays (~ 10 K), this warm dust is strongly correlated with the presence of active star formation and the presences of luminous stars. The λ Ori star-forming region is

R.D. Mathieu (✉)
Department of Astronomy, University of Wisconsin, Madison, WI, USA
e-mail: mathieu@astro.wisc.edu

© Springer-Verlag Berlin Heidelberg 2015   147
C.P.M. Bell et al. (eds.), *Dynamics of Young Star Clusters and Associations*,
Saas-Fee Advanced Course 42, DOI 10.1007/978-3-662-47290-3_11

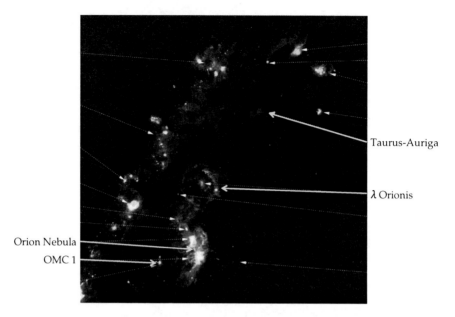

Fig. 11.1 IRAS 100 μm map of the Galactic plane in the direction of the Orion molecular cloud. The Orion molecular cloud extends from λ Orionis to the lower edge of the image. Several star-forming regions are labelled, including the Taurus-Auriga clouds, the Orion Nebula and the OMC 1. Figure adapted from Bally (2008)

located at one end of the Orion giant molecular cloud. The region comprises several OB associations as identified by Adriaan Blaauw, with the hot and luminous Orion Nebula region particularly prominent at 100 μm. The region also includes numerous embedded star clusters and is littered with young low-mass stars throughout. For comparison, the Taurus-Auriga region is also identified. As discussed in Chap. 12, this region has proven seminal for our understanding of star formation, but Fig. 11.1 shows it to be a minor player on the Galactic scale.

Our exploration of the λ Ori region begins with a census of its contents. λ Ori itself is an O8 III star with a mass of roughly 10–15 $M_\odot$. λ Ori has a lower mass B0 V companion at a separation of 4.4 arcsec (or 1900 au in projection). Nine other B stars join λ Ori in a dense clump at the centre. Surrounding λ Ori is an ionisation-bounded HII region, 50 pc in diameter. Just beyond is a ring of neutral hydrogen and associated molecular clouds. Within this ring lie a large number of low-mass young stars, both classical T Tauri stars (with actively accreting circumstellar discs) and weak-lined T Tauri stars (with low accretion rates, traced by emission lines such as Hα, or no evidence at all for a circumstellar disc). FU Orionis is a member of the λ Ori region, and the namesake of a class of very-active young stellar objects characterised by high accretion rates. Finally there are several Herbig-Haro objects and quite a few jets and outflows. So there is star formation activity still underway, although as we shall see it is spatially limited by the recent dispersal of molecular gas and none is occurring near λ Ori itself. (For a more comprehensive discussion of the region see Mathieu

2008; for an overview of young stellar objects and important physical properties see Hartmann 1998.)

11.3 Dust and Molecular Gas

Currently the interstellar dust and molecular gas of the λ Ori region are located largely in the 60 pc diameter ring. Figure 11.2 shows this ring from three observational perspectives—dust extinction, thermal dust emission and molecular gas emission from rotational lines of carbon monoxide: the three closely trace each other.

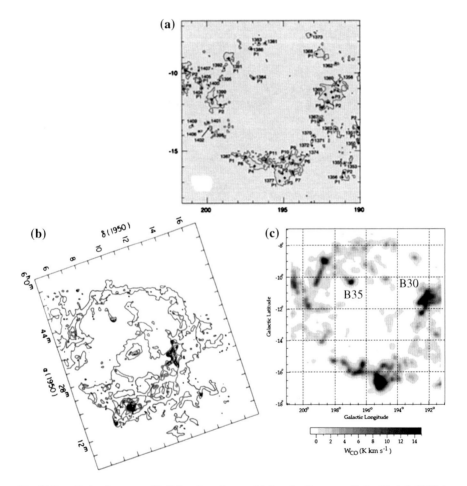

Fig. 11.2   a Extinction map of λ Ori region; *dots* are high extinction cores Dobashi et al. 2005. **b** IRAS 100 μm map. **c** CO (1-0) intensity map (Lang et al. 2000)

An early detection of the ring was made by Barnard, who identified numerous dark clouds in the region (Barnard 1919). The left, middle and right panels of Fig. 11.2 show an extinction map, an IRAS 100 μm map and a CO (1-0) intensity map of the λ Ori region from Dobashi et al. (2005), Maddalena and Morris (1987), Lang et al. (2000) respectively. Two dark clouds that will be particularly significant to our story, Barnard 30 and Barnard 35, are identified. The total molecular mass in the ring now is about 10^4 M$_\odot$, comparable to a giant molecular cloud at the smaller end of the mass range for these objects. The current mass is of course a lower limit on the original molecular gas mass. Interestingly there have been suggestions that the ring is expanding (Maddalena and Morris 1987), but at this point the kinematics remain uncertain.

11.4 Massive Stars

Analysis of the massive star population requires techniques of classical stellar astrophysics. Dolan and Mathieu (2001) used Strömgren photometry for 20 OB stars in the region to place them on a theoretical Hertzsprung-Russell diagram, shown in Fig. 11.3. The Strömgren technique uses a reddening-independent colour ratio to

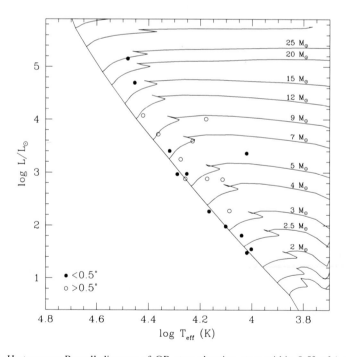

Fig. 11.3 Hertzsprung-Russell diagram of OB stars showing stars within 0.5° of λ Ori (*filled circles*) and more than 0.5° away (*open circles*). Also shown are evolutionary tracks from Schaller et al. (1992). Figure from Dolan and Mathieu (2001)

estimate the spectral type (thus permitting an estimate of extinction from comparison of expected colours to those observed), as well as constraints on surface gravity from measurement of Stark broadening from the Hβ index. The filled circles represent the stars in the central clump, with the two most luminous being λ Ori and its companion HD 36822; the open circles represent the more distributed stellar population.

Twelve of the 20 stars fit well to the theoretical main-sequence, suggestive of membership in the association. The two most luminous stars—λ Ori and HD 36822—are presumed members that have evolved off of the main-sequence, providing a clock with which to date the region as discussed below. The remaining 6 stars lie above the main-sequence by more than can be easily explained by binarity or evolution in a coeval stellar population. They may simply be foreground stars, although one of them—HD 36881—lies amidst the clump immediately around λ Ori. For the high-mass stars we have a good census of the products of this star-forming region, with the exception of yet more massive stars that have gone supernova.

11.5 Low-Mass Stars

Thus far, I have avoided using the phrase 'OB association' to describe the λ Ori star-forming region. With our ever increasing ability to discover young low-mass stars, it is clear that OB stars are simply the highly visible high-mass end of an IMF that almost always extends also to very-low stellar masses. I now introduce several ways that we can find low-mass stars in star-forming regions, using the λ Ori region as a case study.

11.5.1 Hα Emission and Objective Prism Surveys

Finding young low-mass stars through their strong Hα emission dates back to Haro and Herbig in the 1950s, long before it was known that the emission results from accretion from discs onto the pre-main-sequence (pre-MS) stars (see e.g. Haro 1953; Herbig 1954). Duerr et al. (1982) completed an objective prism survey of the entire λ Ori region. The filled circles in Fig. 11.4 show the union of classical T Tauri stars discovered both by these authors and Haro in an earlier study.

Interestingly, from the point of view of present and future synoptic studies, the vast majority of the stars found as Hα emitters are also known variable stars in the region. T Tauri stars as a class were identified in part through their variability by Joy (1945). Large amplitude irregular (and in some cases periodic rotation-related) accretion variability (>0.1 mag in the visible) is an efficient way to search for low-mass young stars. Lower amplitude periodic variability can also be used to measure rotation periods for young (and old) stars due to cool spots rotating on the stellar surface.

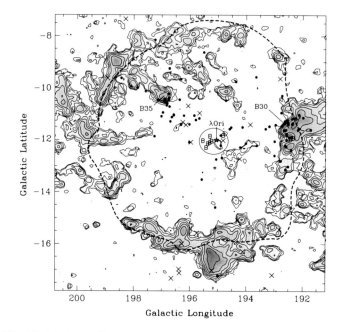

Fig. 11.4 The *filled circles* are Hα sources cataloged by Duerr et al. (1982). The *contours* are the CO (1-0) map of Lang and Masheder (1998), the *dashed line* delimits the HII region, and the 'B' symbols denote B stars near λ Ori. Figure from Dolan and Mathieu (1999)

Even though Duerr et al. (1982) surveyed the entire region, they only found classical T Tauri stars in a roughly linear domain extending from B35 to B30, about 50 pc in length. At about the same time, radio astronomers using millimetre wave molecular line observations were discovering that giant molecular clouds also tended to be elongated at about this scale in size. As a result, Duerr et al. (1982) conjectured that they were looking at a fossil giant molecular cloud within which the group of OB stars and low-mass stars had formed.

11.5.2 Lithium Absorption and Multi-Object Spectroscopic Surveys

In terms of obtaining a complete stellar census, a problem with Hα surveys is that they primarily reveal the classical T Tauri stars, whereas star-forming regions also have substantial populations of young low-mass stars that are not actively accreting. So how does one find all young low-mass stars in a systematic and unbiased way?

A characteristic property of young low-mass stars is presence of strong lithium (Li) absorption lines. Li is present in the photospheres of only young stars in cosmic abundances because it requires temperatures above $\simeq 3 \times 10^6$ K in order to be processed through nuclear reactions. Note this appears before central temperatures of $\simeq 10^7$ K are reached through pre-MS contractions (the onset of the main-sequence where hydrogen burning is initiated). As young low-mass stars are nearly completely convective, the surface Li is brought deep below to the high temperature zones, burnt, and depleted from the surface photospheric abundances. Thus the resonant Li feature at 6708 Å is a diagnostic of youth, if one has the means to obtain a large number of intermediate-resolution spectra ($R > 3000$). Such spectra also provide estimates of surface gravity and heavy element abundance in addition to spectral type (i.e. temperature). If high enough resolution is obtained ($R > 10,000$) they also provide radial velocities, adding an additional kinematic dimension to membership determination.

This survey approach became feasible with the advent of large-N multi-object spectrographs. To give a sense of scale, the WIYN Observatory 3.5-m telescope with the Hydra multi-object spectrograph permits placement of nearly 100 optical fibres on stars within a $1°$-diameter field-of-view. This proved a powerful capability for the WIYN Li survey to be discussed, but even so, the λ Ori region is of order $10°$ in diameter on the sky. A few multi-object spectrographs with larger fields-of-view and higher multiplex advantage have been commissioned in recent years.

Technically, a spectroscopic resolution of 20,000, or 15 km s^{-1}, suffices. Working in the 6640 Å spectral region provides Li 6708 Å, Hα, and a rich array of metal lines blueward of Hα for measuring radial velocities ($\sigma < 1$ km s^{-1}). Spectra with signal-to-noise yielding equivalent-width detection limits of order 0.1 Å suffice. None of

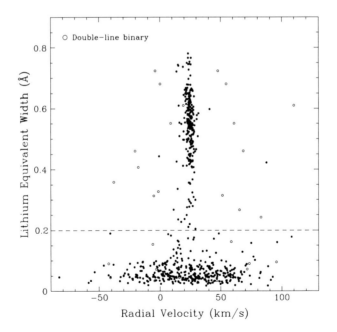

Fig. 11.5 Li line strength and radial velocity cleanly discriminate the low-mass pre-MS stars in the λ Ori star-forming region. Those stars with Li equivalent widths greater than 0.2 Å are taken to be members. These stars have a mean velocity of 24.5 km s^{-1}, similar to that of the Orion molecular cloud, and a velocity dispersion of only 2.3 km s^{-1}. Figure from Dolan and Mathieu (2001)

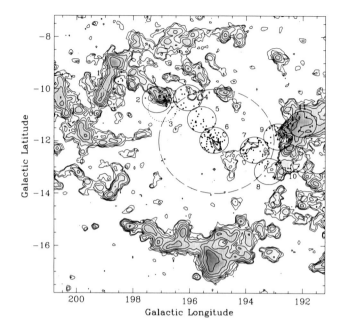

Fig. 11.6 *Smaller circles* are the target fields for the WIYN Li survey for young stars. The *filled dots* mark stars with strong Li absorption and radial velocities commensurate with the Orion region, and thus members of the λ Ori star-forming region. Compare to the Hα survey results in Fig. 11.4. Figure from Dolan and Mathieu (2001)

these specifications are particularly challenging; it is the wide-field large-N multi-object spectrograph that is the key.

Figure 11.5 shows a plot of Li equivalent width versus radial velocity for an unbiased population of late-type stars in the immediate vicinity of λ Ori. The separation between members of this star-forming region and the field is very clean in both dimensions: any star above 0.2 Å Li equivalent width appears to be a member and most have radial velocities consistent with Orion. In addition, if multi-epoch data are available the short-period pre-MS binaries can also be discerned. The 266 pre-MS stars revealed in Fig. 11.5 required about 4200 spectra of 3600 stars ($R < 16$ mag) obtained over the course of approximately 10 nights. The reward for this effort is shown in Fig. 11.6. The 1°-diameter regions surveyed with WIYN follow the linear extension revealed by the Hα surveys, but reveal nearly four times as many young low-mass stars. This unbiased (representative, but not complete) survey yields an estimate of the IMF and ultimately the star formation history of the region.

11.5.3 Photometric Surveys

Photometric surveys for young stars in associations rely on the over-luminosity of pre-MS stars compared to main-sequence stars. Figure 11.7 shows the deep $(RI)_c$

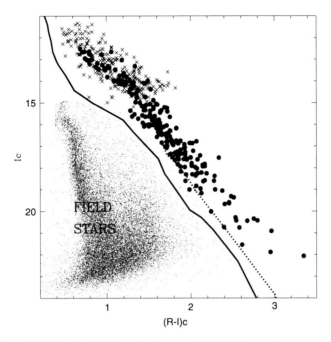

Fig. 11.7   Deep CMD for the field immediately around λ Ori. *Crosses* denote spectroscopically identified members, whereas *filled circles* represent photometrically identified candidate members. The *solid line* is the ZAMS and the *dotted line* is a 5 Myr isochrone both projected to the estimated distance of the region. Figure from Barrado y Navascués et al. (2004)

photometry of Barrado y Navascués et al. (2004) in the immediate vicinity of λ Ori. The solid line represents the zero-age main-sequence (ZAMS) at the Orion distance, while the dotted line is a 5 Myr isochrone; stars above the ZAMS are candidate association members. The separation of most field stars from the candidate members is good, but even so for any one candidate member another diagnostic of membership or youth is in order (e.g. follow-up spectroscopy). Another complication is the effect of interstellar reddening, making all objects appear redder and fainter than they are. In this CMD, the reddening vector is nearly parallel to the isochrones. Thus the age estimate may be valid, but the mass of the object is degenerate with the assumed stellar mass (in the absence of follow-up spectra). The beauty of this particular survey is its depth, reaching down to young brown dwarfs.

Wide-field photometric surveys are necessary to both obtain a global IMF for the region and to explore the spatial distribution of past star formation. Dolan and Mathieu (2002) performed a *V R I* photometric survey over the entire association, the results of which are shown in Fig. 11.8. In this study, after CMD selection a further statistical approach was taken to remove remaining field contamination, so the dots in Fig. 11.8 are 'proxy stars' representing over-densities of pre-MS stars. Despite

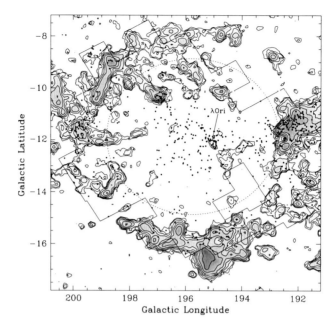

Fig. 11.8 Distribution of 'proxy' pre-MS stars based on optical photometry. The *solid outline* defines the domain of the survey. Note that fields outside the *dashed circle* are projected on the molecular clouds, so the surface densities are likely artificially enhanced by reddening. Figure from Dolan and Mathieu (2002)

surveying the entire region, there is very little evidence for prior star formation outside the linear region extending from B30 to B35.[1]

Very productive wide-field flux-limited surveys have also been performed at X-ray wavelengths. Young stars tend to have very active chromospheres, so that large X-ray luminosities (relative to the bolometric luminosities) are also an excellent diagnostic of youth. The ROSAT All-Sky Survey was able to detect young FGK stars within a couple of hundred parsecs while XMM-*Newton* and *Chandra* observations can detect much lower mass young stars as well as probe star-forming regions at greater distances (Barrado et al. 2011).

11.6 Analysis of a Star-Forming Region

Here we are using the λ Ori region as a tutorial case for a broad analysis of a star-forming region. Every region is unique and so no one region is truly an exemplar. And given the wonderful diversity of astrophysical phenomena in these regions no analysis can be comprehensive. Even so, this region is a good example of how surveys and analyses of star-forming regions are conducted.

---

[1] These photometric data were used to select out field stars from the target sample for the multi-object Li spectroscopy in Figs. 11.5 and 11.6.

11.6.1 Distance

The first thing we need is a reliable distance measurement. For the λ Ori region the photometric distance, which is to say main-sequence fitting to the more massive stars (see Fig. 11.3), remains the best measure at 450 ± 50 pc. This is comparable to the Orion molecular cloud, and so in conjunction with the kinematic (radial velocity) agreement an association of λ Ori with the much larger Orion molecular complex is likely. A more direct estimate of the distance to the Orion Nebula Cluster using VLBI astrometric measurements of masers with known velocities has yielded a distance consistent with the above. The OB stars are a bit far away for accurate and precise *Hipparcos* distances, although the measurements are formally consistent with the photometric distance. *Gaia* should finally provide excellent distances and hopefully three-dimensional structure.

11.6.2 Spatial Distribution of Star Formation

The λ Ori region is marked by the central clump of OB stars, with a radius of only about 2 pc. The compact spatial distribution of this clump is a very important clue to the evolution of the star-forming region. Whether determined from the associated low-mass stars or from *Hipparcos* proper motions, the velocity dispersion of the OB stars is $\simeq 2$ km s^{-1} in one dimension. This is typical of the internal motions of molecular clouds, but currently the only mass present is in the stars, and their mass is much too small to bind the system.[2] Thus on kinematic grounds alone this clump cannot have been unbound for more than $\simeq 1$ Myr.

The other clue from the spatial distribution is the roughly linear distribution of the lower-mass stars. They are not distributed throughout the ring, which implies that the ring was made *post facto*. We also have surface density enhancements of optically visible low-mass stars at B35 and B30, as well as several embedded sources in B35 indicating on-going star formation. As pointed out by Duerr et al. (1982), this spatial distribution of the low-mass stars suggests that they map out a previous giant molecular cloud.

11.6.3 Initial Mass Function

Young regions like λ Ori enable an almost 'pure' measurement of the IMF covering a broad range in mass and needing only modest corrections for the effects of stellar evolution for only the most massive stars. The IMF of the λ Ori region is interesting in several important ways. Figure 11.9 shows the IMF in the immediate vicinity of

---

[2] Just for a quick comparison, the old open cluster NGC 188 (see Chap. 8) has a total mass of about 1,000 M$_\odot$ and a similar core radius, yet an internal velocity dispersion of only 0.4 km s^{-1}.

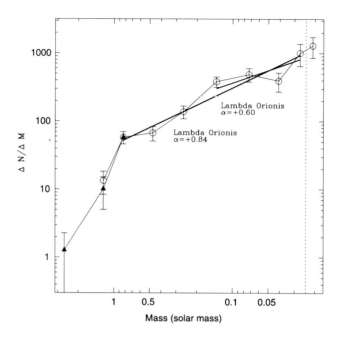

Fig. 11.9 IMF for the λ Ori star-forming region. *Filled triangles* denote data from Dolan and Mathieu (2001), whereas *open circles* represent data from Barrado y Navascués et al. (2004). The *vertical dashed line* represents the completeness limit of the Barrado y Navascués et al. study. The two power-law fits correspond to fits across different mass ranges, namely the stellar/substellar boundary (0.03–0.14 $M_\odot$; $\alpha = +0.6$) and the full sample (0.03–1 $M_\odot$; $\alpha = +0.86$). Figure from Barrado y Navascués et al. (2004)

λ Ori from Barrado y Navascués et al. (2004). Broadly speaking there is reasonable agreement between the IMF local to λ Ori and the field.

However, a detailed analysis shows important differences between local and global IMFs for the λ Ori region. Globally there are 107 stars throughout the region with masses between 0.4 and 0.9 $M_\odot$. Using the IMFs of Miller and Scalo (1979) or Kroupa et al. (1993) one would thus predict 10 or 19 OB stars in the region, respectively. In fact there are 16; so globally the IMF is as expected if drawn from that characterising the field (see e.g. Bastian et al. 2010). Interestingly, however, in the vicinity of λ Ori itself the number of low-mass stars is low by roughly a factor of two compared to the field, whereas away from λ Ori in the B30 and B35 regions the low-mass stars are over-populated by roughly a factor of three. Thus the global IMF of the λ Ori star-forming region resembles the field, while the local IMF appears to vary across the region. Perhaps it is the integration of the star formation process over the entire region that produces the field star IMF.

11.6.4 Total Stellar Population

The global stellar census yields a total mass in the region of all objects greater than $0.1\,M_\odot$ of 450–$650\,M_\odot$. The current total molecular cloud mass is $10^4\,M_\odot$. Using this ratio, the global star formation efficiency in the region is very low, only a few percent. This is typical of star-forming regions. As discussed in Chap. 4, it is the local star formation efficiency which is most critical for determining whether a grouping of young stars is bound. Regrettably, we have no idea what was the mass of the dense core of the natal molecular cloud within which the OB stars likely formed, and so have no direct measure of the local star formation efficiency to compare with the global measure. However, combining information from spatial distributions and measured kinematics can place some constraints.

11.6.5 Accretion Disc Evolution

This topic is not central to the dynamics of star-forming regions, but is nonetheless an interesting outcome of the observations relevant to the initial conditions of planet formation. Furthermore, stellar dynamics might influence the structure and evolution of circumstellar discs. Recall that in the Li survey the spectra also included the Hα line, which for emission equivalent widths above $10\,\text{Å}$ is taken as a proxy for active disc accretion. The presence of classical T Tauri stars with active accretion is evident near the dark clouds B35 and B30. On the other hand, there is a marked lack of Hα emission from the low-mass stars around the OB clump. Many of the Hα stars associated with B30 and B35 have ages similar to pre-MS stars found in the cluster near λ Ori. Dolan and Mathieu (2001) suggest that the absence of Hα emission from the central pre-MS stars is the result of an environmental influence linked to the luminous OB stars (for example external photoevaporation of the circumstellar gas in discs or dynamical interactions driving more rapid disc evolution). Similar results have been found for the λ Ori region with other accretion disc diagnostics (see e.g. Sacco et al. 2008; Barrado et al. 2011). Yet Hernández et al. (2009) find from deep *Spitzer* data that the frequency of accretion discs increases with lower stellar mass and argue for an evolutionary effect whereby the discs around lower-mass stars evolve more slowly than discs around higher-mass stars (cf. Hillenbrand et al. 1998; Carpenter et al. 2009).

11.6.6 Age Distribution

One of the reasons we embarked on a study of this star-forming region was to look for evidence of sequential star formation. For example, did a supernova from a star in the OB star clump accelerate star formation in the molecular clouds B30 and B35?

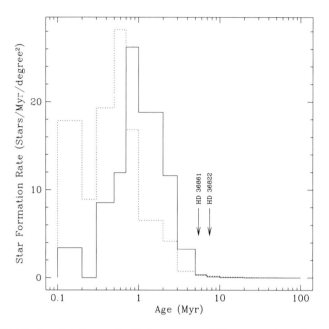

Fig. 11.10 Distribution of stellar ages for the low-mass pre-MS stars as estimated using the models of D'Antona and Mazzitelli (1998). The individual histograms are composed of stars within a 2° radius of λ Ori (*solid line*) and outside 2° (*dashed line*). The *arrows* mark the ages of the two evolved massive stars. Figure from Dolan and Mathieu (2001)

Strömgren photometry of the OB stars allows fairly precise age-dating of the two evolved OB stars, λ Ori and HD 36822. Pre-MS evolutionary tracks allow dating of the young solar-type stars, although the systematic uncertainties between sets of models are substantial. Dolan and Mathieu (2001) show the stellar age distributions for three sets of pre-MS models; for demonstration purposes one of these is shown Fig. 11.10, in which the D'Antona and Mazzitelli (1998) models were used to estimate individual stellar ages. For a detailed discussion of the pros and cons of various age-dating techniques the reader is referred to Soderblom (2010).

In the context of these models, there was a general onset of star formation in the region of the current OB star clump roughly 6–7 Myr ago. Both λ Ori and HD 36822 formed rather early in the story, and then there was a rapid rise in lower-mass star formation in the λ Ori vicinity. About 1 Myr ago the star formation in the λ Ori vicinity drops off precipitously. In the B30 and B35 regions the formation started later, peaked more recently (approximately a million years ago) and is still underway (as also evidenced by currently embedded stars in the clouds). The story is much the same using other models, but the time scales are different. As the OB stars and the solar-type stars are being dated using quite different astrophysics, it is not clear the absolute scales are comparable: it is difficult to rule out quantitatively one scenario or another. In any event, the evidence for sequential star formation in the region is not compelling. Yet, the evidence for star formation ending recently around the OB clump while still continuing in the remaining molecular clouds is stronger. Recall that kinematic arguments suggest that the OB clump could not have lost its binding gas more than 1 Myr ago.

11.7 The Star Formation History of λ Ori

What story can be told for the λ Ori star-forming region inspired by this analysis? A possible scenario is presented schematically in Fig. 11.11. It begins with an elongated molecular cloud having several large-scale density enhancements. Roughly 6 Myr ago numerous OB stars were born in the central cloud but not elsewhere. Low-mass star formation began about the same time in all of the density enhancements, with a birth rate increasing gradually over several million years.

1 Myr ago a massive star in the OB clump went supernova, shredding the central cloud and thus unbinding the central stellar population. A ring of gas was pushed out from the central region, forming much of the ring seen today. (Importantly, Lang et al. 2000 argue that several massive clouds in the ring, such as B223, likely formed near their current locations.) Remnants of the adjacent dense clouds remain as B35 and B30; stars that had been formed within the forward edges of these dense clouds were exposed by the blast wave. Today we have star formation continuing in the remaining gas but it has ceased in the vicinity of the supernova epicentre.

Can we make any predictions concerning the future? The termination of star birth in B35 is 'imminent', as is the escape of the OB stars from their central position. All of the stars will disperse into the field over the next 10 Myr or so. The λ Ori

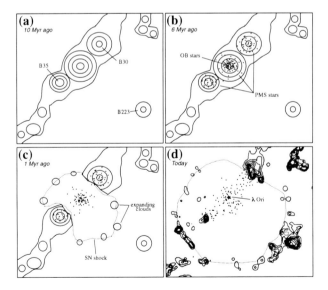

Fig. 11.11 Schematic history of the λ Ori star-forming region showing conditions at 10, 6 and 1 Myr ago, as well as a map of the molecular clouds today. Figure from Dolan and Mathieu (2002)

star-forming region will contribute field stars with an IMF similar to the current estimates, even though the IMF may have been inhomogeneous across the region. If *Gaia*'s successor observes this region 1 Myr from now, perhaps it will identify a remnant association of B stars with a paucity of lower-mass stars. It may also detect a more distributed complex of low-mass stars with similar space velocities, similar to what is known as the Ursa Majoris stream. Eventually the shear of the Galaxy will tear apart the moving group and the λ Ori star-forming region will have completed its contributions to the Galactic stellar population.

11.8 Final Thought

In terms of the dynamics and evolution of star-forming regions, the long-standing question of whether most stars form in OB associations or in T associations or in embedded clusters or in open clusters rests in part on distinctions derived from historical observational techniques. All stars are forming within giant molecular clouds that eventually disperse, leaving behind bound and unbound ensembles of stars across the entire IMF.

At the same time, there remain fundamental questions of star formation that are beyond the purview of this manuscript. I have suggested why star formation ended in the λ Orionis region, but have made no mention of why it started. The observations in hand are largely silent about the star formation rate and what determines it. While I have suggested that the IMF is not universal locally within the giant molecular cloud, nothing has been said about what in the star formation process determines stellar mass. The impact of these issues on the subject of this book remain to be seen.

References

Bally, J. 2008, Handbook of Star Forming Regions, Volume I: The Northern Sky, ed. B. Reipurth, ASP Monograph Publications, 459
Barnard, E. E. 1919, ApJ, 49, 1
Barrado, D., Stelzer, B., Morales-Calderón, M., et al. 2011, A&A, 526, 21
Barrado y Navascués, D., Stauffer, J. R., Bouvier, J., Jayawardhana, R., & Cuillandre, J.-C. 2004, ApJ, 610, 1064
Bastian, N., Covey, K. R., & Meyer, M. R. 2010, ARA&A, 48, 339
Carpenter, J. M., Bouwman, J., Mamajek, E. E., et al. 2009, ApJS, 181, 197
D'Antona, F. & Mazzitelli, I. 1998, in Astronomical Society of the Pacific Conference Series, Vol. 134, Brown Dwarfs and Extrasolar Planets, ed. R. Rebolo, E. L. Martin, & M. R. Zapatero Osorio, 442
Dobashi, K., Uehara, H., Kandori, R., et al. 2005, PASJ, 57, 1
Dolan, C. J. & Mathieu, R. D. 1999, AJ, 118, 2409
Dolan, C. J. & Mathieu, R. D. 2001, AJ, 121, 2124
Dolan, C. J. & Mathieu, R. D. 2002, AJ, 123, 387
Duerr, R., Imhoff, C. L., & Lada, C. J. 1982, ApJ, 261, 135

Haro, G. 1953, ApJ, 117, 73

Hartmann, L. 1998, Accretion Processes in Star Formation, ed. L. Hartmann, Cambridge University Press

Herbig, G. H. 1954, ApJ, 119, 483

Hernández, J., Calvet, N., Hartmann, L., et al. 2009, ApJ, 707, 705

Hillenbrand, L. A., Strom, S. E., Calvet, N., et al. 1998, AJ, 116, 1816

Joy, A. H. 1945, ApJ, 102, 168

Kroupa, P., Tout, C. A., & Gilmore, G. 1993, MNRAS, 262, 545

Lang, W. J. & Masheder, M. R. W. 1998, PASA, 15, 70

Lang, W. J., Masheder, M. R. W., Dame, T. M., & Thaddeus, P. 2000, A&A, 357, 1001

Maddalena, R. J. & Morris, M. 1987, ApJ, 323, 179

Mathieu, R. D. 2008, Handbook of Star Forming Regions, Volume I: The Northern Sky, ed. B. Reipurth, ASP Monograph Publications, 757

Miller, G. E. & Scalo, J. M. 1979, ApJS, 41, 513

Sacco, G. G., Franciosini, E., Randich, S., & Pallavicini, R. 2008, A&A, 488, 167

Schaller, G., Schaerer, D., Meynet, G., & Maeder, A. 1992, A&AS, 96, 269

Soderblom, D. R. 2010, ARA&A, 48, 581

Chapter 12
Overview of Star-Forming Regions

Robert D. Mathieu

12.1 Introduction

The dynamical evolution of young star clusters and associations is an evolving interplay of the gravitational potential, as determined by the spatial distribution of gas and stars, and the motions of the gas and stars. Arguably, more investigation has been put into the content and structure of star-forming regions than any other aspect of star formation, far more than can be comprehensively presented here. Rather, the goal is to introduce a few nearby star-forming regions which are important for our discussion of dynamics. These regions are not characteristic in any systemic sense. Their significance derives from the extensive studies that have been applied to them over the years, and thus their role in laying the foundation for our understanding of star formation.

12.2 Taurus-Auriga

An appropriate place to start a discussion of the Taurus-Auriga[1] star-forming region is the spectacular photograph of the Taurus dark clouds taken by Barnard et al. (1927; see Fig. 12.1). The filamentary structure, the dense cores, even some of the young stellar objects are all evident in this image. At the time there was still discussion about whether these dark areas were intervening clouds or holes in the stellar distribution.

In Fig. 12.2 is shown a star count map and a J–H colour map based on the 2MASS all-sky survey data. Even in the near-infrared, the clouds remain defined by reduction in stellar surface density. Furthermore there is a close association of enhanced

---

[1] A detailed overview of the Taurus-Auriga star-forming region and its constituents can be found in Kenyon et al. (2008).

R.D. Mathieu (✉)
Department of Astronomy, University of Wisconsin, Madison, WI, USA
e-mail: mathieu@astro.wisc.edu

© Springer-Verlag Berlin Heidelberg 2015
C.P.M. Bell et al. (eds.), *Dynamics of Young Star Clusters and Associations*,
Saas-Fee Advanced Course 42, DOI 10.1007/978-3-662-47290-3_12

Fig. 12.1 Photographic image of the Taurus dark clouds (Barnard et al. 1927)

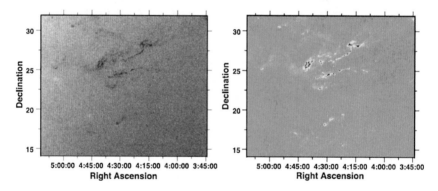

Fig. 12.2 Maps of near-infrared star counts (*left*) and J–H colours (*right*) in the Taurus-Auriga region, based on the 2MASS survey. Figure adapted from Kenyon et al. (2008)

reddening with low stellar surface density: evidently the Taurus-Auriga dark clouds are in fact dust clouds.

Figure 12.3 shows a CO $J = 1$-0 map of the Taurus-Auriga region, showing yet again the characteristic tilted U-shape distribution of the material in the region. The dust (see Fig. 12.2) and molecular gas are intimately associated. The total amount of molecular gas—the primary diffuse constituent—is about 10^4 $M_\odot$, so in fact the Taurus-Auriga clouds are a rather small star-forming region made important only by their proximity of 140 pc from the Sun.

Of particular importance is the evident clumpiness on smaller size scales (see Fig. 12.4). Sites of particular high extinction have long been known since they were

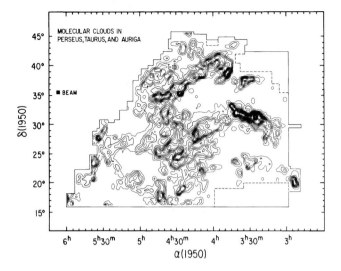

Fig. 12.3 Map of velocity-integrated CO emission. Figure from Ungerechts and Thaddeus (1987)

Fig. 12.4 Positions of cores in the Taurus region on a $C^{18}O$ integrated intensity map. Figure from Onishi et al. (1998)

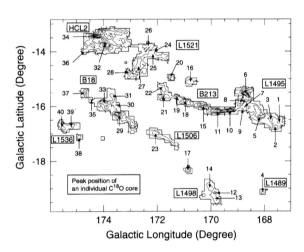

catalogued by Barnard and later Lynds. These dense cores have been explored in depth over the last several decades, beginning with the pioneering studies by Myers and Benson with high-density molecular tracers ($> 10^5$ cm^{-3}) that showed the cores to be sites of 1–10 M$_\odot$ of dense molecular gas roughly 0.1–1 pc in diameter.

With the launch of the IRAS satellite the close association of these dense molecular cores with low-mass star formation was evident in the correlation of infrared source positions with many of these cores. On the other hand, the X-ray survey missions such as ROSAT were effective at finding the chromospherically bright, often older, pre-main-sequence stars scattered throughout the region and not necessarily closely associated with the current distribution of gas and dust.

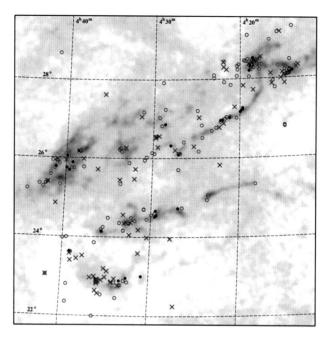

Fig. 12.5 Positions of Class 0/I (*black filled circles*), II (*red open circles*) and III (*blue crosses*) objects in the Taurus star-forming region superimposed on a greyscale extinction map. Figure from Luhman et al. (2010)

Figure 12.5 gives an overview of the young star distribution in the Taurus-Auriga clouds. Here Class 0/I objects have rising near- to far-infrared spectral energy distributions (deeply embedded protostars), Class II objects are T Tauri stars with active accretion discs ('classical T Tauri stars'), and Class III objects are pre-main-sequence stars suffering modest extinction with minimal or no circumstellar discs. The essential point of Fig. 12.5 is that there is an extremely tight correlation of the Class 0/I objects with the molecular gas distribution, while the Class III objects can be found both associated with gas or not. This evolving spatial distribution with evolutionary state (and, loosely, to astrophysical age) of young stars reflects the dissolution of this star-forming region.

The Class 0/I objects are highly spatially correlated, or clumped, with a mean separation of about 0.3 pc, which is on the order of dense core radii: the clumpiness in the molecular gas shows up in the stellar distributions, with surface densities of 20–30 pc^{-2} (e.g. Gomez et al. 1993).

Even so, these stellar surface densities are not sufficient to bind the groups. Indeed, other than the multiple star systems such as binaries and triples, there likely is no stellar structure in Taurus-Auriga that is itself bound. When the gas is dispersed—and how that happens is an open question given the lack of OB stars—the very loosely bound young stars of Taurus-Auriga will simply diffuse into the field star populations of the Milky Way.

12.3 Orion Molecular Cloud

The Orion Molecular Cloud (OMC) is the nearest giant molecular cloud, comprised of two sub-units, and is unremarkable in terms of its size or mass. Orion A is associated with the Orion Nebula Cluster (ONC) and the L1641 dark cloud. Orion B is associated with the Flame (NGC 2024) and Horsehead Nebulae (also NGC 2023, 2068, and 2071). Figure 12.6 shows the entirety of the cloud in 100 μm dust emission, and also places it in the Galactic context of other nearby star-forming regions. Detailed discussions of the stellar content as well as the gas and dust distributions in the region can be found within several chapters of the Handbook of Star Forming Regions Vol. II: The Southern Sky by Reipurth (2008).

Any discussion of the evolution of the several OB associations in the vicinity of the OMC must start with Adriaan Blaauw. Figure 12.7 is an optical image of the OMC region, identifying the associations studied by Blaauw in the early 1960s. For a sense of scale, at the top of the image is the λ Orionis association whose evolution was discussed in Chap. 11. Note that the size-scale of the clump of 10–12 OB stars at the core of the λ Ori region is comparable to the ONC and there are a comparable number of OB stars in the two regions. The essential difference is that the ONC is still associated with its natal gas, while in the older λ Ori region the natal gas has been dispersed, likely by a supernova.

Moving on to the classic OB associations of Blaauw, two of his essential findings were that the ages of the associations covered a range of ∼10 Myr and correlated inversely with their physical extent (as can be seen in Fig. 12.7). He further used proper motions to show that the so-called nuclear turn-off ages (from post-main-

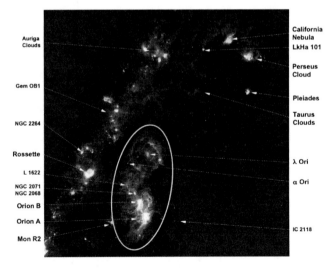

Fig. 12.6 100 μm map of the Galaxy in the direction of Orion. The Orion molecular cloud is shown within the oval. Figure adapted from Bally (2008)

Fig. 12.7 Hα map of the
OMC region, with classical
OB associations and ages
identified. Image courtesy
of W.J. McDonald

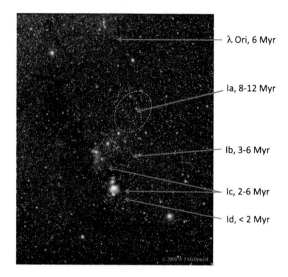

λ Ori, 6 Myr

Ia, 8-12 Myr

Ib, 3-6 Myr

Ic, 2-6 Myr

Id, < 2 Myr

sequence evolution of the most massive stars) correlated roughly with the dynamical
(expansion) ages of the associations. Put simply, Orion was a laboratory to study the
evolution and dissolution of OB associations.

Though Blaauw focused on the OBA stars, in fact low-mass star formation is
also happening throughout the region, both within the OB associations and beyond.
Of course this was also known by Blaauw and his contemporaries, given the rich
population of T Tauri stars discovered through their variability and by Hα surveys.
A modern near-infrared survey for variability is shown for Orion A in Fig. 12.8,
showing the broad distribution of low-mass young stars along the north-south ridge.
Given typical internal motions of 1 km s^{-1}, the crossing time of even the denser
Orion Nebula region exceeds the typical stellar ages of 1 Myr. Again, the structure of
the gas tends to dictate the initial spatial distribution and dynamics of star-forming
regions.

The highest density of near-infrared variables is associated with the ONC, the
densest region of current star formation in Orion to which we will turn next. The
essential point here from Fig. 12.8 is that star formation is a spatially continuous
process throughout molecular clouds. We tend to classify and thereby segregate
conceptually. But from the point of view of star formation, there is no clean boundary
between the distributed star formation in L1641 and the rich cluster in the ONC (cf.
Allen and Davis 2008). The distinction between T associations and OB associations
is largely historical and observational. Every i.e. 'OB association' in the OMC also
has an associated population of low-mass T Tauri stars.

Following the insight of Blaauw, let us now focus on a canonical case that illustrates the dynamical evolution of associations. The ONC (see Fig. 12.9; also known
as the Trapezium Cluster) is the most massive, compact and youngest collection of
young stars within the OMC. While an accounting depends on choice of boundary, the stellar population includes several thousand stars. The typical ages are 1–
2 Myr, and others remain embedded along the line-of-sight. The IMF is log-normal

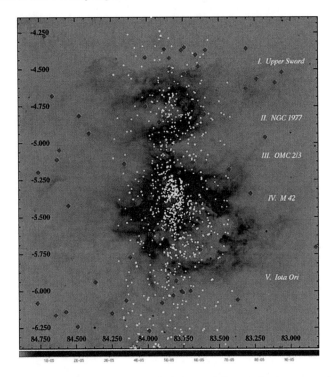

Fig. 12.8 The distribution of young stars toward Orion A. *Yellow circles* denote near-infrared variable stars, whereas the *blue diamonds* and *red squares* represent the OB members of the Orion OB 1c and OB1d associations respectively. Note that the highest surface density of the *yellow circles* corresponds to the position of the Orion Nebula region. The reverse greyscale image is an MSX 8 μm image. Figure from Muench et al. (2008)

Fig. 12.9 The Orion Nebula Cluster at three wavelengths—optical (*left*), near-infrared (*middle*) and X-ray (*right*). Figures from Lada and Lada (2003) and Lada (2010)

(with some uncertainty at the very-lowest stellar masses; cf. Andersen et al. 2011; Da Rio et al. 2012), much like that derived from the field (Muench et al. 2002).

It is illustrative to consider the ONC as an open cluster as done by Hillenbrand and Hartmann (1998). King model fits yield a core radius of only 0.2 pc. Note that the short crossing time of order 0.1 Myr supports the use of a dynamical equilibrium model, at least for the core. With 2200 stars within this radius, the core density becomes $2.1 \times 10^4\,M_\odot\,pc^{-3}$. The core radius is much smaller than typical for open clusters, and the core density much higher, by an order of magnitude.

However, the crossing time for stars in the halo of the cluster approaches or exceeds the ages of the stars. In fact, outside the core the cluster is clearly elongated in the north-south direction (see e.g. Fig. 1 in Hillenbrand and Hartmann 1998). Even if the gravitational potential is no longer dominated by the gas, most of these stars have not traversed the new potential; certainly much of the system has not completed violent relaxation (cf. the cold collapse scenario of Allison et al. 2010). Finally, the dynamical mass (i.e. derived from stellar motions) for the cluster is $4800\,M_\odot$ while the observed mass is $1800\,M_\odot$, which leaves open the possibility that the cluster may be unbound and expanding depending on the distribution of the gas mass and its contribution to the potential. With the possible exception of the core, the ONC will almost certainly evolve on dynamical timescales. Numerous authors have noticed that the ONC is mass segregated, at least for the most-massive stars (see e.g. Fig. 6 in Hillenbrand and Hartmann 1998). Whether this is the consequence of the star formation process or dynamical relaxation is a key question whose answer is not yet entirely clear. The two-body relaxation time for the core is of order 1 Myr, and dynamical friction on the most massive stars makes the equipartition time shorter still.

Perhaps the most interesting question for us is whether the ONC is destined to emerge as a bound open cluster or dissolve as the other OB subgroups in Orion. The comparison of the observed and dynamical masses suggests that the cluster is unbound. This seemingly straightforward conclusion is somewhat complicated by uncertainty about location of the gas. The Orion Nebula is well-known to be a blister on the front face of the OMC. We do not know whether the several thousand solar masses of gas along the line-of-sight are actually very near the cluster, dynamically associated with it and perhaps binding it together, or whether this material is well behind the cluster. In either case its tidal impact must also be considered in the cluster evolution. The answer to whether the cluster is bound or unbound may be 'yes' (as in a bit of both). As shown by N-body simulations (Lada et al. 1984; Kroupa and Boily 2002), after gas-loss a marginally bound system may leave behind a bound core surrounded by a dispersing halo. This may be the fate of the ONC. It certainly will not be a massive open cluster such as M67 or even the Pleiades, so we continue to search for examples of the progenitors of such systems. What is certain is that the current ONC will expand from its current state as a result of the loss of gas mass.

12.4 Young Embedded Clusters

With the advent of surveys with large format, sensitive, near-infrared images, it became evident that many and perhaps most young stars in giant molecular clouds

are formed in embedded clusters[2] containing dozens to several hundreds of stars with
typical radii of <0.5 pc. Lada and Lada (2003) estimate about 200 embedded clusters
within 2.5 kpc of the Sun, implying of order 10,000 such clusters in the Galaxy at the
moment. Of course, determining membership for any given star in an infrared image
is challenging. This challenge includes field stars in foreground projection, other stars
forming in the molecular cloud but not associated with a particular spatial grouping,
and background stars observed through patchy extinction. True members can be
separated from field stars through signatures of youth, such as infrared excesses and
chromospheric X-ray emission. These diagnostics do not distinguish other young
stars in the molecular cloud, nor are such contaminants kinematically distinct.

Lada and Lada (2003) established commonly used dynamical criteria for embedded clusters. First, they require the stellar grouping to be stable against the tidal fields
of the Galaxy and interstellar clouds. Lada (2010) notes that this requires of order 8–
10 stars of 0.5 $M_\odot$ stars within a 1 pc radius. Second, they also require that, if bound,
the two-body relaxation time not be shorter than the typical lifetime of open clusters
of 100 Myr. This criterion requires 30–40 stars. Finally, the cluster should be partially or entirely embedded in its natal gas cloud. Importantly, there is no requirement
that embedded clusters actually be bound by their stellar mass, but only that these
criteria are satisfied should the cluster be bound. The mass spectrum of embedded

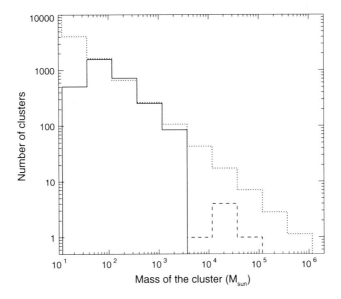

Fig. 12.10 The mass spectrum of embedded clusters for the Galaxy. The *solid line* shows the masses
of clusters within 2.5 kpc (Lada and Lada 2003), while the *dashed line* shows massive embedded
clusters from Ascenso (2008). The *dotted line* represents a spectral index of $\alpha = -1.7$. Figure from
Lada (2010)

---

[2] A detailed review can be found in Lada and Lada (2003).

clusters is shown in Fig. 12.10. In the local Galaxy, the slope of the mass spectrum is reasonably represented as a power-law with spectral index α of between -1.7 and -2.0. Whether this distribution holds all the way to the most-massive young clusters, such as Westerlund 1, is not clear. If it does, more such clusters remain to be found. At the other extreme, the low-mass turnover in Fig. 12.10 seems to be real, but each of these groupings includes very few stars (e.g. Porras et al. 2003). In any case, many of the embedded clusters in the local Galaxy have very small masses. The spectral index of α with values of between -1.7 and -2.0 is interesting with respect to the question of in what environment is a star most likely to form. A slope of $\alpha = -2.0$ represents equal mass per decade: most stars forming in embedded clusters are just as likely to form in clusters with masses between 10 and 100 $M_\odot$ as in clusters between $10^{4-5} M_\odot$.

Wide-field infrared surveys of local molecular clouds indicate that most young stars are found in embedded clusters, perhaps as high as 70–90 % (see review in Lada and Lada 2003). This implies that this is the dominant mode of star formation in the Milky Way. In this context, it is interesting to compare the 'typical' star-forming region, in terms of total mass and central stellar density, to constraints on the birth site of our Solar System (Adams 2010). However there appears to be a continuum of embedded cluster properties, at least in stellar surface density, from the richest regions to the low-density aggregates (Meyer et al. 2008; Bressert et al. 2010). Clearly, the dynamics of young clusters and associations are intimately related (see e.g. Fig. 12.12) and their dissolution is a primary mechanism of populating the field. Because associations are thought to represent up to 90 % of the outcomes of star formation in the Galactic disc, this suggests that associations originate in embedded clusters.

Turning to the age distribution of young clusters, in a survey of young clusters Leisawitz et al. (1989) found that only clusters younger than 5 Myr were associated ($d < 25$ pc) with massive ($>2 \times 10^4 M_\odot$) molecular clouds. Leisawitz et al. (1989) conclude that during the 5 Myr after the formation of a cluster containing massive stars the interstellar environment changes dramatically, and by 10 Myr little remains of the natal giant molecular cloud. (This timescale is consistent with our detailed discussion of the λ Ori above; see Chap. 11). This provides an upper limit on the timescales for clusters to separate from their parent molecular clouds and perhaps molecular cloud lifetimes (if the onset of star formation is rapid once clouds form). The ages of the embedded clusters are more typically 1–3 Myr. Given that this is comparable to their internal age spreads, the definition of an age distribution for embedded clusters becomes a bit problematic. Somewhat better defined is their birth rate of 2–4 kpc^{-2} Myr^{-1} (Lada and Lada 2003). This is ten times the formation rate derived from the population of currently bound open clusters.

Regarding the dynamics of young clusters, the survival rate is perhaps the most important issue. Figure 12.11 compares the expected age distribution given constant production and survival of embedded clusters to the observed distribution of cluster ages. By 10 Myr most embedded clusters have already disappeared, a timescale very similar to that for separation from their natal clouds. Evidently, the vast majority of

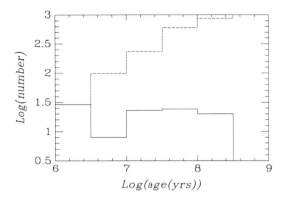

Fig. 12.11 Observed frequency distribution of ages for open and embedded clusters (*solid line*) compared with that predicted for a constant rate of cluster formation based on the embedded clusters (*dotted line*). By 3 Myr there is a discrepancy between number observed and predicted, suggesting that embedded clusters soon dissolve after leaving their natal molecular clouds. Figure from Lada and Lada (2003)

embedded star clusters do not survive emergence from their molecular clouds, and that they must dissolve in a way still to be determined.

At this point, a few remarks are in order. First, we tend to use the phrase 'bound' with regard to star-forming events carelessly: it is essential that we clearly define with the location, size-scale and the content of a region whose boundedness is under discussion. Location is important because given the irregular structure of star-forming regions the boundedness will vary. Size matters because even within regular structures cores may be bound while halos are unbound. Total content is critical because the role of gas in determining the gravitational potential within which young stars move can be dominant and transitory. A stellar system may be bound in the presence of gas and unbound with its loss.[3] Second, the internal structure of embedded clusters will play a role in the nature of their dissolution, and their boundedness. Lada and Lada (2003) suggest that there tend to be two qualitatively different structures. Of order 60 % tend to be compact, centrally condensed clusters, of which the Trapezium cluster is a more massive example. These can be fit with dynamical equilibrium models (e.g. King models). Others have very extended, irregular multiple peaks, such as the partially embedded cluster NGC 2264 (see Chap. 13, Fig. 13.4) to which fitting a dynamical equilibrium model would be truly nonsensical. In addition, most embedded clusters are elongated with aspect ratios as large as 2, much as found for the ONC (Allen et al. 2007). Figure 12.12 shows two regions—IC 348 and IC 2391—along with their dynamical times. Interestingly, the cluster structure does not correlate well with the dynamical (crossing) times. Despite an age of 20 dynamical times, the IC 2391 region is highly structured, looking more like an aggregate of subclusterings (each possibly in dynamical equilibrium). These regions may have

---

[3]It is worth reminding that gas is also subject to external pressure. As such, the boundedness of a gas cloud may not be a measure of the gravitational potential alone.

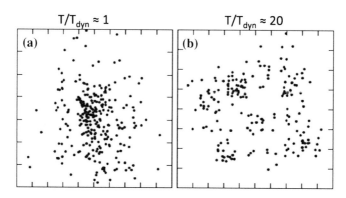

Fig. 12.12 Spatial distribution of low-mass stars in the regions of the IC 348 star-forming region (**a**) and the open cluster IC 2391 (**b**). Also shown are the ratios of cluster (stellar) ages to dynamical times. Figure adapted from Cartwright and Whitworth (2004)

Fig. 12.13 Optical image of B59, a dense core at the intersection of an array of filaments and the site of an embedded cluster. Figure from Lada (2010)

experienced 'cold collapse' (a version of violent relaxation) described in detail in Chap. 6.

Finally, let me draw your attention to the beautiful image of the Barnard dark cloud B59 in Fig. 12.13. Myers (2009) argues that the filamentary structure of dark cloud complexes play a major role in funnelling gas toward dense molecular cores, and earlier in this volume (see Clarke Chap. 6) we were shown several theoretical simulations suggesting the same. It is striking that where the filaments in Fig. 12.13 come together there is a small embedded cluster (e.g. Covey et al. 2010). This image is telling us vividly that we must let go of our simplifications ('the horse is a sphere'): there is a great deal of structure and complexity in these star-forming regions that may be crucial to the emergence of young clusters or associations.

References

Adams, F. C. 2010, ARA&A, 48, 47

Allen, L. E. & Davis, C. J. 2008, Handbook of Star Forming Regions, Volume I: The Northern Sky, ed. B. Reipurth, ASP Monograph Publications, 621

Allen, L., Megeath, S. T., Gutermuth, R., et al. 2007, Protostars and Planets V, ed. B. Reipurth, D. Jewitt & K. Keil, University of Arizona Press, 361

Allison, R. J., Goodwin, S. P., Parker, R. J., Portegies Zwart, S. F., & de Grijs, R. 2010, MNRAS, 407, 1098

Andersen, M., Meyer, M. R., Robberto, M., Bergeron, L. E., & Reid, N. 2011, A&A, 534, A10

Ascenso, J. 2008, PhD thesis, Centro de Astrofísica da Universidade do Porto

Bally, J. 2008, Handbook of Star Forming Regions, Volume I: The Northern Sky, ed. B. Reipurth, ASP Monograph Publications, 459

Barnard, E. E., Frost, E. B., & Calvert, M. R. 1927, A Photographic Atlas of Selected Regions of the Milky Way

Bressert, E., Bastian, N., Gutermuth, R. et al. 2010, MNRAS, 409, L54

Cartwright, A., & Whitworth, A. P. 2004, MNRAS, 348, 589

Covey, K. R., Lada, C. J., Román-Zúñiga, C., et al. 2010, ApJ, 722, 971

Da Rio, N., Robberto, M., Hillenbrand, L. A., Henning, T., & Stassun, K. G. 2012, ApJ, 748, 14

Hillenbrand, L. A., & Hartmann, L. W. 1998, ApJ, 492, 540

Kenyon, S. J., Gómez, M., & Whitney, B. A. 2008, Handbook of Star Forming Regions, Volume I: The Northern Sky, ed. B. Reipurth, ASP Monograph Publications, 405

Kroupa, P., & Boily, C. M. 2002, MNRAS, 336, 1188

Lada, C. J. 2010, Royal Society of London Philosophical Transactions Series A, 368, 713

Lada, C. J., & Lada, E. A. 2003, ARA&A, 41, 57

Lada, C. J., Margulis, M., & Dearborn, D. 1984, ApJ, 285, 141

Leisawitz, D., Bash, F. N., & Thaddeus, P. 1989, ApJS, 70, 731

Luhman, K. L., Allen, P. R., Espaillat, C., Hartmann, L., & Calvet, N. 2010, ApJS, 186, 111

Meyer, M. R., Flaherty, K., Levine, J. L., et al. 2008, Handbook of Star Forming Regions, Volume I: The Northern Sky, ed. B. Reipurth, ASP Monograph Publications, 662

Muench, A. A., Lada, E. A., Lada, C. J., & Alves, J. 2002, ApJ, 573, 366

Muench, A., Getman, K., Hillenbrand, L., & Preibisch, T. 2008, Handbook of Star Forming Regions, Volume I: The Northern Sky, ed. B. Reipurth, ASP Monograph Publications, 483

Myers, P. C. 2009, ApJ, 700, 1609

Onishi, T., Mizuno, A., Kawamura, A., Ogawa, H., & Fukui, Y. 1998, ApJ, 502, 296

Porras, A., Christopher, M., Allen, L., et al. 2003, AJ, 126, 1916

Reipurth, B. 2008, Handbook of Star Forming Regions, Volume II: The Southern Sky, ed. B. Reipurth, ASP Monograph Publications

Ungerechts, H., & Thaddeus, P. 1987, ApJS, 63, 645

Chapter 13
Kinematics of Star-Forming Regions

Robert D. Mathieu

13.1 Introduction

The previous chapters on star-forming regions have focused on spatial distributions of gas and stars. Here we focus on the internal kinematics, the motions of the young stars within the star-forming regions.

The Taurus-Auriga star-forming region, both the molecular gas and the stars, provides an excellent example of the kinematic scales of star-forming regions. This region has been studied extensively, including millimetre-wavelength maps, stellar proper motions and stellar radial velocities. Table 13.1 provides several measures of the internal kinematics.

The velocity dispersion between nearby dense molecular cores, that is the dispersion amongst the velocity centroids of dense molecular cores within groups, is very small, of order 0.5–1.0 km s^{-1}. The global gas velocity dispersion over the entire region is larger, 1–2 km s^{-1}. Figure 13.1 shows the position-velocity plot for the gas. Ignoring for the moment the separate Perseus cloud east of 3^h30^m, one is struck by how narrow the distribution is in velocity, showing the small local velocity dispersion. That it is aligned vertically indicates that the global velocity dispersion also is small. So this is a very quiet dynamical system. This is fairly characteristic for those giant molecular clouds in which there are no OB stars at the moment, which is to say no major global energy input. Indeed the Orion cloud as a whole is a very quiet system, away from the regions where there are embedded OB stars.

Furthermore, the mean velocity difference between the gas and the stars is consistent with zero (though it is difficult to reconcile the absolute velocity scales to better than 0.5 km s^{-1}; see e.g. Cottaar et al. 2014). The stellar velocity dispersions within the clumps of young stars are also small. The global stellar velocity dispersion of the

R.D. Mathieu (✉)
Department of Astronomy, University of Wisconsin, Madison, WI, USA
e-mail: mathieu@astro.wisc.edu

© Springer-Verlag Berlin Heidelberg 2015   179
C.P.M. Bell et al. (eds.), *Dynamics of Young Star Clusters and Associations*,
Saas-Fee Advanced Course 42, DOI 10.1007/978-3-662-47290-3_13

Table 13.1 Kinematic scales in Taurus-Auriga

Kinematic diagnostic	km s^{-1}
Local gas velocity dispersion	0.5
Global gas velocity dispersion	1–2
Local stellar velocity dispersion	$\lesssim$1–2
Global stellar velocity dispersion	2
$v_{star} - v_{gas}$	0.2 ± 0.4

References: Jones and Herbig (1979), Hartmann et al. (1986) and Ungerechts and Thaddeus (1987)

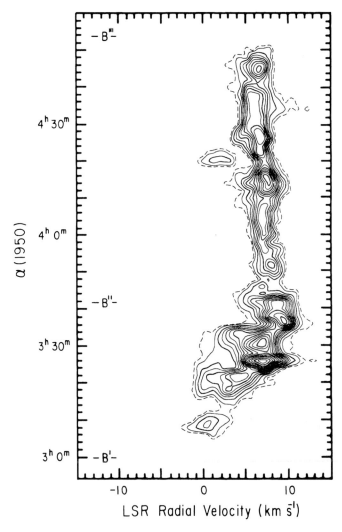

Fig. 13.1 Space-velocity diagram of CO (1-0) line temperature along a cut across the Taurus-Auriga and Perseus molecular cloud complexes. Figure from Ungerechts and Thaddeus (1987)

stars is about 2 km s^{-1}, although this is a somewhat difficult measurement to make. So these are also very quiet stellar systems. This makes the observational study of the internal motions of the young stars in star-forming regions a technical challenge.

13.2 OB Associations After *Hipparcos*

The *Hipparcos* mission was a wonderful bookend to the career of Adriaan Blaauw and in particular to his studies of OB associations, for his classic papers in the early 1960s were the first major studies of the internal kinematics of star-forming regions. At the time the major observational question was the expansion rate of OB associations, building from the earlier theoretical work arguing that they must be unbound. The proper motion studies of Blaauw demonstrating such expansion remain classics in our field.

The contribution of *Hipparcos* to the field was primarily in the domain of kinematic membership of these associations, extending down to B, A and F stars. Nearby associations are distributed over large areas of the sky. At the same time they are no longer associated with gas and dust, and thus confusion with field stars limits secure identification of members of lower mass than the short-lived OB stars. Figure 13.2 shows a proper motion map of the Scorpius-Centaurus association, in this case showing 532 members selected from 4156 *Hipparcos* stars in this region of the sky. The co-movement of the association is very clear.

As Blaauw pointed out 50 years ago, if the sub-associations are unbound and expanding then the sequence of different surface densities evident in Fig. 13.2 represents a sequence of ages. (Here 'age' means the time since the natal gas was removed and the system became unbound.) From this perspective, Lower Centaurus is the oldest part of this system and Upper Scorpius is the youngest. At higher Galactic

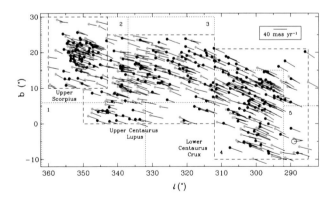

Fig. 13.2 *Hipparcos* positions and proper motions for 532 members of the Scorpius-Centaurus association, with three sub-associations identified. Figure adapted from de Zeeuw et al. (1999)

longitude, just off the figure, is the Ophiuchus molecular cloud in which active low-mass star-formation continues today. Presumably at one time this entire region was a giant molecular cloud, of which the Ophiuchus cloud is the last vestige. Indeed this region is often cited as an example of sequential star formation across a giant molecular cloud, with the star formation in each sub-region ended by a supernovae which then generated star formation in the adjacent region. Another possibility, of course, is that for reasons in the nature of the cloud itself that we do not understand this sequence of formation happened without any particular causal connection. Even so, each region might still be sequentially unbound by supernovae at different eras. We had hoped to obtain internal velocity dispersions, i.e. the differences between the vectors in Fig. 13.2, from *Hipparcos*, but the available precision only placed upper limits of $1-1.5 \, \mathrm{km \, s^{-1}}$. It is in this area that *Gaia* will shine.

13.3 Kinematics in Star-Forming Associations

We begin again with the λ Orionis region as a case study. As a reminder, there are 11 OB stars in a central clustering. Figure 13.3 shows *Hipparcos* proper-motion vectors for the brighter of the OB stars, shown for motions over the past 3 Myr. Given the current physical proximity of these OB stars, clearly they were not unbound 3 Myr ago—indeed, with these proper motions their current radius suggests that they were bound as recently as 1 Myr ago. Given their ages of order 7 Myr, the stars were orbiting in a bound clump of stars and gas for most of their lifetimes. Likely they were actually even more concentrated than presently.

Now with the lower-mass stars in the vicinity of the clump, precise radial velocity measurements with precisions of about $0.7 \, \mathrm{km \, s^{-1}}$ have been achieved, especially for those stars that are not actively accreting. The active mass accretion leads to veiled spectra and emission lines which make it more difficult to obtain high precision.[1] Fortunately, whether the OB stars accelerated their disc evolution or because normal evolution has simply depleted the discs around many of these stars, most of the low-mass stars in the λ Ori region are not classical T Tauri stars. We find a radial velocity dispersion of about $2.2 \, \mathrm{km \, s^{-1}}$ for the low-mass stars in the vicinity of the OB stars. And indeed, the dispersion of the proper motion measurements for the OB stars is also $2.5 \, \mathrm{km \, s^{-1}}$.

This region is also identified as the open cluster Collinder 69. But is it in fact bound? We can use a very generalised version of the viral theorem

$$\frac{1}{2}\sigma^2 = \frac{GM_{\mathrm{bound}}}{R}, \tag{13.1}$$

---

[1] Recent large internal surveys, such as the Sloan Digital Sky Survey III APOGEE IN-SYNC Project, have been able to reach internal consistency of $<0.3 \, \mathrm{km \, s^{-1}}$ (see e.g. Cottaar et al. 2014). This requires a comprehensive modelling effort, a very large database of observations to control for most systematic issues, and sophisticated data reduction efforts.

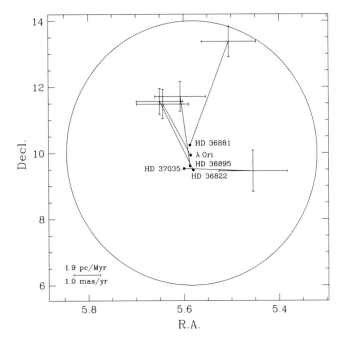

Fig. 13.3 Positions 3 Myr ago of 5 OB stars in the central clump of the λ Ori association, based on *Hipparcos* proper motion data. The error bars are derived from proper motion errors. The circle schematically indicates the position of the ionisation front at the present epoch. Figure from Dolan and Mathieu (1999)

in addition to values for the velocity dispersion and radius of the region ($\sigma \simeq 2.2\,\mathrm{km\,s^{-1}}$ from above and $R \simeq 2\,\mathrm{pc}$) to estimate that a mass of $\sim 1000\,M_\odot$ is required to bind the system. There are about 10 OB stars, representing about $100\,M_\odot$ or so. There are about 70 solar-type stars and about 200 lower-mass stars that have been found (see e.g. Bayo et al. 2011). The required mass definitely is not present in the stars.[2] Clearly, that a substantial mass of gas, more than the stellar mass, was needed to hold this system together for several million years. And despite being clustered, these stars do not a bound cluster make. Today, all of these stars are expanding away from each other, soon to disperse into the Galactic field.

We next turn to the star-forming region NGC 2264 (see Fig. 13.4), long included in lists of both open clusters and star-forming regions, depending on one's background. The stars are typically 2–5 Myr old. While not an old system, nonetheless it is not as

---

[2] We note that a more careful estimate of the virial parameter for the region would require an assessment of the density distribution as a function of radius leading to a factor η between 6–11 (e.g. Portegies Zwart et al. 2010) making the region less bound. In addition, potential differences in the velocity dispersion as a function of stellar mass should be considered. A typical approach is to consider a two-mass-bin sample approximation with high-mass stars having a lower velocity dispersion and lower-mass stars having a higher velocity dispersion (e.g. Binney et al. 2008) accounting for possible mass segregation. Even with these complexities in the analysis.

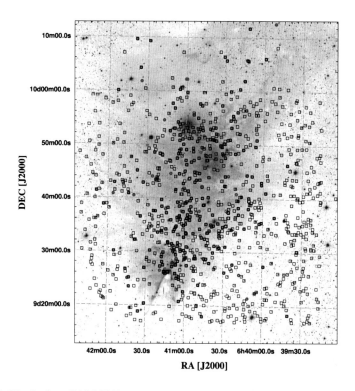

Fig. 13.4 Distribution of NGC 2264 target stars in the radial velocity survey of Fűrész et al. (2006). The *filled circles* are classical T Tauri stars, and show the clumpy and asymmetric spatial structure of the young stars in this star-forming region. Figure from Fűrész et al. (2006)

young as the Orion Trapezium region (although there are still embedded T Tauri stars, so apparently star formation continues). While often called a young open cluster in the literature, even a cursory examination shows the spatial distributions of both the stars and the gas to still be very clumpy, with the stars and gas spatially associated and not dynamically relaxed. The total cluster size is about 8 pc, with two predominant clumps each about 4 pc in radius. Given a velocity dispersion of 1–2 km s^{-1}, the crossing time is 2–4 Myr. As stressed previously, with star-forming regions we are looking at systems with comparable crossing times and ages, and so we should not be surprised that these systems are irregular and not well-mixed.

This dynamical youth can also be seen in the radial velocity distribution for the region (see Fig. 13.5). The essential clues are the asymmetry in the velocity distribution and the unusually large velocity dispersion, $\simeq$3.5 km s^{-1}. More careful examination reveals large-scale systematic trends in the region (see Fig. 13.6). Locally, the typical dispersion is again about 1–2 kms^{-1}. But globally not only do we see spatial organisation but also kinematic organisation across the region (also seen in other regions such as Orion discussed below).

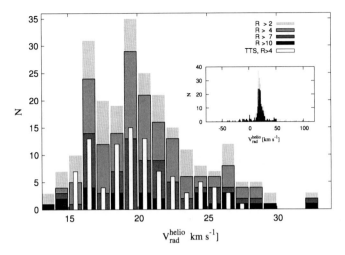

Fig. 13.5 Radial velocity distribution for 344 stars in NGC 2264. The distribution is unusually wide and non-Gaussian. Shading indicates measurement quality, improving with increasing value of R. The white bars show a selection of stars with strong $H\alpha$ emission, showing little difference between the kinematics of stars with and without active accretion discs. The insert is an expanded velocity range to show how clearly the cluster members form a peak in the velocity space. Figure from Fűrész et al. (2006)

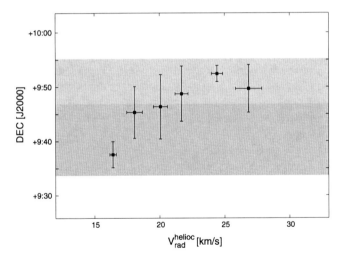

Fig. 13.6 North-south velocity gradient of young stars in NGC 2264, shown by plotting the mean declination values in 2 km s^{-1}-wide radial velocity bins against the mean radial velocity of the bins. The error bars represent the rms of declinations and radial velocities of stars in a given velocity bin. The shaded areas show the declination ranges for the two main condensations of $H\alpha$ stars. Figure from Fűrész et al. (2006)

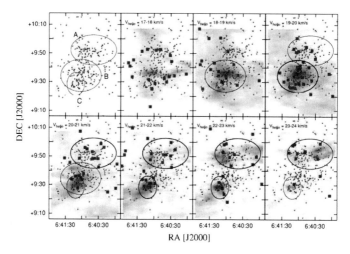

Fig. 13.7 The *red dots* are the Hα-emission stars in the NGC 2264 regions. The *red squares* are young stars with measured radial velocities. The *blue shading* shows the ^{13}CO emission from molecular gas. Three sub-clumps delineated by the molecular emission are identified. Figure from Fűrész et al. (2006)

Figure 13.7 will be familiar to readers who are experts in radio astronomy. It is the equivalent of a channel map. The red dots are the Hα-emission stars in the region, with the red squares being stars with measured radial velocities. The blue shading shows the ^{13}CO emission from molecular gas. The density of red squares tends to correlate with the amount of blue shading in each velocity channel. There is an evident close association of the stars with the gas. But these stars are not embedded in the gas; extinction measurements show that most of the gas is behind the stars, so it is not entirely obvious what is the situation in the third dimension. Perhaps these large structures in the stellar velocity distribution reflect large-scale gradients in the natal molecular cloud.

What can we say about the NGC 2264 star-forming region based on these kinematic studies? First, given a dynamical timescale for the region comparable to the stellar ages, the continued presence of clustering—both spatial and kinematic—of young stars in the region is not surprising given the clumpy gas distribution. Second, the scale of the motions are gravitational. The total mass in this system—gas and stars—is about 4,000 $M_\odot$. Given a radius of about 4 pc, we expect motions of 2 km s^{-1}. The gas motions far exceed thermal (see also Clarke Chap. 1) and so are responsive to the gravitational field and pressure. The stars of course do not respond to pressure, and so the close association of gas and stars in both spatial and kinematic dimensions suggests that both are responding in large part to the gravitational potential on a crossing timescale.

13.4 Stellar Kinematics in Young Star Clusters

For the purposes of studying the kinematics in young star clusters, we will choose the Orion region as our case study. If one peruses the literature, however, one finds reference to the Trapezium Cluster, the Orion Nebula Cluster and the Orion Nebula region, all centred on the Trapezium stars but with different extents. Why? Because nature does not provide clear demarcations around the structures in star-forming regions, and it is not evident where are the boundaries of the ever larger stellar systems around the Trapezium OB stars.

In any case, the size-scale of the stellar clustering in this region is around 2 pc diameter. Again for a rough estimate, if we adopt a 2 km s^{-1} velocity dispersion, we find a crossing time of a couple Myr. The oldest stars in these regions have a derived age of about 2 Myr, with most stars being much younger.[3] It is likely that few stars have completed more than one orbit in the local gravitational potential. In addition, the stellar distribution shows significant structure, especially being elongated in the north-south direction.

In 1988 two excellent proper motion studies of this region were published, one of very-high precision (van Altena et al. 1988) and one covering a wider field at excellent precision (Jones and Walker 1988). van Altena et al. (1988) measured a one-dimensional velocity dispersion of 1.5 km s^{-1} for the core of the region. Jones and Walker (1988) found a velocity dispersion of 2.5 km s^{-1}, possibly with some anisotropy further from the core and some mass dependence. But perhaps most importantly, they make the dynamical argument that, 'On the basis of the velocity dispersion, and the stellar and gas masses, we conclude that we are not seeing a protocluster but a system in disruption.'[4]. This statement was based on a global analysis; it remains unclear whether the core of the region (the so-called Trapezium Cluster) is unbound. Nonetheless, the words of Jones and Walker were prescient in recognising that clusterings of young stars were not necessarily destined to be open clusters.

More recently, this region has been the subject of a wide-field, high-precision radial velocity study (Tobin et al. 2009; see also Fűrész et al. 2008). Again, the region shows structure in the stellar spatial and velocity distributions. Figure 13.8 shows a position-velocity diagram for the region. The mode of the stellar velocity distribution at each declination is delineated by the blue line, superimposed upon the ^{13}CO distribution. The spatial extent of Fig. 13.8 is 16 pc, and the location of

---

[3]There are some that appear to be much older, which is somewhat confusing—perhaps they are part of the Orion 1C association that is projected in front of it. Alternatively, we may not understand the ages of pre-main-sequence stars as well as we hope.

[4]As a technical aside, historically many of the proper motion studies of the Trapezium region focused on detection of expansion. Classically this was difficult in measuring proper motions from plates, for the slightest differences in plate scale would mimic systemic radial motions (or alternatively, an expansion term was a parameter determined from the stars on the plates). Thus neither of these more modern studies spoke to the expansion of the cluster. With a well-determined independent frame of reference, *Gaia* should finally accomplish the goal of the first astrometrists who studied this star-forming region.

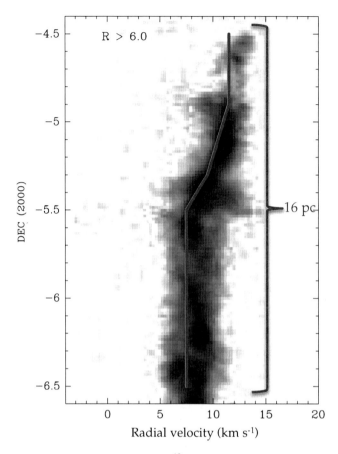

Fig. 13.8 Position-velocity diagram for stars and ^{13}CO molecular gas emission around the Trapezium. The *blue line* represents the mode of the stellar velocity distribution for kinematic members at each declination. Figure adapted from Tobin et al. (2009)

the Trapezium is shown. Most notable is the significant systemic velocity shift of the stellar velocities, following the north-south ridge of moderate-density molecular gas in this region. Specifically there is a 2.5 km s^{-1} shift over a scale of only 2 pc or so, compared to the 16 pc extent of the entire region shown in Fig. 13.8. Equally importantly, the kinematics of the stars follows that of the gas. Finally, this gradient happens very near to the Trapezium and the associated clustering of stars.

Thus, as with NGC 2264, the scale of the systematic signatures in the velocities suggest on-going gravitational effects. Tobin et al. (2009) and Proszkow et al. (2009) use these data in an attempt to back out the initial conditions of the cluster formation. These authors argue that the systematic velocity shift is the result of infall of an elongated, sub-virial, collapsing filament associated with the formation of the Trapezium cluster. Whether this interpretation is correct or not is perhaps not the key point; we are very early in the game. Many of us remember the ambiguities and multiple interpretations in the study of molecular line widths as we tried to understand whether clouds were rotating or collapsing. More important is that in terms of the internal kinematics of star-forming regions, we are entering into the domain of detailed kinematic maps as compared to simply velocity dispersions. *Gaia* observations combined with sub-km s^{-1} precision radial velocities will soon allow us to examine in detail the systematic flows within star-forming regions driven by gravity and internal motions.

References

Bayo, A., Barrado, D., Stauffer, J., et al. 2011, A&A, 536, A63

Binney, J. & Tremaine, S. 2008, Galactic Dynamics: Second Edition, ed. J. Binney & S. Tremaine, Princeton University Press

Cottaar, M., Covey, K. R., Meyer, M. R., et al. 2014, ApJ, 794, 125

de Zeeuw, P. T., Hoogerwerf, R., de Bruijne, J. H. J., Brown, A. G. A., & Blaauw, A. 1999, AJ, 117, 354

Dolan, C. J. & Mathieu, R. D. 1999, AJ, 118, 2409

Fűrész, G., Hartmann, L. W., Szentgyorgyi, A. H., et al. 2006, ApJ, 648, 1090

Fűrész, G., Hartmann, L. W., Megeath, S. T., Szentgyorgyi, A. H., & Hamden, E. T. 2008, ApJ, 676, 1109

Hartmann, L., Hewett, R., Stahler, S., & Mathieu, R. D. 1986, ApJ, 309, 275

Jones, B. F. & Herbig, G. H. 1979, AJ, 84, 1872

Jones, B. F. & Walker, M. F. 1988, AJ, 95, 1755

Portegies Zwart, S. F., McMillan, S. L. W., & Gieles, M. 2010, ARA&A, 48, 431

Proszkow, E.-M., Adams, F. C., Hartmann, L. W., & Tobin, J. J. 2009, ApJ, 697, 1020

Tobin, J. J., Hartmann, L., Furesz, G., Mateo, M., & Megeath, S. T. 2009, ApJ, 697, 1103

Ungerechts, H. & Thaddeus, P. 1987, ApJS, 63, 645

van Altena, W. F., Lee, J. T., Lee, J.-F., Lu, P. K., & Upgren, A. R. 1988, AJ, 95, 1744

Chapter 14
Pre-main-sequence Binaries

Robert D. Mathieu

14.1 Introduction

In the early 1980s there were rapid advances in our understanding of the formation of single stars, based on both observational and theoretical progress in answering important questions concerning protostellar accretion discs.

It has been long clear that binaries were a common outcome of the star formation process. After all, Abt and Levy (1976) found that at least half of the stars in the field were binaries. Even so, forming a single star was challenging enough, and so studies of binary formation were largely postponed. At many star formation meetings there would be a session on binary stars, with Hans Zinnecker showing high-angular-resolution images of his favourite pre-main-sequence (pre-MS) binaries, Mike Simon reporting on his lunar occultation work (the moon conveniently passing across the Taurus-Auriga and Ophiuchus star-forming regions), and the author speaking about the implications of orbital solutions for pre-MS spectroscopic binaries, and closing with a theorist—often Cathie Clarke—explaining it all! Then Andrea Ghez finished her dissertation work, a speckle imaging[1] survey of pre-MS stars at Palomar (Ghez et al. 1993) and Christoph Leinert and his colleagues completed their speckle survey of Taurus-Auriga at Calar Alto (Leinert et al. 1993). Both found very high binary frequencies, even higher than the field, and binary stars became too exciting to postpone any further (see e.g. Mathieu 1994). Here we will focus on binary populations among pre-MS stars, the evolutionary phase between protostars and the main-sequence. The

---

[1]Speckle imaging takes advantage of very short exposures in an attempt to freeze the seeing of large telescopes so that potentially diffraction-limited images are less sensitive to atmospheric turbulence. The image centroids of bright point sources in the field-of-view are registered and the individual exposures are co-added to search for fainter nearby sources. Because in such short exposures one can be limited by the noise of the detector rather than the background sky emission, the sensitivity is much lower than modern adaptive optics assisted imaging.

R.D. Mathieu (✉)
Department of Astronomy, University of Wisconsin, Madison, WI, USA
e-mail: mathieu@astro.wisc.edu

© Springer-Verlag Berlin Heidelberg 2015   191
C.P.M. Bell et al. (eds.), *Dynamics of Young Star Clusters and Associations*,
Saas-Fee Advanced Course 42, DOI 10.1007/978-3-662-47290-3_14

mass range will be, roughly, 0.1 to $2\,M_\odot$. The samples will include both 'classical T Tauri stars' (CTTS) with active circumstellar accretion discs and 'weak-lined T Tauri stars' (WTTS) which may have circumstellar discs with low accretion rates, discs in transition between optically-thick and optically-thin or no discs at all. The binary orbital periods will range from 1 day to greater than 10,000 days.

Several aspects of pre-MS binaries are also covered in the recent review of Duchêne and Kraus (2013).

14.2 Pre-main-sequence Binary Frequency

14.2.1 Definition

Perhaps the most fundamental measurable—binary frequency—is subject to significant ambiguities of definition. Here we will count each single, binary, triple, etc. as one 'stellar system'. As in Chap. 9, the multiple system frequency (which we will mix with the term binary frequency) is the number of binary, triple and higher-order systems divided by the total number of stellar systems. It is also critical to specify the separation (or period) domains, within which one is citing a frequency, as well as the mass ratio range to which one is sensitive, as rarely does an observational study include the entirety of these domains.

14.2.2 Frequency as a Function of Star-Forming Region

The most striking finding is the very high frequency of wide binaries in Taurus-Auriga, relative to the field population. This is confirmed in many low-density star-forming regions. Taurus-Auriga was found to have a frequency a factor of two larger than the field for binaries with separations between 10 and 150 au, which was especially notable given that the field binary frequency is of order 50 %. Such a high frequency was never evident among the closer (spectroscopic) binaries. It also seemed to be independent of the presence of discs. At the same time work in young star clusters, such as the Orion Nebula Cluster (ONC), seemed to show that at roughly the same separations (50–150 au) there was no evidence of an overabundance in pre-MS binaries at the factor of two level (Padgett et al. 1997; Liu et al. 2003). In fact, cluster populations were very field-like in terms of binary frequencies. And so there was a sense that perhaps environmental effects were at play in the formation of binary stars. King et al. (2012) have recently performed a 'second-generation' study, considering the regions of Taurus, Chamaeleon, and Ophiuchus, the young cluster IC 348, and the ONC. They have performed an analysis of the existing data with particular care regarding separation domains, primary star mass range, dynamic ranges for detecting secondary stars, etc. Figure 14.1 shows their derived multiplicity fractions against the local number density of stars. Importantly, they find very

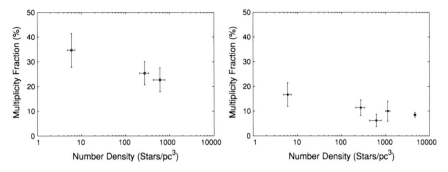

Fig. 14.1 Plots of multiplicity fraction versus number density in five star-forming regions. *Left panel*: Results for Taurus, Chamaeleon I and Ophiuchus (from *left* to *right*, respectively) for binary separations of 18−830 au. *Right panel*: Also includes the higher-density IC 348 cluster and the ONC for binary separations of 62−620 au. Figure adapted from King et al. (2012)

little environmental dependence over several decades of number density. Only Taurus at wider binary separations shows a possible overabundance (see left panel of Fig. 14.1). Perhaps importantly, Taurus here represents the region with the lowest local number density. It is important to keep in mind though that the sample sizes of pre-MS binaries remain small. We are not yet in a position to detect anything but large frequency differences, of order a factor two.

14.2.3 Frequency as a Function of Orbital Period

Pre-MS binary populations have not been comprehensively studied across all orbital periods, compared to solar-type stars in the field. Nonetheless we can ask whether the shape of the pre-MS orbital period distribution is consistent with the field. Figure 14.2 compares the Taurus binary separation distribution to that in two nearby open clusters as well as that of the field. Considering only the shape of the observed frequency distributions, not the absolutely frequencies (i.e. the vertical scaling), every indication is that the Gaussian-like shape in the field orbital period distribution is established very early during formation, with peaks at very similar periods. (See also field and pre-MS comparisons in Mathieu 1994 and Melo 2003, both of which include spectroscopic binaries.)

Currently we have very little knowledge regarding the frequency of the very widest binaries that are found in the field. The issue here is not angular resolution— they are easily detected in nearby star-forming regions. The challenge is establishing dynamical association. In the field, these systems are identified as common proper motion pairs. In star-forming regions identifying common proper motion pairs is technically challenging because of the small internal motions and higher-than-field ambient density. More fundamentally, it is physically challenging to determine association when the orbital periods and the dissolution times of clusterings of pre-MS stars are comparable.

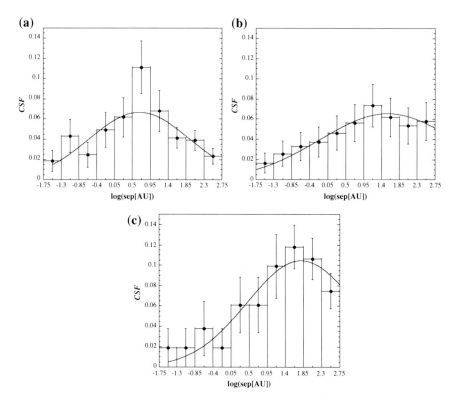

Fig. 14.2 Comparison of companion frequencies for three samples. (**a**) α Per and Praesepe open clusters, (**b**) the field as defined by Duquennoy and Mayor (1991), and (**c**) Taurus pre-MS stars. Figure adapted from Patience et al. (2002)

14.2.4 Higher-Order Multiplicity

Bo Reipurth and Hans Zinnecker have pointed out since the very earliest days that there are a large number of higher-order multiples among pre-MS binaries. Table 3 from Sterzik et al. (2005) gives a sense of the nature of such multiples. In the field higher-order multiples represent 10 % of the stellar systems. Whether higher-order multiples are formed at an even higher frequency and then lose the most loosely bound companions through subsequent dynamical evolution remains to be established observationally. Interestingly, Sterzik et al. (2005) suggested that higher-order systems may be necessary to form tightly bound binaries, through internal evolution. (See also Reipurth and Clarke 2001 for interesting views concerning this involving brown dwarfs.) On the other hand (see also Clarke Chap. 5) the evolution of wide binaries might provide a signature of the early dynamical state and evolution of star-forming regions. It appears the evolution of higher-order multiple systems is ripe for further exploration, both observationally and theoretically.

14.2.5 Protobinaries

In 2000 at IAU Symposium 200 on the formation of binary stars, I said in my closing remarks that I thought the next discovery domain would be the protostellar binaries, that is to say, embedded "Class I" systems. At the time the only such protobinary known was L1551, discovered by Rodríguez et al. (1998) at 7 mm wavelength observations with the VLA. This discovery was yet another case—including T Tauri itself—where a canonical case in the development of single-star formation theory, and indeed the discovery object for bipolar outflows, turns out to be a binary star. We are still in the discovery phase for protobinaries, which should very soon explode with ALMA. One difficulty in comparing these results to the field will be to assess the mass ratio range to which one is sensitive in the protostellar phase where it is difficult to translate infrared (or sub-millimetre) flux ratios into mass ratios for the central protostellar objects (not to mention uncertainty on what the final masses will be). While we do not yet have a major systematic survey, Duchêne, Bouvier and their colleagues have been taking high-resolution infrared imaging of Class I sources, with results shown in Fig. 14.3. The essential point of Fig. 14.3 is that even at the protostellar phase there is appreciable evidence for binarity having already being established, at least at intermediate separations. Whether all of the pairs will remain dynamically bound after the loss of the natal material remains an unknown and interesting question.

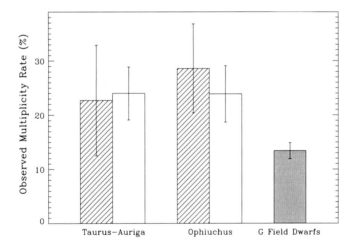

Fig. 14.3 Observed multiplicity rates in the projected separation range 110−1400 au for Class I protostars (*hatched histograms*) and T Tauri stars (*open histogram*) in the Taurus-Auriga and Ophiuchus molecular clouds. The field G-dwarf multiplicity rate is from Duquennoy and Mayor (1991). Figure from Duchêne et al. (2007)

14.3 Pre-main-sequence Binaries and Disc Evolution

The past three decades have been marked by tremendous theoretical and observational progress in our understanding of protostellar accretion discs (see e.g. the review of Dullemond and Monnier 2010). In fact many of the canonical protostellar accretion discs surround binary stars, and so they become even more interesting. It has been long known that a companion embedded within a disc will dynamically clear a gap, or perhaps a central hole depending on the separation. Figure 14.4 shows two results of smoothed particle hydrodynamics (SPH) simulations of binaries surrounded by coplanar discs (Artymowicz and Lubow 1994). The resonances of the binary are effective at clearing a gap on very-short (near dynamical) timescales. Consequently the concept of a single disc must be discarded, to be replaced by a circumbinary disc and (possibly) two circumstellar discs; one circumprimary and one circumsecondary. For very-wide binaries where a typical disc size of, say, 100 au is much smaller than the binary separation, it is quite possible that normal single-star circumstellar disc evolution processes largely hold true. At the other extreme of binary separations of, say, 1 au or less, then such a typical disc is essentially circumbinary. For intermediate binary separations, significant circumbinary and circumstellar discs both may be present.

Indeed, not only does a binary clear gaps but it was long thought that these same resonances would halt the inflow of circumbinary material. There is a balance at the Lindblad resonances of the outward torques being driven by the binary and the inward torques being driven by the processes causing the accretion flow. Thus replenishment of the circumstellar discs would not happen. Were that the case, a key question is: if the circumstellar discs are actively accreting onto the stellar surfaces, how do the discs survive? With typical surface accretion rates of $10^{-7} - 10^{-8}\,M_\odot\,yr^{-1}$, unreplenished circumstellar discs would rapidly exhaust themselves except in the widest binaries.

The classical T Tauri binary UZ Tau is a favourite case study for accretion discs in a binary environment. Should you return to one of George Herbig's early papers

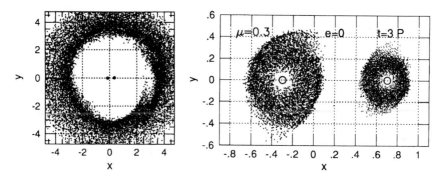

Fig. 14.4 Examples of a circumbinary disc (*left panel*) and circumstellar discs (*right panel*) in young binary systems. The discs are dynamically bounded by resonances from the binaries. Figure adapted from the SPH calculations of Artymowicz and Lubow (1994)

and look at a list of T Tauri stars, you will find UZ Tau East noted as one of the most active. Indeed Herbig (1977) called it an 'eruptive T Tauri star'. Adding modern data, UZ Tau E shows all of the diagnostics for a high-accretion-rate disc—strong Hα emission, ultraviolet excess, spectral veiling indicative of a surface accretion rate of 10^{-7} M$_\odot$ yr^{-1}, large irregular photometric variability, infrared excess with a power-law spectral energy distribution, chromospheric activity, an outflow of 10^{-8} M$_\odot$ yr^{-1}, and a microjet. All within a relatively massive disc of 0.024 M$_\odot$. Each of these diagnostics has since been explained within the paradigm of an accretion disc around a single star, and indeed UZ Tau E is often used as a canonical case. However, UZ Tau is an object of Joy (1945), who noted that the system was a 'double' of 3.7 arcsec separation. The pair known as UZ Tau East and West are at a separation of 500 au in projection. Speckle observations showed that UZ Tau W is itself a binary, 50 au in projection. Finally, it was discovered that UZ Tau E is a spectroscopic binary with a period of 19.0 days, an orbital eccentricity of 0.24, and a mass ratio of 0.29 (Prato et al. 2002): the system is a quadruple! Millimetre interferometric observations show that both binaries have massive discs (see Fig. 14.5), and

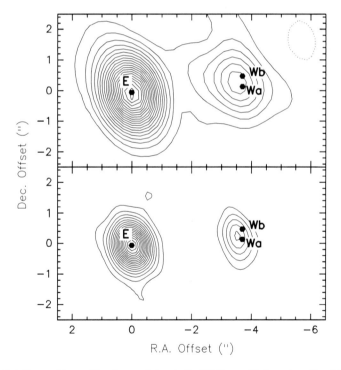

Fig. 14.5 Relative positions of the discs and stars in the UZ Tau multiple system. *Top*: 2.7 mm map, contour step 0.9 mJy/beam (35 mK, 2σ). *Bottom*: 1.3 mm map, contour step 5 mJy/beam (140 mK, 4σ). The emission around UZ Tau W is not resolved, with the offset centroid suggesting circumstellar discs. Figure from Guilloteau et al. (2011)

thus within UZ Tau are examples of all three binary-disc configurations described earlier.

The case of UZ Tau E is remarkable. The periastron separation is only $\simeq 0.1$ au. Thus the picture we get of the system includes a massive circumbinary disc surrounding the binary, and a large accretion flow onto at least one of the stellar surfaces. Clearly there is no room for circumstellar discs to be a significant reservoir for the surface accretion flow. It is difficult to escape the conclusion of a flow of material from the circumbinary disc, across the cleared region, onto the stellar surfaces, possibly with very tiny circumstellar discs. (And let us not forget the significant role of the stellar magnetic fields at these separations.) Despite two stars coming within several hundredths of an au of each other every 19 days, the star shows every accretion diagnostic typically attributed to an extended, massive protostellar accretion disc around a single star. How do we understand and describe the flow from the circumbinary disc to the surfaces?

Another favourite pre-MS binary of mine is DQ Tau, which may be providing us with at least a start on the answer to this question. DQ Tau is actually a binary very similar to UZ Tau E—orbital period of 15.8 days, eccentricity of 0.56 and a mass ratio near unity (Mathieu et al. 1997). DQ Tau is not as active as UZ Tau E, but nonetheless is a classical T Tauri star with all the indications of accretion onto at least one stellar surface and a substantial disc, presumably circumbinary. (Interestingly, though, the spectral energy distribution of DQ Tau does not show signs of a gap; Mathieu et al. 1997.) The secrets of DQ Tau were revealed serendipitously. Gibor Basri and an undergraduate student, Keivan Stassun, analysed a photometric time series on DQ Tau, looking for spot modulation as an indicator of the rotation period (of a single star). Instead, they found the brightness to spike regularly with a period of $\simeq 16$ days. Not every cycle, but clearly periodic. When we compared notes (one cannot underestimate the human factors such as friendship in scientific discovery), it became clear that DQ Tau was brightening at many (but not all) periastron passages. This was subsequently also seen in emission lines diagnostic of accretion such as Hα (Basri et al. 1997; Mathieu et al. 1997).

At about the same time, Pawel Artymowicz and Steve Lubow were performing SPH simulations of circumbinary discs around binary stars. Surprisingly, they found that under certain disc conditions pulsed accretion streams would flow from the inner edge of the circumbinary disc to the domains of the stars (Artymowicz and Lubow 1996). (The SPH resolution did not permit examining the actual flow onto the stellar surfaces.) For a binary much like DQ Tau, these streams would form at apastron and the material would fall to the stars roughly at periastron. Subsequently Günther and Kley (2002) found much the same using a grid-based code, with an example outcome shown in Fig. 14.6, and in recent years numerous numerical simulations also have found such streams (see e.g. Shi et al. 2012). Although the conditions necessary for them to occur remain to be well established, such accretion streams may be the key to understanding how mass accretion occurs in the formation of binary stars and thereby formation in most stars.

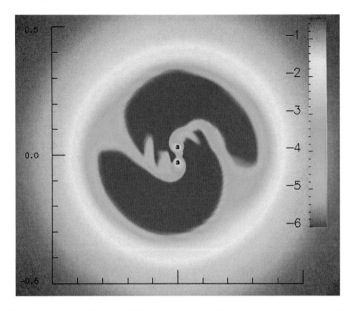

Fig. 14.6 Grid-based numerical simulation of disc accretion flows from the circumbinary disc to the stars within the DQ Tau system. The length scales are in au and the colour-coding is in log surface mass density. Figure from Günther and Kley (2002)

Of course, one would like to observe the DQ Tau accretion stream directly. John Carr, Joan Najita, and colleagues (including the author) used the Keck/NIRSPEC to look for the fundamental overtone emission of CO in the thermal infrared. CO is a tracer of hot dense gas, but the key is that it is optically thick at typical disc densities. So if you detect it, then there is likely an optically-thin region in the disc, arguably in the gap (or disc atmosphere). We did detect the fundamental overtone emission (see Fig. 14.7), and this provides some ability to do a kinematic map of where the gas is located. We found the emission to derive from the region where a gap is expected. The total mass in the gap is about 10^{-10} $M_{\odot}$, which is optically thin and somewhat less than expected for the stellar accretion rates. We do not know in detail whether there are accretion streams. More recently, Boden et al. (2009) have used near-infrared interferometry to detect a resolved infrared feature on the same size-scale as the binary itself. They suggest that the static visibility offset over many orbital periods indicates a system in quasi-equilibrium with material inflow replenishing dissipated material in the binary region. As an aside, we note that the stars in DQ Tau are close enough in their periastron approach that their stellar magnetospheres likely interact, as suggested by Basri et al. (1997). Periodicity has been found at millimetre wavelengths (Salter et al. 2010) and enhanced X-ray emission at a periastron passage (Getman et al. 2011), suggesting magnetic flaring activity. The connections, if any, between the dynamical and magnetospheric processes have not been addressed.

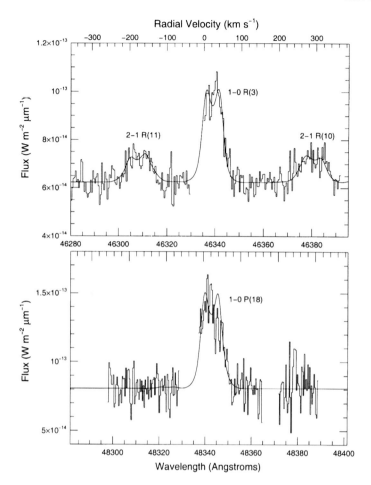

Fig. 14.7 Near-infrared spectra of CO fundamental rovibrational emission lines from DQ Tau. Overplotted is a model of a disc gap having a mass surface density of 5×10^{-4} g cm^{-2}, an inner radius of 0.1 au, an outer radius of 0.5 au, and a temperature gradient of $770(R/0.5\,\text{au})^{-0.5}$. Figure from Carr et al. (2001)

Since the studies of DQ Tau others have looked for similar phenomena in other binaries. Jensen et al. (2007) examined UZ Tau E. It too shows photometric variation on the orbital period consistent with SPH modelling, although the periodicity is not as evident in the Hα emission. The very short-period (and nearby) pre-MS binary V4046 Sgr shows changes in the structure of Balmer lines with the orbital period, which Stempels and Gahm (2004) suggest may be due to flows. And of course there is always a case that causes difficulties. Alencar et al. (2003) did not find any periodic photometric variability in the binary AK Sco (orbital period of 13.6 days, eccentricity of 0.47), but do find periodic features in Hα emission which are not consistent with

SPH predictions. So there remains much work to be done before our understanding of accretion in pre-MS binaries is on more solid ground.

14.4 Concluding Thought

The essential finding of the discovery phase two decades ago was that binaries are frequent among pre-MS objects. A primary goal here is to impress upon you that when you are observing pre-MS stars, when you are modelling young stars, and when you are seeking to understand the properties of a star-forming region, do not ever forget that there is very likely another star in the story. Ignoring the companions will lead you astray, especially if your ideas are based in the current theory of single-star formation. Unfortunately, this theory is likely not entirely appropriate for as many as half of your objects of interest. We have yet to connect, in a deep way, the nature of these binary populations and the dynamical evolution of young star clusters and associations. However, we can be sure that there is one.

References

Abt, H. A. & Levy, S. G. 1976, ApJS, 30, 273
Alencar, S. H. P., Melo, C. H. F., Dullemond, C. P., et al. 2003, A&A, 409, 1037
Artymowicz, P. & Lubow, S. H. 1994, ApJ, 421, 651
Artymowicz, P. & Lubow, S. H. 1996, ApJ, 467, L77
Basri, G., Johns-Krull, C. M., & Mathieu, R. D. 1997, AJ, 114, 781
Boden, A. F., Akeson, R. L., Sargent, A. I., et al. 2009, ApJ, 696, L111
Carr, J. S., Mathieu, R. D., & Najita, J. R. 2001, ApJ, 551, 454
Duchêne, G., Delgado-Donate, E., Haisch, Jr., K. E., Loinard, L., & Rodríguez, L. F. 2007, Protostars and Planets V, ed. B. Reipurth, D. Jewitt & K. Keil, University of Arizona Press, 379
Duchêne, G. & Kraus, A. 2013, ARA&A, 51, 269
Dullemond, C. P. & Monnier, J. D. 2010, ARA&A, 48, 205
Duquennoy, A. & Mayor, M. 1991, A&A, 248, 485
Getman, K. V., Broos, P. S., Salter, D. M., Garmire, G. P., & Hogerheijde, M. R. 2011, ApJ, 730, 6
Ghez, A. M., Neugebauer, G., & Matthews, K. 1993, AJ, 106, 2005
Guilloteau, S., Dutrey, A., Piétu, V., & Boehler, Y. 2011, A&A, 529, 105
Günther, R. & Kley, W. 2002, A&A, 387, 550
Herbig, G. H. 1977, ApJ, 217, 693
Jensen, E. L. N., Dhital, S., Stassun, K. G., et al. 2007, AJ, 134, 241
Joy, A. H. 1945, ApJ, 102, 168
King, R. R., Parker, R. J., Patience, J., & Goodwin, S. P. 2012, MNRAS, 421, 2025
Leinert, C., Zinnecker, H., Weitzel, N., et al. 1993, A&A, 278, 129
Liu, M. C., Najita, J., & Tokunaga, A. T. 2003, ApJ, 585, 372
Mathieu, R. D. 1994, ARA&A, 32, 465
Mathieu, R. D., Stassun, K., Basri, G., et al. 1997, AJ, 113, 1841
Melo, C. H. F. 2003, A&A, 410, 269
Padgett, D. L., Strom, S. E., & Ghez, A. 1997, ApJ, 477, 705
Patience, J., Ghez, A. M., Reid, I. N., & Matthews, K. 2002, AJ, 123, 1570
Prato, L., Simon, M., Mazeh, T., Zucker, S., & McLean, I. S. 2002, ApJ, 579, L99

Reipurth, B. & Clarke, C. 2001, AJ, 122, 432
Rodríguez, L. F., D'Alessio, P., Wilner, D. J., et al. 1998, Nature, 395, 355
Salter, D. M., Kóspál, Á., Getman, K. V., et al. 2010, A&A, 521, 32
Shi, J.-M., Krolik, J. H., Lubow, S. H., & Hawley, J. F. 2012, ApJ, 749, 118
Stempels, H. C. & Gahm, G. F. 2004, A&A, 421, 1159
Sterzik, M. F., Melo, C. H. F., Tokovinin, A. A., & van der Bliek, N. 2005, A&A, 434, 671

Part III
From Whence the Field?

Chapter 15
Galactic Demographics: Setting the Scene

I. Neill Reid

15.1 Introduction

Star clusters and associations are the agents for change in galactic environments.
They mark locations where the density of the interstellar medium (ISM) was sufficiently high that self-gravity overcame pressure, inducing collapse at multiple locations. Once formed, nuclear processes within the stars generate energy and transform
the interior chemical composition. Mass-loss, through winds and more violent phenomena, returns processed material to the ISM, enriching the heavy metal content,
generating shocks within neighbouring interstellar clouds that stimulate further star
formation and, in some cases, leading to breakout galactic fountains that send material far into the halo and intergalactic medium.

The present series of chapters has three main strands: an examination of the
detailed processes involved in how gas within an interstellar cloud redistributes
itself to form stars and star systems; an exposition of the dynamical evolution of
clusters and associations, paying particular attention to the role of binary and multiple systems; and, finally, a discussion of the general properties of the field population
within the Galactic disc, and how those properties can provide insight into the past
history of cluster formation within the Milky Way and other galactic systems. To shift
metaphors, these three topics form a Russian doll, moving from the spatially compact, short timescale star formation process through medium-scale cluster evolution
and dissipation to integration within the large-scale field population, constituting the
mix-mastered residue from the long past history of formation and dispersal of star
clusters and associations.

My chapters tackle the larger scales. This introductory chapter aims to provide a
broad context for the discussion by laying out the basic properties of the Milky Way
galaxy and of its component stellar populations. These are wide-ranging topics that

I.N. Reid (✉)
Space Telescope Science Institute, Baltimore, MD, USA
e-mail: inr@stsci.edu

© Springer-Verlag Berlin Heidelberg 2015
C.P.M. Bell et al. (eds.), *Dynamics of Young Star Clusters and Associations*,
Saas-Fee Advanced Course 42, DOI 10.1007/978-3-662-47290-3_15

are covered by a broad swathe of the astronomical literature. Rather than trying to provide blanket coverage, my intentions throughout the course are to provide sufficient references to give the interested reader a starting point for further exploration. Apologies in advance to those omitted from explicit citation.

15.2 The Nature of the Milky Way

The Milky Way has been known as a luminous, celestial band for more than 3,000 years. Its popular name derives from the Roman Via Lactea, but it was also the River of Heaven (Al Nahr, Tien Ho, Akash Ganga), a celestial pathway (Waetlinga Straet, Wotan's Way, Winter Street) and, to the Greeks, the Galactic circle or Galaxy. Indeed, a handful of Greek philosophers, including Democritus and, perhaps, Aristotle, even ascribed its diffuse light to a vast congregation of extremely distant stars, an hypothesis that was verified only when Galileo turned his spyglass skywards in 1609.

The original explanation for the congregation of stars known as Milky Way is often ascribed to the Durham clergyman, Thomas Wright, and in his 1750 treatise *An Original Theory or New Hypothesis of the Universe* Wright did suggest that the concentration of stars into a luminous band might reflect geometric projection along a thin, extended distribution. However, Wright envisaged that distribution as a ring-like structure, much like the rings of Saturn. The philosopher Immanuel Kant (1755) and the mathematician, physicist Johann Heinrich Lambert (1761) were the first to independently propose that the Sun lay within an extended disc of stars. Their hypothesis was quantified by William Herschel, who effectively invented the discipline of Galactic astronomy with the star-gaging surveys carried out from Bath in the 1770s. Figure 15.1 shows Herschel's 1785 representation of the local stellar distribution. The strong bifurcated feature is due to the Great Rift in Cygnus; the recognition of the presence of interstellar absorption lay more than a century in the future.

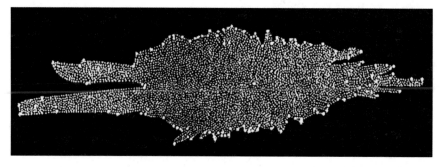

Fig. 15.1 Herschel's model of the Milky Way, deduced from his celestial sweeps and star-gaging. Figure adapted from Herschel (1785)

As part of his celestial sweeps, Herschel encountered numerous diffuse objects and stellar conglomerates. Those nebulae had previously attracted the attention of Edmond Halley, while Charles Messier was in the process of constructing a reference list of the brighter fuzzy objects in support of his main interest of comet hunting. Galileo had argued that closer inspection of such objects would inevitably resolve all as aggregates of faint stars, and such was the case for many systems, notably open clusters like the Pleiades, Praesepe and Messier 67, and globular clusters like Messier 13, 15 and 92. However, even with the development of larger telescopes with greater light grasp and higher resolution, many systems stubbornly resisted resolution.

Herschel constructed his own catalogue of nebulae based on observations with his 20-foot and 40-foot telescopes, eventually compiling a list of 2,500 systems. Herschel's work was taken up and extended to the southern hemisphere by his son, Sir John Herschel, whose General Catalogue included over 5,000 nebulae. By this point, nebulae were classed in three broad categories: gaseous clouds, star clusters and white nebulae. Immanuel Kant had originally suggested that some might prove to be island universes, distant Milky Ways populated by numerous stars. Initially, the elder Herschel subscribed to that viewpoint although his opinions evolved, notably with the discovery of several planetary nebulae. In the latter cases, the diffuse emission was clearly linked to a central star and Herschel came to advocate a close association between nebulae and the star formation process. In contrast, his son fell back on Galileo's suggestion that most, probably all, nebulae would eventually be resolved as star clusters. Herschel's General Catalogue showed a clear deficiency of nebulae within the Milky Way. At the time this was taken as an argument against the island universe hypothesis (where the expectation was a uniform distribution), but this actually reflects dust and absorption in the Galactic Plane.

The General Catalogue was succeeded in 1888 by the New General Catalogue, compiled by John Louis Emil Dreyer, based partly on observations with the 72-inch diameter Leviathan of Parsonstown built by William Parsons, the third Earl of Rosse. During the 1840s, Rosse had used the Leviathan to survey and sketch a number of his nebulae, resulting in the clear detection of spiral structure in several systems, notably Messier 51 (in 1845, the Whirlpool Nebula) and Messier 99 in Coma Berenices (in 1848). The implications were unrecognised, but as photography came to supplant direct visual observations (more of this in Chap. 17), it became evident that many white nebulae shared these morphological characteristics.

The island universe concept received a boost in 1885, with the eruption and subsequent decay of a bright stellar source, S Andromeda, within Messier 31, the Andromeda galaxy. That observation, together with the discovery of novae in other spiral systems, led Heber Curtis of Lick Observatory to espouse that viewpoint. Curtis identified the Milky Way as a relatively small structure, centred near the Sun a concept similar to the model developed by the Dutch astronomer, Jacobus Kapteyn (see Chap. 17). Curtis also noticed that his photographs of spiral systems showed a number with dark bands across the mid-section, which he speculated might be due to dust absorption.

In contrast to Curtis, Harlow Shapley favoured the Big Galaxy hypothesis, envisaging the Sun lying on the outskirts of a single vast system whose centre lay towards

Sagittarius, the centroid of the galactic globular cluster distribution. The two concepts were the subject of the 1919 Great Debate organised by the National Academy of Sciences and designed to address two issues: how large is the Milky Way? and are spiral nebulae island universes? Famously, Shapley and Curtis adopted different speaking styles (populist vs. specialist) and chose to place different emphasis on the two issues, so there was no clear winner at the time. However, Edwin Hubble's subsequent discovery of Cepheid variables in M31 (1923) laid the matter to rest; their distances, estimated using Henrietta Leavitt's period-luminosity relation derived from Magellanic Cloud Cepheids, clearly demonstrated that M31 was not a small stellar aggregate within the confines of even Shapley's Big Galaxy.

Shortly thereafter, the important role played by interstellar absorption within the Milky Way was finally established in a quantitative manner. Dust is largely confined near the Galactic plane. R.J. Trumpler carried out a survey of open star clusters, using the derived Hertzsprung-Russell (H-R) diagrams to estimate distances and hence sizes. Taken at face value, his results suggested that open clusters increased in size with increasing distance, a distinctly non-Copernican result. Trumpler (1930) argued that a more plausible explanation was the existence of material in the line-of-sight that attenuated light from the more distant clusters, giving apparently larger distances. Interstellar absorption was the key ingredient that allowed reconciliation between Shapley's Big Galaxy (which became smaller) and Curtis' island universe.

Island universes come in many forms. As photographic images of galaxies accumulated, morphological patterns started to emerge, leading to the simple tuning-fork classification scheme devised by Edwin Hubble (1926). Galaxies were classed as elliptical (E), spirals (S) and barred spirals (SB). Ellipticals were sub-divided based on their apparent ellipticity (E0 to E7), and spirals as early (Sa/SBa), intermediate (Sb/SBb) or late (Sc/SBc) depending on the relative size of the bulge component (decreasing from Sa to Sc). These regular systems were supplemented by a class of irregular galaxies. Hubble's system has been refined, but still survives as a useful classification and a challenge to galaxy formation models.

Our nearest large neighbor, M31, and its satellite galaxies played a key role in further expanding our understanding of the constituents of the Milky Way, specifically in supporting Walter Baade's development of the stellar population concept. The first clue came from more local observations, as proper motion surveys revealed a handful of stars with extremely high velocities relative to the Sun. Analyses by Lindblad and Oort indicated that the kinematic characteristics were tied to the relative role played by systemic rotation and random motions; high-velocity stars, the subject of Oort's 1926 thesis, were almost exclusively pressure-supported, with negligible rotation.

Further clues came from dwarf galaxies. In 1939, Baade identified Cepheids in the irregular galaxy, IC 1613, placing it at a similar distance to M31, albeit at substantially lower total luminosity. Shortly thereafter, Baade and Hubble identified RR Lyrae variables in the Sculptor and Fornax dwarf galaxies discovered by Shapley. Those systems were similar in size to IC 1613, but the brightest stars were red rather than blue, and they lacked the star-forming regions that were conspicuous in the more distant system. Baade drew explicit comparisons with the Galactic globular clusters.

The clinching data came with Baade's wartime observations of M31 and its companions. Interned as an alien, but retaining his observing privileges on Mt. Wilson as a Carnegie Observatories staff member, Baade took advantage of the blackout conditions to resolve the brightest stars in the central regions of M31 and in its satellites, M32, NGC 195 and NGC 205. Those stars were red, as in the Scuptor and Fornax dwarf galaxies and Galactic globulars, in contrast to the bright blue stars evident in IC 1613 and in M31's spiral arms.

Based on those observations, Baade advanced the concept of distinct stellar populations, namely Population I and Population II. We now know that this distinction represents the dichotomy between an old, evolved stellar population (Population II) and a gas-rich system with on-going star formation, generating short-lived, high-mass stars (Population I). Crucially, Baade demonstrated that this provided a means of characterising the properties of stars within not only the Milky Way, but also other stellar systems. Quoting directly, 'This leads to the further conclusion that the stellar populations of the galaxies fall into two distinct groups, one represented by the well-known HR diagram of the stars in the Solar Neighbourhood (the slow-moving stars), the other by that of the globular clusters. Characteristic of the first group (type I) are highly luminous O- and B-type stars and open clusters; the second (type II), short-period Cepheids (RR Lyraes) and globular clusters. Early-type nebulae (E-Sa) seem to have populations of the pure type II. Both types seem to co-exist in the intermediate and late-type nebulae (Sb-Sc spiral galaxies). The two types of stellar populations had been recognised among the stars of our own Galaxy by Oort as early as 1926.'

Turning to the Milky Way, suggestions that it was itself a spiral galaxy had been made since the late 19th century. Curtis drew an analogy with the spiral nebulae that he photographed from Lick, and that viewpoint gained wider acceptance with identification of Cepheids in the nearby spirals M31 and M33. Baade's results, however, offered a means of settling this question; specifically, the observations that bright OB stars outlined spiral arms suggested that mapping the distribution of such stars in the Milky Way might outline underlying spiral structure. Working with J.J. Nassau, S. Sharpless and D. Osterbrock at Yerkes Observatory, William W. Morgan carried out a photographic survey of most of the northern Milky Way that revealed sections of the features we now know as the Sagittarius arm, the Orion Spur and the Perseus Arm. Presented at the 1951 AAS Christmas meeting, and later supported by HI observations through the nascent radio astronomy program in the Netherlands, the results confirmed the Milky Way as a spiral galaxy and garnered a standing ovation.

15.3 The Milky Way as a Galaxy: Large-Scale Properties

What do modern observations reveal about the overall properties of the Milky Way galaxy? Figure 15.2 shows the all-sky map derived from near-infrared (1.25–3.5 μm) observations made by the Diffuse Infrared Background Experiment (DIRBE) on NASA's Cosmic Background Explorer (COBE) mission. Observations at those wave-

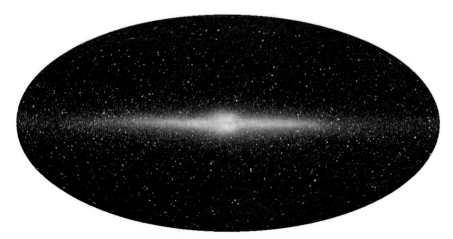

Fig. 15.2 All-sky false-colour near-infrared (1.25, 1.6 and 3.5 μm) map produced by DIRBE/COBE

lengths are dominated by starlight, with some contribution from hot dust. The concentration of stars within the Galactic disc is obvious, as is the boxy/peanut-like central bulge. Detailed star counts towards the bulge strongly suggest the presence of a stellar bar, with consequences for the local kinematics of disc stars (see Chap. 17). The general consensus is that, given an external viewpoint, we would likely classify it as a barred spiral, either type SBc or SBbc. As a comparison, the Andromeda galaxy is also generally classed as Hubble type Sb, while M33 in Triangulum is an Sc galaxy.

The total mass of the Milky Way can be estimated by constructing mass models that take into account constraints imposed by the spatial distribution and kinematics of the Galactic stellar populations and the measured motions of its satellites. The models include the dark matter halo and several baryonic components, including the bulge, the stellar halo and the disc, with the latter usually modelled as two components, thin and thick. Results (Wilkinson and Evans 1999; McMillan 2011) indicate a total mass of $\sim 6 \times 10^{11}$ M$_\odot$ within a radius of ~ 60 kpc and a total virial mass ($r < 300$ kpc) of $\sim 1.3 \times 10^{12}$ M$_\odot$, with uncertainties of at least a factor 2. Interestingly, applying similar models to M31 indicates that our neighbour is similar in mass, perhaps smaller by $\sim 10\,\%$ (Evans and Wilkinson 2000; Watkins et al. 2010). In any event, baryons are a minority constituent within the Milky Way, $\sim 6.5 \times 10^{10}$ M$_\odot$, or less than 5 % of the total mass. The luminosity is estimated as $\sim 2 \times 10^{10}$ L$_\odot$ ($M/L \sim 65$) or $M_V \sim -21$ mag; M31's luminosity is estimated as slightly higher, at $\sim 2.6 \times 10^{10}$ L$_\odot$ (van den Bergh 1999). This gives the Milky Way a luminosity $L \sim 0.8 L_*$, where L_* is the luminosity of a galaxy at the breakpoint in the Schechter (1976) galaxy luminosity function, the transition between a power-law at faint magnitudes and an exponential distribution at bright magnitudes.

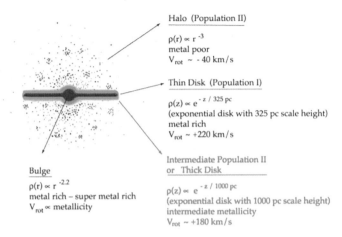

Fig. 15.3 The Milky Way's stellar population. Figure courtesy of S. Majewski

The spatial distribution and gross characteristics of the constituent stellar populations in the Milky Way are outlined schematically in Fig. 15.3 (see also Freeman and Bland-Hawthorn 2002). The least substantial, and most extended, of these populations is the stellar halo, with a mass of $\sim$4 $\times$ $10^8\,M_\odot$ (Bell et al. 2008). Its most prominent constituents are globular clusters, of which approximately 130 are currently known. Halo stars form a non-rotating, pressure-supported system with a near-spheroidal distribution, with heavy-element chemical abundances ranging from one-tenth to less than one-ten thousandth that of the Sun. The halo is essentially gas-free, with no evidence for on-going star formation. As discussed further in Chap. 18, these are the local representatives of Baade's Population II, remnants of the Milky Way's first major star formation episode.

Baade originally identified the Galactic Bulge with classical Population II, an identification that appeared to be confirmed with the identification of RR Lyrae variables within his eponymous window. The scarcity of main-sequence stars more massive than the Sun, with only a relatively small number of even A stars identified, indicates a predominantly old population. However, spectroscopic observations have shown that the metal-poor stars within in the Bulge are a minor constituent, probably representing the innermost halo stars. Most Bulge stars are metal-rich, with a significant tail extending to metallicities a factor of 2–3 higher than the Sun (McWilliam and Rich 1994). Its origins remain unclear.

Bulge stars exhibit significant rotation, possibly correlated with metallicity. The visible extent of the Bulge in the DIRBE image corresponds to a diameter of $\sim$3 kpc, encompassing the stellar bar (see Chap. 17). The stellar mass is estimated as $\sim$1–2 $\times$ $10^{10}\,M_\odot$ (Kent 1992; Dwek et al. 1995—although note that the latter paper assumes a Salpeter IMF and therefore probably overestimates the contribution of low-mass stars, see Chap. 16). There is evidence for on-going star formation (e.g. the Arches cluster near the Galactic Centre), but this may reflect gas being funnelled from the disc into the central regions and the black hole at the Galactic Centre.

The substantial majority of baryonic material in the Milky Way is in the disc, a flattened, extended, rotationally-supported component with a total mass estimated as 4–$5 \times 10^{10}\,M_\odot$ (McMillan 2011). Almost all the gas and dust in the Milky Way (and in other spiral galaxies) lies close to the Galactic mid-Plane. Consequently, the disc is the primary location for on-going star formation, most star clusters lie close to the disc, and disc stars are the product of an extensive star-forming history.

Galaxy discs are generally characterised by the presence of extensive, detailed structure, particularly at blue and ultraviolet (UV) wavelengths where young star-forming regions stand out disproportionately. The underlying mass distribution is more regular. Ken Freeman's (1970) seminal analysis of surface photometry of several nearby spiral galaxies showed that the azimuthally-averaged surface brightness profile is well-characterised by an exponential distribution:

$$I(r) = I_\circ r^{ar}, \tag{15.1}$$

where $I_\circ$ is the (extrapolated) central surface brightness and a (also written as hR) is the scalelength, with values of the latter parameters ranging from ~ 2.5 to 5.5 kpc.

Subsequent analyses, notably by van der Kruit and Searle (1981, 1982), confirmed these results and also indicated that many galaxies showed clear evidence that the radial profile is truncated at 3–5 scalelengths (see Fig. 15.4). This has been inferred as suggesting a cut-off in the star formation process, possibly tied to the gas surface density declining below a critical threshold (van der Kruit and Freeman 2011). However, recent observations, notably with the GALEX satellite, have shown clear evidence for UV light beyond the cut-off radius in some galaxies (e.g. M83, Thilker et al. 2005), in some cases resolved as UV-bright knots. This suggests the presence of on-going star formation, but at a much lower level than within the main body of the disc.

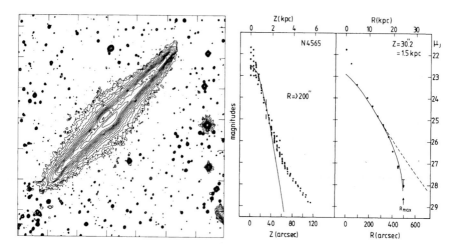

Fig. 15.4 *Left panel*: Surface photometry of the edge-on Sb spiral, NGC 4565. *Right panel*: Derived density profiles along and perpendicular to the disc. Figure adapted from van der Kruit and Searle (1981)

Perpendicular to the disc, many spiral galaxies show evidence for complex density distributions. Close to the mid-Plane, the distribution is well-matched by a simple exponential, but deviations suggestive of the presence of a second component appear at moderate to large heights. This behaviour was noted originally by Burstein (1979), who found that modelling the surface brightness profiles of five S0 galaxies required three components: bulge, disc, and what Burstein termed the 'thick disc'. van der Kruit and Searle (1981, 1982) confirmed that this additional component was required to match some galaxies in their sample (e.g. NGC 4565; see Fig. 15.4), with the additional component generally more prominent in systems with larger bulges.

In the case of the Milky Way, determining the radial density distribution is complicated by our location close to the mid-Plane and the consequent necessity of allowing for interstellar absorption along the line-of-sight both towards and away from the Galactic Centre. Nonetheless, recent analyses suggest that the data are consistent with an exponential scalelength of 2.5–3 kpc (see Table 15.1) and a sharp decline/cut-off in the density distribution $\sim$6 kpc beyond the Solar radius, or $\sim$14 kpc from the Galactic Centre (Robin et al. 1992).

Determining the vertical density distribution of the Galactic disc is a more tractable problem. Starcounts show clear evidence for more stars at distances $z > 1.5$ kpc above the Plane than can be modelled with the single exponential associated with disc stars in the 1960s and 70s. As Fig. 15.5 shows, the distribution can be represented using two exponentials characterised as the thin and thick discs, as suggested originally by (Gilmore and Reid 1983). Succeeding years have seen considerable debate regarding both the parameters that should be associated with a two-exponential fit (specifically, the scaleheight and local normalisation of the thick disc) and whether the thin and thick disc are distinct stellar populations, or subsets of an underlying

Table 15.1 Scalelengths and scaleheights for the thin and thick disc

Method	h_R^{thin} (kpc)	h_z^{thin} (pc)	h_R^{thick} (kpc)	h_z^{thick} (pc)	Thick/thin	References
Photographic starcounts		300		1450	2 %	(1)
SEGUE starcounts			4.1 ± 0.4	750 ± 70		(2)
SDSS starcounts	2.6	300	3.6	900	12 %	(3)
2MASS K giant starcounts	3.0 ± 0.1	270 ± 10		1060 ± 50		(4)
Pioneer X flux measurements	4.5–5					(5)
2MASS starcounts	3.7 ± 1.0	360 ± 10	5.0 ± 1.0	1020 ± 30	$7 \pm 1 \%$	(6)
Spectroscopic survey			3.4 ± 0.7	695 ± 45		(7)

References: (1) Gilmore and Reid (1983); (2) de Jong et al. (2010); (3) Jurić et al. (2008); (4) Cabrera-Lavers et al. (2005); (5) van der Kruit (1986); (6) Chang et al. (2011); (7) Kordopatis et al. (2011)

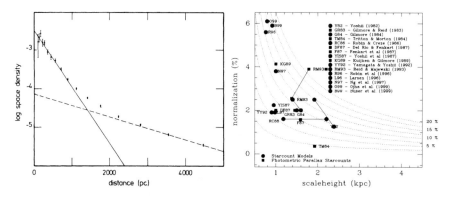

Fig. 15.5 *Left panel*: Two-component fit to star counts towards the South Galactic pole (from Gilmore and Reid 1983). *Right panel*: The various combinations of scale height and local normalisation proposed for the thick disc (from Siegel et al. 2002)

continuum. In other words, is the two-exponential fit simply a mathematical representation, or do these two components have some relation to the underlying physics of galaxy formation? We return to this issue in Chap. 19.

15.4 Star Formation in the Milky Way

Star formation is the key process that drives galaxy evolution. In spiral galaxies like the Milky Way, most star formation is triggered along spiral arms where gas is concentrated by the spiral density wave (illustrated in Fig. 15.6). Gas is shocked and compressed (see Bertin and Lin 1996 for a thorough discussion of this process). The M51 maps show the progression from cold, dense CO clumps, overlying a broader HI distribution, along the inner edge of the arms through the narrow band of $H\alpha$ associated with HII regions to UV light from OB stars along the outer edge of the arms. The narrow $H\alpha$ distribution reflects the relative short lifetimes of the high-mass stars that generate ionising radiation. Spiral arms are long-lived, but nonetheless transient features. Density wave patterns are likely to evolve over time and, if so, spiral structure will evolve in a corresponding fashion.

Star formation manifests its presence at many wavelengths across the electromagnetic spectrum (see Fig. 15.7), providing a number of opportunities to measure the global star formation rate within galactic systems. Hot stars register their presence both directly at UV wavelengths (allowing for interstellar absorption) and indirectly, through processed radiation from hot dust, and mid- and far-infrared wavelengths; emission lines due to hydrogen ($H\alpha$, $P\alpha$) and ionised metals (OII, SII) from ionised gas in HII regions are evident at optical and near-infrared wavelengths; and complex molecular features due to excitation of polycyclic aromatic hydrocarbons (PAHs) are evident at mid-infrared wavelengths. Moving to longer wavelengths, radio emission

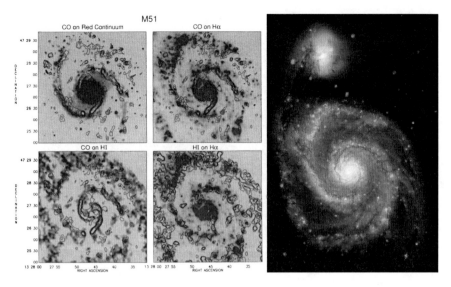

Fig. 15.6 Spiral structure in M51, the Whirlpool galaxy. *Left panel*: Shows cold (CO), warm (HI) and hot (Hα) gas concentrated along the spiral arms (from Rand et al. 1992). *Right panel*: Composite image comprising the following data: *Chandra* X-ray (*purple*), HST optical (*green*), *Spitzer* infrared (*red*) and GALEX UV (*blue*; figure taken from http://thefabweb.com/wp-content/uploads/2012/08/The-Whirlpool-Galaxy-M51-Composite-Image.jpg?25d8db)

Fig. 15.7 Tracing star formation across the electromagnetic spectrum: the spectral energy distribution of the dwarf irregular galaxy, NGC 1705. Figure courtesy of D. Calzetti

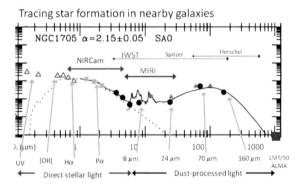

is generated by thermal (free-free emission) and non-thermal processes (synchrotron radiation from cosmic rays generated in supernovae remnants), while, at the other extreme, soft X-rays are generated by thermal emission from gas heated by supernovae and stellar winds from high-mass stars.

Measurements at these wavelengths can be used to estimate the global star formation rates in galactic systems. Calzetti et al. (2009) have reviewed a wide range of methods and summarised the resulting calibrations (their Table 1; reproduced in Fig. 15.8). All of these indicators rely on phenomena associated with *massive* stars, typically exceeding ~ 8–$10\,M_\odot$, while lower-mass stars account for the bulk of mass in star-forming regions. Estimates of the total star formation rate therefore rely on an

Table 1: SFR Calibrations.

Waveband	SFR ($M_\odot$ yr^{-1})	Comments	Reference
X–ray	$(1.7\pm0.3)\times10^{-40}L_{2-10\ keV}$ (erg s^{-1})	SFR<50 $M_\odot$ yr^{-1}	(1), (2)
	$(8.9\pm1.8)\times10^{-40}L_{2-10\ keV}$ (erg s^{-1})	SFR$\gtrsim$50 $M_\odot$ yr^{-1} or young gals	(2)
	$(1.5\pm0.5)\times10^{-40}L_{0.2-2\ keV}$ (erg s^{-1})		(1)
UV	$(8.1\pm0.9)\times10^{-29}$ l_ν (erg s^{-1} Hz^{-1})	0.13-0.26 μm range	(3), (4)
Hα	$(5.3\pm1.1)\times10^{-42}$ L(Hα) (erg s^{-1})		(3)
MIR	1.27×10^{-38} [L(24 μm) (erg s^{-1})]$^{0.885}$	L(24 μm)= ν l(ν)	(5), (6)
	5.3×10^{-42} [L(Hα)$_{obs}$ + 0.031 L(24 μm)]	see text	(5), (7)
FIR	3.0×10^{-44} L(8-1000 μm) (erg s^{-1})	see text	(3)
Radio	4.1×10^{-21} $\nu^{0.1}$ $l_{\nu,T}$ (W Hz^{-1})	$l_{\nu,T}$=thermal emiss.	(8)
	3.5×10^{-22} $\nu^{0.8}$ $l_{\nu,NT}$ (W Hz^{-1})	$l_{\nu,NT}$=non-thermal emiss.	(8)
	4.0×10^{-22} $l_{1.4\ GHz}$ (W Hz^{-1})	$l_{1.4\ GHz}$=obs. radio luminosity	(9), (8)

Note. — References: (1) Ranalli et al. 2003; (2) Persic & Rephaeli 2007; (3) Kennicutt 1998; (4) Salim et al. 2007; (5) Calzetti et al. 2007; (6) Alonso–Herrero et al. 2006; (7) Kennicutt et al. 2007; (8) Schmitt et al. 2006; (9) Yun et al. 2001.

Fig. 15.8 Calibrating global star formation rates. Table 1 taken from Calzetti et al. (2009)

assumed form for the underlying initial mass function (IMF, see Chap. 16). In general, higher luminosity galaxies have higher global star formation rates (see Fig. 15.9).

These calibrations can be applied to estimating the global star formation rate in the Milky Way. As with estimates of the radial density distribution, investigations have to make allowance for the presence of dust obscuration in the mid-Plane. Chomiuk and Povich (2011) have reviewed recent analyses based on radio measurements of free-free radiation, far-infrared measurements of dust emission and star counts of OB stars and young stellar objects. Integrating over the disc, they find values ranging from 0.5 to 2.6 $M_\odot$ yr^{-1}, with an average value of 1.9 $\pm$ 0.4 $M_\odot$ yr^{-1}. As Fig. 15.9 shows, this value is broadly consistent with the estimated luminosity of the Milk Way.

Star formation is distributed along spiral arms, but the large-scale activity is resolved into a series of smaller scale star-forming events. Table 15.2 lists basic characteristics of different stages in this process. The overall scheme is clear; the details, less so. Star formation becomes apparent within molecular clouds as localised density concentrations that evolve to host (generally) multiple stars. Concentrations of these star-forming clumps are characterised as embedded clusters. As winds from high-mass stars and supernovae clear the remaining gas, the denser embedded clusters emerge as open clusters and the cloud complex as a whole takes on the characteristics of an extended OB association. Clusters dissipate and dissolve with time through gravitational interactions, and globular clusters represent the residuals of the densest star-forming regions from the earliest epochs of galaxy formation. This thumbnail sketch obviously omits a many complications; much more thorough discussions of the physical processes of the star formation and the evolution from embedded clusters to associations and open clusters are given in the chapters presented elsewhere in this volume (see Clarke and Mathieu).

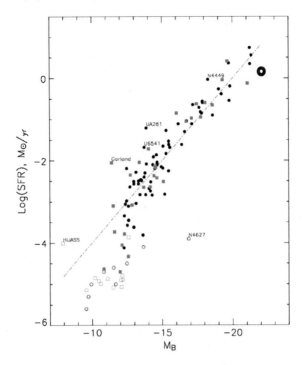

Fig. 15.9 The global star formation rate in spiral galaxies. The Milky Way's location is indicated by the large circle. Figure adapted from Kaisin and Karachentsev (2008)

Table 15.2 Properties of star-forming regions and star clusters

	Embedded cluster	OB association	Open cluster	Globular cluster
Size	Few–10 pc	20–500 pc	Core radius $\sim$2 pc	10–40 pc
Mass	100–1000 $M_\odot$	20–80 OB stars	100–1000 $M_\odot$	10^4–10^6 $M_\odot$
Density*	Few stars pc^{-3}	0.1 stars pc^{-3}	$\sim$10–100 stars pc^{-3}	10^3 stars pc^{-3}
Gravitationally bound?	?	No	Yes	Yes
Age	<10 Myr	2–15 Myr	Typically <250 Myr	10–13 Gyr
Numbers		12 within 650 pc	$\sim$3000	$\sim$150

* The star density in these systems can vary substantially; thus, while the average star density in the Orion Nebula Cluster is $\sim$100 stars pc^{-3}, the core density exceeds 10^4 stars pc^{-3} (Hillenbrand and Hartmann 1998)

15.5 Stellar Abundances

The chemical abundance distribution within galactic systems is driven by the star formation process. Stellar nucleosynthesis transforms hydrogen, helium and light elements to heavy elements, which are returned by mass-loss and winds to the ISM where they contribute to the next generation of star formation. The expectation, therefore, is that the average metallicity of a galaxy increases with time, as more generations of star formation add their nucleosynthetic products to the ISM.

Supernovae are particularly important sources of heavy metals, since that process provides the only means of generating elements heavier than iron. Supernovae come in two main flavours: Type Ia SNe, which are generally believed to be the result of a white dwarf in a binary system accreting sufficient material to exceed the Chandrasekhar mass; and Type II SNe, generated by core collapse in a high mass ($M \gtrsim 7 \, M_\odot$) stars. The two processes occur on different timescales and generate ejecta with different abundance distributions: most white dwarf progenitors were intermediate-mass stars, with lifetimes >1 Gyr, and, primarily, they generate elements close to the iron peak; in contrast, massive stars can undergo core collapse within 10–100 Myrs of their formation and have more diverse products, with a high proportion of α-elements (notably O, Ne, Mg, Si, S, Ca), some iron peak, s-process and r-process elements. Since the type II SNe evolve faster, the first few generations of recycled materials include a higher α/Fe abundance ratio than later generations (Matteucci and Greggio 1986), and this becomes evident when one compares the detailed abundance distributions of halo and disc stars (see Fig. 15.10). The overwhelming majority of disc stars, including the Sun, have α-abundances that are a factor of 3–4 lower than in halo stars. Recognising that iron abundance is a chronometer, it becomes clear that the Milky Way enriched its metallicity to close to the solar value within the first 1–2 Gyr of its existence as a star-forming entity.

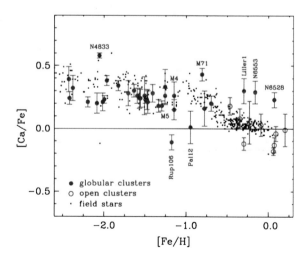

Fig. 15.10 The evolution of α/Fe abundance ratio as characterised by measurements of calcium abundance in a range of stellar systems. [Fe/H] is the logarithmic abundance of iron relative to hydrogen, scaled to 1 ([Fe/H] = 0 dex) at the solar abundance. Figure adapted from Gratton et al. (2004)

Stellar abundance variations manifest themselves as changes in the relative strength of spectral features. Metallicities are generally measured either through direct analysis of line strengths from spectra, or using narrowband photometric indices designed to sample specific spectral features. In the former case, the spectroscopic line strength measurements are matched against theoretical predictions such as curve of growth analyses or spectral synthesis models. The photometric indices are calibrated empirically, using stars with spectroscopic metallicity determinations. Spectroscopic measurements are more precise than photometric estimates, but are limited to brighter objects by the necessity of acquiring a high signal-to-noise spectrum.

Stellar abundances are measured relative to the Sun and usually given in the form [M/H], where M represents heavy elements (often Fe) and the measurements are in a logarithmic scale. Thus, [Fe/H] $= -1$ dex indicates a stellar abundance of iron that is one-tenth the abundance in the Sun. Extensive observations exist for stars in the vicinity of the Sun, both spectroscopic and photometric (primarily using Strömgren photometry). The results (see Fig. 15.11) reveal an asymmetric distribution that peaks close to the solar value, with an extended tail towards lower abundances. Approximately 40 % of local stars are more metal-rich than the Sun, while less than 5 % have abundances [Fe/H] < -0.5 dex.

Metallicities can also be determined for gaseous nebulae, notably HII regions, using measurements of emission lines produced by neutral and ionised oxygen, carbon and nitrogen. The measurements are in terms of absolute abundances, usually expressed in a logarithmic scale where the abundance of hydrogen is set equal to 12, and provide a particularly effective means of identifying spatial variations in abundance within galaxies. The results indicate that metallicity increases towards

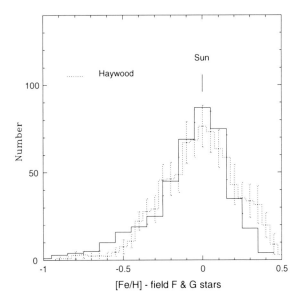

Fig. 15.11 The abundance distribution of field F and G stars in the vicinity of the Sun (from Reid et al. 2007). Note that the distribution peaks close to the Sun's metallicity. Also plotted is the distribution derived by Haywood (2002)

the central regions of the Milky Way and other spirals, with typical gradients of
0.05–0.07 dex kpc^{-1}. Some galaxies (e.g. M33: Cioni 2009) show evidence for a
flattening in the radial variation at large radii. Within the Milky Way, observations of
HII regions or OB stars (both sampling current star-forming regions) are generally
consistent with a slope of $\sim$0.1 dex kpc^{-1} that may flatten at radii beyond $\sim$12 kpc
(Smartt et al. 2001; Rudolph et al. 2006). There are sometimes mismatches between
the average abundance of gas and the stars at the same radius, perhaps reflecting
stellar migration (discussed further in Chaps. 17 and 20).

We should highlight an interesting complication regarding the solar metallicity:
oxygen is the third most abundant element in the Sun; despite that fact, considerable uncertainty remains over the exact value of the solar oxygen abundance. Until
recently, the standard value was [O] = 8.83 dex, as given by Grevesse and Sauval
(1998). However, detailed line analysis of solar spectra led to proposals reducing that
value by almost two-thirds to 8.66 dex (Asplund 2005). A subsequent re-analysis
leads to a slightly higher value, [O] = 8.69 dex (Asplund et al. 2009). This revision is
not without further implications, since the lower abundance leads to lower opacities
at the base of the convective envelope. That, in turn, leads to sound speeds, density
profiles and helium abundances that are in conflict with helioseismology analyses
(Serenelli et al. 2011). There is, however, possible reconciliation in sight, and we will
return to this issue in Chap. 21. This uncertainty clearly complicates tying the stellar
abundance scale in general, and the Sun's metallicity in particular, to ISM metallicities, which are usually determined by measuring the absolute oxygen abundance in
HII regions.

15.6 The Sun's Place in the Milky Way

The second chapter in this series will concentrate on the properties of the stars, dust
and gas (mainly the stars) within the few tens of parsecs that define the immediate
Solar Neighbourhood. Before focussing in on that scale, it is useful to consider the
Sun's location within the Galaxy from a broader perspective.

The Sun's location vertically within the Galactic Plane is surprisingly well-
determined. Matching starcounts towards the North and South Galactic Poles indicates that the Sun lies somewhat towards the North Pole, offset by 20 ± 0.35 pc
(Humphreys and Larsen 1995). Alternatively, one can map the distribution of young
objects, which are expected to be closely confined in narrow distributions centred on
the mid-Plane. Observations of Cepheids indicate an offset of 26 ± 3 pc (Majaess et al.
2009); OB stars give 19.6 ± 2.1 pc (Reed 2000); and measurements of open clusters
indicate 22.8 ± 3.3 pc (Joshi 2005). Overall, the results are remarkably consistent,
placing the Sun $\sim$20 pc above the mid-Plane of the Galactic disc.

The Sun's distance from the Galactic Centre, the Solar Radius or R$_\circ$, has been
determined using a variety of techniques. Observations of distance indicators such
as RR Lyraes and Type II Cepheids allow estimates of the distance to the centroid of
the Bulge, giving values of 8.1 ± 0.6 and 7.7 ± 0.7 kpc, respectively (Majaess 2010;

Majaess et al. 2009). Alternatively, proper motion measurements of objects near or at the Galactic Centre can be used to estimate the distance, since the apparent motion reflects the Sun's orbital velocity around the Galactic Centre. Measurements of OH masers in high-mass star-forming regions give a value of 8.24 ± 0.55 kpc. Direct measurement of the radio source Sgr A* give values of 8.0 ± 0.4 kpc (Ghez et al. 2008) and 8.33 ± 0.35 kpc (Gillessen et al. 2009). Overall, the various indicators indicate that $R_\circ \sim 8.0$ kpc, with an uncertainty of 5 %. As noted above, the stellar density distribution in the disc shows a sharp decrease approximately 6 kpc beyond the Sun's orbit, implying a total radial extent of $\sim$14 kpc.

Focusing on the local environment, the Sun lies in an interarm region relatively close to the star-forming feature known as the Orion Spur, which itself lies between the inner Sagittarius spiral arm and the outer Perseus arm (see Fig. 15.12). Zooming in to a scale of a few hundred parsecs (see Fig. 15.13), it becomes apparent that the Sun is in a quiescent region. The nearest active star-forming regions (ρ Ophiuchus, Chamaelon, Taurus and the Scorpius-Centaurus association) and their associated molecular clouds are more than 150 pc from the Sun. The nearest open clusters are the Hyades (age $\sim$ 650 Myr, distance $\sim$ 50 pc), the Pleiades (age $\sim$ 130Myr, distance $\sim$ 130 pc) and Praesepe (age $\sim$ 650Myr, distance $\sim$ 180 pc). There are relatively few stars younger than the Pleiades in the immediate vicinity of the Sun. Nonetheless, with that caveat, the stars populating the Solar Neighbourhood represent a fair sampling of the stellar content of the Galactic disc, a subject pursued further in the next chapter.

Fig. 15.12 The Sun's location within the Milky Way. The major spiral arm features and the Galactic Bar are labelled. Figure from Momany et al. (2006)

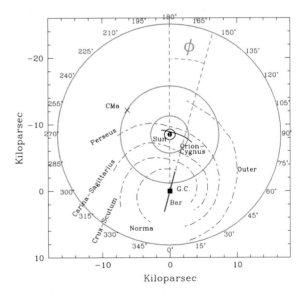

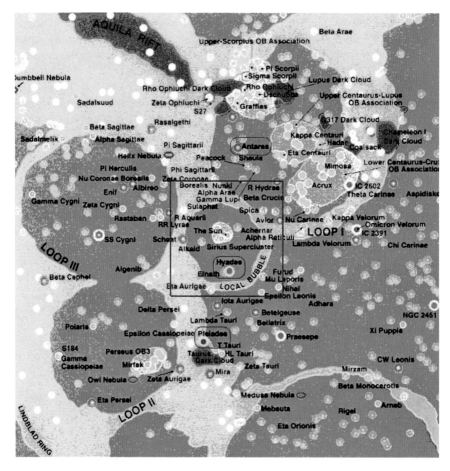

Fig. 15.13 A map of the ~400 × 400 pc region centred on the Sun. Shaded areas map higher gas density, star-forming regions. The Galactic Centre lies at ~12 o'clock in this diagram, while the Hyades cluster lies in the direction of the Galactic anticentre. Figure from Henbest and Couper (1994)

References

Asplund, M. 2005, ARA&A, 43, 481

Asplund, M., Grevesse, N., Sauval, A. J., & Scott, P. 2009, ARA&A, 47, 481

Bell, E. F., Zucker, D. B., Belokurov, V., et al. 2008, ApJ, 680, 295

Bertin, G. & Lin, C. C. 1996, Spiral Structure in Galaxies: A Density Wave Theory, ed. Bertin, G. and Lin, C. C., MIT Press.

Burstein, D. 1979, ApJ, 234, 829

Cabrera-Lavers, A., Garzón, F., & Hammersley, P. L. 2005, A&A, 433, 173

Calzetti, D., Sheth, K., Churchwell, E., & Jackson, J. 2009, in The Evolving ISM in the Milky Way and Nearby Galaxies, ed. K. Sheth, A. Noriega-Crespo, J. Ingalls, & R. Paladini.

Chang, C.-K., Ko, C.-M., & Peng, T.-H. 2011, ApJ, 740, 34

Chomiuk, L. & Povich, M. S. 2011, AJ, 142, 197

Cioni, M.-R. L. 2009, A&A, 506, 1137

de Jong, J. T. A., Yanny, B., Rix, H.-W., et al. 2010, ApJ, 714, 663

Dwek, E., Arendt, R. G., Hauser, M. G., et al. 1995, ApJ, 445, 716

Evans, N. W. & Wilkinson, M. I. 2000, MNRAS, 316, 929

Freeman, K. C. 1970, ApJ, 160, 811

Freeman, K. & Bland-Hawthorn, J. 2002, ARA&A, 40, 487

Ghez, A. M., Salim, S., Weinberg, N. N., et al. 2008, ApJ, 689, 1044

Gillessen, S., Eisenhauer, F., Trippe, S., et al. 2009, ApJ, 692, 1075

Gilmore, G. & Reid, N. 1983, MNRAS, 202, 1025

Gratton, R., Sneden, C., & Carretta, E. 2004, ARA&A, 42, 385

Grevesse, N. & Sauval, A. J. 1998, Space Sci. Rev., 85, 161

Haywood, M. 2002, MNRAS, 337, 151

Henbest, N. & Couper, H. 1994, The Guide to the Galaxy, ed. Henbest, N. and Couper, H., Cambridge
    University Press.

Herschel, W. 1785, Royal Society of London Philosophical Transactions Series I, 75, 213.

Hillenbrand, L. A. & Hartmann, L. W. 1998, ApJ, 492, 540

Hubble, E. P. 1926, ApJ, 64, 321

Humphreys, R. M. & Larsen, J. A. 1995, AJ, 110, 2183

Joshi, Y. C. 2005, MNRAS, 362, 1259

Jurić, M., Ivezić, Ž., Brooks, A., et al. 2008, ApJ, 673, 864

Kant, I., 1755, Allgemeine Naturgeschichte und Theorie des Himmels (Königsberg and Leipzig,
    Germany)

Kaisin, S. S. & Karachentsev, I. D. 2008, A&A, 479, 603

Kent, S. M. 1992, ApJ, 387, 181

Kordopatis, G., Recio-Blanco, A., de Laverny, P., et al. 2011, A&A, 535, 107

Lambert, J. H., 1761, Cosmologische Briefe uber die Einrichtung des Waltbaues (Augsburg, Ger-
    many)

Majaess, D. 2010, AcA, 60, 55

Majaess, D. J., Turner, D. G., & Lane, D. J. 2009, MNRAS, 398, 263

Matteucci, F. & Greggio, L. 1986, A&A, 154, 279

McMillan, P. J. 2011, MNRAS, 414, 2446

McWilliam, A. & Rich, R. M. 1994, ApJS, 91, 749

Momany, Y., Zaggia, S., Gilmore, G., et al. 2006, A&A, 451, 515

Rand, R. J., Kulkarni, S. R., & Rice, W. 1992, ApJ, 390, 66

Reed, B. C. 2000, AJ, 120, 314

Reid, I. N., Turner, E. L., Turnbull, M. C., Mountain, M., & Valenti, J. A. 2007, ApJ, 665, 767

Robin, A. C., Creze, M., & Mohan, V. 1992, ApJ, 400, L25

Rudolph, A. L., Fich, M., Bell, G. R., et al. 2006, ApJS, 162, 346

Schechter, P. 1976, ApJ, 203, 297

Serenelli, A. M., Haxton, W. C., & Peña-Garay, C. 2011, ApJ, 743, 24

Siegel, M. H., Majewski, S. R., Reid, I. N., & Thompson, I. B. 2002, ApJ, 578, 151

Smartt, S. J., Venn, K. A., Dufton, P. L., et al. 2001, A&A, 367, 86

Thilker, D. A., Bianchi, L., Boissier, S., et al. 2005, ApJ, 619, L79

Trumpler, R. J. 1930, PASP, 42, 214

van den Bergh, S. 1999, JRASC, 93, 200

van der Kruit, P. C. 1986, A&A, 157, 230

van der Kruit, P. C. & Freeman, K. C. 2011, ARA&A, 49, 301

van der Kruit, P. C. & Searle, L. 1981, A&A, 95, 105

van der Kruit, P. C. & Searle, L. 1982, A&A, 110, 61

Watkins, L. L., Evans, N. W., & An, J. H. 2010, MNRAS, 406, 264

Wilkinson, M. I. & Evans, N. W. 1999, MNRAS, 310, 645

Chapter 16
The Solar Neighbourhood

I. Neill Reid

16.1 Introduction: Act Locally, Think Globally

Observations of an individual star cluster provide insight into the circumstances and characteristics of one particular star-forming event. The stars within the general field near the Sun represent the ensemble product of numerous star-forming events over the history of the Galactic disc. Our goal is to determine the statistical properties of stars in the Galactic thin disc with a view to probing that past history and reconstructing the most influential events.

Achieving that goal requires a sample of stars that is representative of the diverse characteristics of the disc. This demands that the sample be selected in a fair, unbiased manner; that it has sufficient size to provide adequate representation for different subsets of disc stars; and that the members of the sample are well characterised on an individual basis. There is a natural tension among these requirements that must inevitably lead to compromise.

Ideally, one should select a reference sample based on criteria that are completely independent of any of the parameters that are under analysis. However, intrinsic factors can limit the ability to apply the same selection criteria across the board. As an example, O and B stars are visible across most of the Galaxy; however, their short evolutionary lifetimes restrict them to active star-forming regions, where measurements can be hindered by uncertainties in their distance and by localised reddening and absorption. In contrast, most low-mass M dwarfs and brown dwarfs have such low luminosities that their detection becomes problematic at distances of more than 10–20 pc from the Sun. These objects are much brighter at young ages, but relatively few such regions are accessible to observations, while calibrating those observations against reliable theoretical models adds further complications. Regardless, any global analysis needs to forge reliable connections between different samples.

I.N. Reid (✉)
Space Telescope Science Institute, Baltimore, MD, USA
e-mail: inr@stsci.edu

© Springer-Verlag Berlin Heidelberg 2015
C.P.M. Bell et al. (eds.), *Dynamics of Young Star Clusters and Associations*,
Saas-Fee Advanced Course 42, DOI 10.1007/978-3-662-47290-3_16

Despite these reservations, one can learn a lot about the global properties of the Galactic disc by focusing on the Sun's immediate neighbours. The scarcity of very-young systems is inevitable. However, as summarised in Chap. 17, stellar orbits evolve with time due to both random gravitational interactions with massive bodies and to systematic orbit migration. Consequently, stars that now reside within the immediate vicinity of the Sun originated from radically different locations within the disc. The nearest, and best-studied, stars therefore provide a clear opportunity to investigate the formation history and the large-scale characteristics of the stellar populations in the disc.

The present chapter describes how the stars in the Solar Neighbourhood can be used to constraint Galactic parameters. Section 16.2 summarises our current knowledge of the local stellar populations; Sect. 16.3 uses those stars to construct the stellar luminosity function, the number of stars as a function of intrinsic luminosity. Section 16.4 comments on the fraction of binary and multiple star systems locally; Sect. 16.5 discusses the current constraints on the stellar mass function, the number of stars that form as a function of mass, and Sect. 16.6 considers the current evidence for the universality of that function. Section 16.7 provides a final summary, while the Appendix gives a more detailed summary of some of the techniques used to derive these statistical quantities.

16.2 The Local Volume and Local Samples

The opening chapter in this series included a broad review of the large-scale structure of the Milky Way. Figure 16.1 focuses on the Sun's immediate environment, providing a schematic map of features within approximately 100 pc. As noted previously, there are no active star-forming regions within this volume, and the $\sim$650 Myr-old Hyades is the only open star cluster within this volume. However, over the last two decades, astronomers have identified tens of young (5–15 Myr) stars and brown dwarfs that appear to form moving groups and associations which are passing through the Solar Neighbourhood (Zuckerman and Song 2004). These young interlopers are discussed in more detail in Chap. 20.

The Sun lies close to the centre of a region known as the Local Bubble, whose extents are sketched out in Fig. 16.1. The feature has a diameter of $\sim$80 pc, and encloses a low-density region of the interstellar medium (ISM). In the lowest-density regions, the particle density within the Bubble is $\sim$0.005 atoms cm^{-3}, or almost 100 times lower than the average in the ISM, and the gas temperature approaches 10^6K (Cox and Reynolds 1987; Frisch 2007). The Sun itself is merely passing through this region, with a relative velocity of $\sim$25 km s^{-1} directed towards Cygnus, at approximately 10 o'clock in Fig. 16.1. At present it lies within a higher-density feature known as the Local Cloud or Local Fluff, where the particle density approaches $\sim$0.25 atoms cm^{-3}.

The Local Bubble was likely formed some 20–60 million years ago, probably by shocks generated by the rapid expansion of supernova ejecta. At that time, the Sun

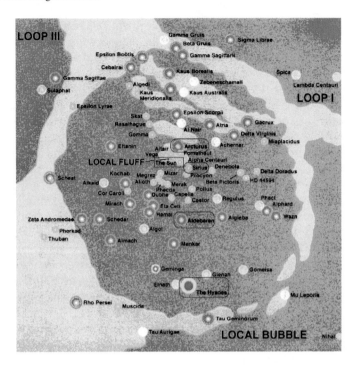

Fig. 16.1 A schematic representation of the Sun's immediate environs. As in Fig. 15.13 of Chap. 15, the Galactic Centre lies at ~12 o'clock. Figure from Henbest and Couper (1994)

lay far from this region. A relative velocity of $1\,\mathrm{km\,s^{-1}}$ corresponds to motion of $1\,\mathrm{pc}$ in 10^6 years. Thus, the Sun's relative motion place it at least $450\,\mathrm{pc}$ distant from the event that triggered the formation of the Local Bubble.

These relative motions are the key to why we might expect local stars to represent more than a history of local star-forming events. Even with a peculiar motion of $0.5\,\mathrm{km\,s^{-1}}$, a star will drift $50\,\mathrm{pc}$ from its formation site in the 10^8 years required for a single Galactic rotation at Solar Radius. As discussed in the following chapter, random gravitational encounters with other objects will add to the peculiar motion, leading to significant mixing within the disc. The net result is that the stars currently close to the Sun may have originated at Galactic radii spanning $4\text{--}10\,\mathrm{kpc}$. This is fortunate, because the lowest-luminosity stars are simply too faint to observe in any kind of detail beyond the confines of the immediate Solar Neighbourhood.

A clear understanding of the degree of completeness is crucial in the analysis of any statistical sample. In the case of nearby stars, this translates to effective means of both identifying candidates and determining reliable distance estimates. Large-scale astrometric surveys can deal with candidate selection by observing all targets brighter than a given magnitude: as examples, the *Hipparcos* satellite obtained milli-arcsecond astrometry, and correspondingly accurate trigonometric parallaxes, for all stars brighter than 9th magnitude; *Gaia* will extend coverage to 19–20th magnitude, with astrometric uncertainties ranging from ∼24 μas at 16th magnitude to ∼300 μas at the faint limit. Without such data, other techniques need to be applied to identify candidate nearby stars. The most effective and widely used are proper motion and photometric parallax. Both techniques have limitations.

The angular motion of a star depends linearly on the tangential velocity relative to the Sun, and inversely on the distance. The stellar velocity distribution for disc stars is relatively narrow (see Chap. 17); thus, stars with high proper motions tend to be close to the Sun. However, since the distribution is narrow, there is a finite probability that a nearby star has a low velocity relative to the Sun, and therefore a low proper motion. Those stars need to be accounted for in any statistical analysis.

Photometric parallax estimates rely on the correlation between surface temperature and luminosity that holds over several stages of a star's evolution, notably during its tenure on the main-sequence. Figure 16.2 shows optical M_V, $B-V$ and near-infrared M_J, $I-J$ colour-magnitude diagrams that highlight the strengths and limitations of this method of distance estimation. Clearly, photometric parallaxes are ineffective for evolved stars. Along the main-sequence, the dispersion about the mean colour-magnitude relations, stemming from age and metallicity variations among the local stars, introduces significant uncertainty in the absolute magnitude estimates for individual stars. Moreover, unless appropriate adjustments are applied, both the dispersion and changes in slope along the main-sequence have the potential to introduce systematic bias in statistical parameters derived from these data, as discussed further in the Appendix to this chapter.

Building on these techniques, several reference samples of nearby stars have been compiled over the past 50 years or more. The more widely used examples include:

- **5 pc sample**: The Dutch astronomer Peter van de Kamp was one of the pioneers in nearby star surveys, and maintained a catalogue of the nearest stars throughout

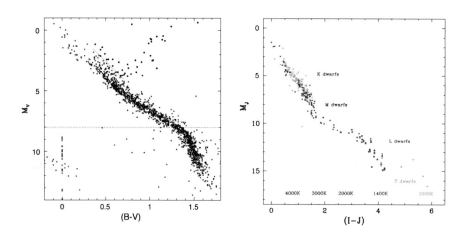

Fig. 16.2 Colour-magnitude diagrams outlined by nearby stars with accurate trigonometric parallax measurements. *Left panel*: The M_V, $B-V$ colour-magnitude diagram (figure from Reid et al. 2002). *Right panel*: The M_J, $I-J$ colour-magnitude diagram which extends to much later spectral types

his career. His initial compilation (van de Kamp 1930), based on photographic astrometric parallax measurements for high proper motion stars, built on previous work by Hertzsprung and included 36 stars ranging in brightness from Sirius to Barnard's star, Wolf 359 and Proxima Centauri. His final compilation (van de Kamp and Lippincott 1975) encompassed 60 stars in 45 systems within a 17 light-year distance limit. The current 5 pc sample includes 66 stars (including 4 white dwarfs) and 6 brown dwarfs in 50 systems. It can be considered complete for hydrogen-burning stars, but not for brown dwarfs, as evidenced by the recent discovery of two brown dwarf systems, a binary (Luhman 2013) and an isolated dwarf (Luhman 2014), no more than 2 pc from the Sun. The latter object is the coolest known dwarf yet discovered, with an effective temperature of ∼250 K.

- **The PMSU survey**: While van de Kamp focused on within 5 pc, Wilhelm Gliese cast a wider net. His Nearby Star Catalogues included photometry, astrometry and spectroscopic information on all stars suspected of lying within 25 pc, with the final Third Catalogue (CNS3) including data for 3803 stars in 3264 systems (Jahreiß and Gliese 1993). Drawn from a variety of heterogeneous sources, those data and the derived distances have a correspondingly broad range of reliability, particularly for lower-mass stars. In the 1990s, Suzanne Hawley, John Gizis and I embarked on the PMSU spectroscopic survey, obtaining optical spectra for the late-K and M dwarfs in the CNS3 (Reid et al. 1995; Hawley et al. 1996). Unsurprisingly, analysing the results for completeness shows that the effective distance limit scales inversely with absolute magnitude, running from ∼20 pc for early-type M dwarfs to 8 pc for the latest-type dwarfs.

- **The 8 pc sample**: Drawn from the PMSU catalogue, the survey volume is a factor 4 larger than the 5 pc sample. The initial sample was limited to regions accessible from the northern hemisphere ($b > -30°$) and included 151 stars in 106 systems; the current sample, expanded to cover the full sky, includes 191 stars and brown dwarfs in 139 systems. With only marginal changes over the last decade, this sample is effectively complete for hydrogen-burning stars.

- **The RECONS 10 pc sample**: Extending coverage from 8 to 10 pc doubles the sample volume. Led by Todd Henry, the RECONS program is focused on that goal, using trigonometric parallax measurements to establish reliable distances (e.g. Riedel et al. 2011). The program has added valuable new data for numerous red dwarfs within 25 pc of the Sun, particularly for stars in the less well-surveyed southern hemisphere. The current 10 pc catalogue includes 357 stars and brown dwarfs in 259 systems within the 10 pc limit, an increase of 64 objects and 46 systems over the 2000 census, but still likely incomplete at the ∼10–15 % level.

- **The 20 pc ultracool dwarf sample**: Low-mass stars and brown dwarfs have extremely cool surface temperatures and radiate only a small fraction of their flux at optical wavelengths. The development of near-infrared sky surveys, such as 2MASS and DENIS, has therefore been crucial in surveys for these objects. Ultracool dwarfs, defined as those with spectral types M6.5 or later, can be identified using near-infrared colours alone, and distances can be estimated based on spectroscopic parallaxes. Adding Luhman's binary, the latest census includes 95 systems with distances within 20 pc of the Sun (Reid et al. 2008).

- **The** *Hipparcos* **25 pc sample**: At the other extreme, the *Hipparcos* catalogue remains the best resource for defining a complete sample of nearby AFGK stars. The catalogue is essentially complete to $M_V = 8.5$ mag within 25 pc of the Sun (Jahreiß and Wielen 1997). On that basis, we identified 805 systems within that distance limit with $M_V = 8.5$ mag, including 760 disc main sequence stars, 4 subdwarfs and 41 evolved stars (Reid et al. 2002, PMSU4). The sample includes 230 known binary or multiple systems; this implies a relatively low multiplicity fraction, as discussed further below, and it is likely that almost as many remain to be discovered.

In analysing these datasets, it is important to have a clear understanding of the likely level of incompleteness, and the associated uncertainties that brings to the analysis. Moreover, while the parameters measured for each star represent the best estimate on an individual basis, there are often systematic corrections that need to be applied when an ensemble of data. Those systematic corrections include Malmquist bias, Lutz-Kelker bias and the potential for systematic in calibrating relations. These issues are discussed further in the Appendix.

The nearby star samples, either singly or in combination, and often supplemented by data for star clusters, provide a means of exploring statistical properties of the disc, including:

- The luminosity function, $\Phi(M_i)$, the number of stars per unit absolute magnitude (or luminosity) per unit volume; integrating this quantity gives the contribution to the total luminosity from stars with different spectral types.
- The mass function, $\Psi(M)$, the number of stars per unit mass per unit volume; this quantity describes how a molecular cloud redistributes its mass to form stars; integrated and matched against the integrated luminosity function, it gives the mass-luminosity ratio for a stellar population.
- Stellar multiplicity, the proportion of single stars, binaries, triples and higher-order multiple systems; this quantity is tied to the granularity and angular momentum within star-forming cloud cores, and the variation of those properties with total mass; the multiplicity of stellar systems evolves with time as systems undergo a variety of gravitational encounters in the general field.
- Stellar kinematics, mapping the cumulative effect of gravitational interactions due to both random encounters and stochastic migration.
- Stellar metallicities, probing the dispersion of chemical composition within localised cloud complexes, and/or mixing within the disc.

The remaining sections of this chapter focus on the luminosity function, stellar multiplicity and the mass function. Stellar kinematics and the metallicity distribution are discussed further in Chaps. 17 and 19, respectively.

16.3 The Stellar Luminosity Function

The luminosity function is a purely observational parameter that can be expressed in terms of wavelength specific or bolometric absolute magnitudes, or, indeed as a function of spectral type. Figure 16.3 shows the luminosity function at visual wavelengths, as derived by combining data from the 8 pc sample and the *Hipparcos* 25 pc sample (from PMSU4). The luminosity function peaks at $M_V \sim 12$ mag ($\sim 0.25\, M_\odot$) and the substantial contribution from subsolar-mass stars is clear: A and F stars are poorly represented, G dwarfs account for approximately 15 % of the nearby stars, while M dwarfs contribute over 70 %. In short, $\Phi(M_V)$ has the classic Anne Elk (Miss) shape: thin at one end, much, much thicker in the middle, and then thin again at the far end.

Figure 16.4 extends coverage to lower masses by adding data for L dwarfs from the 2MASS Ultracool Survey and for T dwarfs from the online database http:// DwarfArchives.org. T dwarfs have extremely low luminosities, and are correspondingly difficult to locate, and the latter sample is known to be incomplete. Almost all of these systems are brown dwarfs which are unable to support hydrogen fusion, and therefore cool monotonically with time. These substellar-mass objects evolve fairly rapidly through spectral types M and L, but the cooling rate slows with decreasing temperature, leading to the increase in number densities from early- to late-type T dwarfs. Brown dwarf characteristics are discussed further in Chap. 20.

Integrating over the luminosity function gives the overall space density of stellar objects. Taking the 8 pc sample, the 139 systems and 191 objects break down into the following categories: 7 brown dwarfs, of which 2 are isolated and 5 are companions

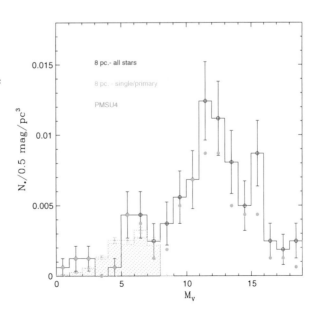

Fig. 16.3 The luminosity function at visual magnitudes for nearby stars from the 8 pc and *Hipparcos* 25 pc (PMSU4) samples. The error bars represent the formal Poisson uncertainties

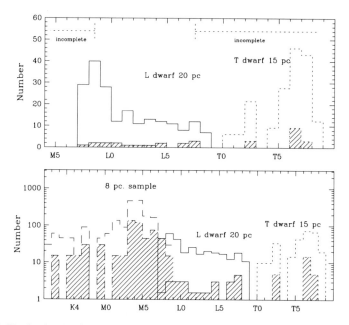

Fig. 16.4 The luminosity function across the substellar boundary. *Upper panel*: Plots L dwarf data from the 2MASS Ultracool Survey (Reid et al. 2008) and T dwarf statistics from the online T dwarf database http://DwarfArchives.org. *Lower panel*: Matches those data against the 8 pc sample, highlighting the contribution from companions in binary and multiple systems. Note that the *lower panel* is scaled logarithmically on the *y*-axis. Figure from Reid et al. (2008)

in stellar systems; 12 white dwarfs, 7 isolated and 5 companions; and 172 hydrogen-burning stars, 92 are isolated, 38 primaries in binary or multiple systems, and 47 companions. This corresponds to an overall space density of 0.065 systems pc^{-3}, 0.086 stars pc^{-3} including 0.0056 white dwarfs pc^{-3}, an overall multiplicity of 27 % (a total of 38 binary/multiple systems).

These number densities are smaller by ~30 % than one would predict based on the 5 pc sample statistics: the 46 systems and 60 objects in that sample translate to predictions of 184 systems and ~240 objects within 8 pc. Arguments have been made that these deficits reflect incompleteness in the 8 pc sample. However, closer inspection raises some questions since the discrepancies between the two samples are driven by the number of early- and mid-type M dwarfs, not the lowest luminosity later-type M and L dwarfs. The implication is that the 'missing' stars in the 8 pc sample have apparent magnitudes in the range 12–14 mag, a domain that has been well surveyed by even photographic proper motion surveys. It is difficult to see how more than 20 such systems could have evaded detection over the last half century.

An alternative hypothesis is that the difference in number density between the 5 and 8 pc surveys provides a measure of the level of structure within interarm regions. Viewed purely as a sampling procedure (i.e. placing a sample volume at a series of position on the Galactic disc), the density of stellar systems derived from the 5 pc sample has a formal uncertainty of ~15 %; thus, the difference with respect to the 8 pc sample is a 2σ discrepancy. *Gaia* will, of course, have the last word on this topic.

16.4 Stellar Multiplicity

Figure 16.4 clearly shows the contribution that companions make to the local stellar systems. Understanding how those systems form is a key challenge for star formation theory, which is why binarism/stellar multiplicity crops up as an important topic in all three sets of chapters in this book. Clearly, one needs to account for binary and multiple systems in computing quantities like the luminosity function and mass function. Indeed, one needs to assess whether those quantities should be constructed on an object-by-object basis, or whether some subset of combinations is appropriate. The frequency and properties of the nearby binary systems, the distribution of mass ratios and separations as a function of spectral type/mass, can provide insight on such hypothetical processes (see e.g. Reid 1991).

Binary systems span a wide range of physical and angular separations; as a consequence, a variety of techniques must be employed to obtain a thorough census. Direct imaging can be used to find companions at wide separations, while techniques such as speckle imaging and adaptive optics, often combined with coronagraphic light suppression, can be used at smaller separations. Spatially unresolved companions are identified through the reflex motion of the primary, measured either by accurate astrometry or, more usually, radial velocity monitoring.

Reliable multiplicity statistics require a well-defined sample. The reference work in this field is the (Duquennoy and Mayor 1991) analysis (hereafter DM91) of the multiplicity of 164 solar-type stars from the then-current second version of Gliese's Nearby Star Catalogue. Their analysis included 13 years of high-precision radial velocity monitoring. Based on their observations, 44 % of the sample stars were identified as having companions. Allowing for incompleteness, DM91 estimated that only ~45 % of stellar systems are single, ~45 % binary and ~10 % multiple with an overall multiplicity of ~55 %. The implication is that majority of solar-type stars are found in binary systems.

The DM91 analysis pre-dated the release of *Hipparcos* parallax data, but re-analysing the sample using *Hipparcos* distance measurements to set the completeness limits leaves the observed binary fraction is essentially unchanged, at $41 \pm 7\%$ (PMSU4). Raghavan et al. (2010) have recently re-analysed a larger sample of 454 FGK stars with *Hipparcos* parallax measurements $\pi > 40$ mas (i.e. $d < 25$ pc), combining spectroscopic, high-resolution imaging and wide-field searches for common proper motion companions. They find a similar observed multiplicity of ~44 %. Including all possible companions, they find that $54 \pm 2\%$ of systems are single stars, $34 \pm 2\%$ are binary, $9 \pm 2\%$ are triple and $3 \pm 1\%$ have four or more components, for an overall multiplicity of 46 %. Moreover, the new observations set significantly stronger constraints on as-yet undiscovered stellar companions, with only a handful likely to emerge from future observations. Raghavan et al. (2010) conclude that, contrary to the received opinion since DM91, a (slight) majority of solar-type systems are single.

These investigations have been extended to lower-mass stars. Contemporaneously with DM91, Fischer and Marcy (1992) and Henry and McCarthy (1993) compiled astrometric/imaging and radial velocity information, respectively, to derive multiplicity statistics for samples of nearby M dwarfs stars, primarily earlier-type dwarfs. Both analyses derived lower overall binary frequencies for M dwarfs than for solar-type stars, with measurements of 34 % and 42 ± 9 % respectively. Those values are consistent with the multiplicity of 34 % (27 % binary, 7 % multiple) for stars in the 8 pc sample (Reid and Gizis 1997).

With the discovery of numerous brown dwarfs over the last decade, it has been possible to extend multiplicity surveys to later-type M dwarfs and the even cooler L and T dwarfs (more on this in Chap. 20). Binary surveys are largely limited to direct imaging, but the low luminosities and relatively small distances of the targets render the observations (primarily with the *Hubble* Space Telescope) capable of identifying binaries at separations as small as 2–3 au. The results indicate that the binary fraction continues to decline, with typical values of 12–17 % (Burgasser et al. 2007; Reid et al. 2008). Figure 16.5 summarises the overall variation as a function of the spectral type of the primary.

What drives this trend to lower binarity among lower-mass stars? At least part of the answer is captured in Fig. 16.6, which shows the distribution of separations of known binary systems as a function of the total system mass (Burgasser et al. 2003, 2007). The parent sample is drawn from observations of stars within the general Galactic field, and therefore represents a predominantly old (> 1 Gyr) population. It is clear that wide binaries become less and less common among lower-mass systems. This strongly suggests a gravitationally-based effect, either affecting the formation process or post-formation dynamical evolution. Surveys of young associations and star-forming regions have revealed a number of wide (few hundred to $\sim$1000 au) separation low-mass binaries in star-forming regions (e.g. Luhman 2004—see further below), suggesting that dynamical evolution play a strong role in the scarcity of such systems in the general field.

An alternative approach is to consider the distribution in binding energy of binary systems, the energy required to disrupt the system. The binding energy is given by $E = -(G M_1 M_2)/2a$, where M_1 and M_2 are the component masses, a is the orbital semi-major axis, and G is the gravitational constant. The results (see Fig. 16.7) show that the widest low-mass binaries tend to be factors of 10–20 more tightly bound than higher-mass systems. Close et al. (2003) have suggested the presence of a break in the minimum binding energy at $M_{\rm tot} \sim 0.3 \, M_\odot$, corresponding to the transition from the exponential distribution to the power-law in Fig. 16.6. This has been taken as evidence for a different formation mechanism for brown dwarf binaries or even brown dwarfs in general. However, it should again be noted that wide very-low-mass (VLM) systems with lower binding energies are found in active star-forming regions, demonstrating that such systems form although not with high frequency. It is possible

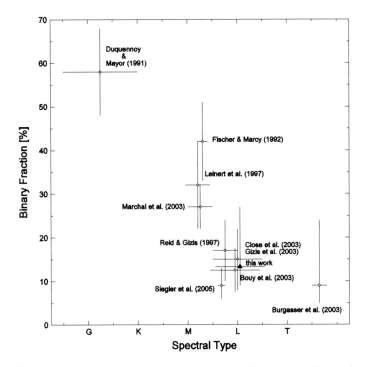

Fig. 16.5 Binary frequency as a function of the spectral types of the primary. Figure adapted from Bouy et al. (2005)

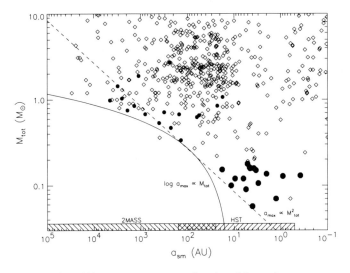

Fig. 16.6 The separation of binary components as a function of the total system mass; *solid points* mark systems with ultracool dwarf components and the *dotted lines* outline exponential and power-law representations of the maximal separation. Figure from Burgasser et al. (2003)

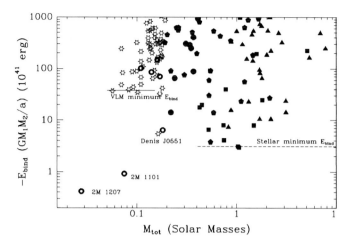

Fig. 16.7 The binding energy of binary systems as a function of the total mass. The minimum binding energy limits outlines in the diagram correspond to escape velocities of $\sim$3.8 km s^{-1} for VLM binaries and $\sim$0.6 km s^{-1} for higher mass stars. Figure adapted from Close et al. (2003)

that the observed distribution results from dynamical evolution acting on a relatively sparse initial sample of VLM systems.

Mass ratio, $q = M_2/M_1$, is the second key parameter that characterises a binary system. There has been considerable discussion over the decades as to whether there is a preference for equal-mass systems, or whether the components in binary systems are essentially drawn at random from the same underlying mass distribution (Trimble 2008). Figure 16.8 shows that this question does not have a simple one-dimensional answer. The data plotted outline the mass ratio distribution as a function of linear separation for companions to solar-type stars within 25 pc of the Sun, combining radial velocity analyses of spectroscopic binaries with imaging data for resolved systems. The companion masses are derived either from the radial velocity curve, assuming an average value of $\sin^3 i = 0.679$ where the inclination in unknown, or from photometry of resolved systems. It is clear that the distribution changes as a function of the separation: within $\sim$50 au, the distribution is essentially bimodal, with mass ratios either tightly concentrated near a value of unity (i.e. close to equal mass) or below 0.01 M$_\odot$, in the planetary regime; beyond $\sim$50 au, there are as yet no planetary-mass detections, but the mass distribution of stellar companions is significantly broader. A similar situation holds for stars in clusters; for example, there is a clear deficit of low-mass companions to solar-type stars in the Pleiades (Reggiani and Meyer 2011).

The deficit of brown dwarf companions to solar-type stars has been highlighted extensively in the radial velocity surveys for exoplanets as the 'brown dwarf desert' (e.g. Marcy and Butler 2000). This is not a selection effect since higher-mass brown dwarfs produce larger reflex motions in the parent star, and are therefore much easier to detect in high-precision radial velocity surveys. Figure 16.8 clearly demonstrates

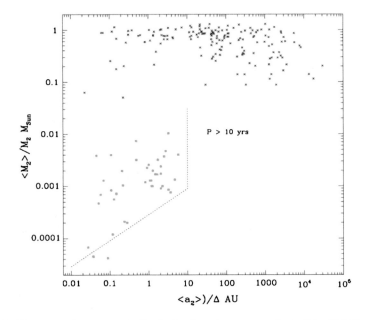

Fig. 16.8 The companion mass ratio distribution for solar-type stars in the Solar Neighbourhood, including exoplanets. Much has been made in the literature on the scarcity of brown dwarf companions at small separations (the 'brown dwarf desert'), but note that the scarcity extends through M dwarfs to K dwarfs i.e. there is a strong preference for near-equal mass systems at small separations. Figure from Reid and Hawley (2005)

that the 'desert' extends well into the stellar regime, with only a bare handful of K and M dwarf companions within $\sim$50 au of the parent star but significant numbers at larger separations. Brown dwarfs are found as companions at wider separations, as are low-mass stellar companions. In other words, there is a strong tendency towards equal-mass binary systems at small separations among solar-type stars, with a significantly flatter companion mass distribution at larger separations.

Observations of M dwarf and brown dwarf binaries show a similar trend towards equal-mass systems at small separations. In the case of brown dwarfs, one cannot estimate masses from the photometric properties since brown dwarfs evolve along very similar cooling tracks, irrespective of mass; however, if one assumes that the two objects are coeval, then the different in luminosity can be used to infer the mass ratio of the system. Some systems clearly have low mass ratios (e.g. the young brown dwarf binary, 2MASS 1207AB; Chauvin et al. 2004), but the overall distribution peaks strongly near unity, indicating an overwhelming majority of near equal-mass systems (Burgasser et al. 2007; Reid et al. 2008).

How does one interpret these results? One possibility is to consider two modes of binary formation. At small separations, competitive accretion within the original star-forming core may dictate that only high-q systems survive: binary systems only survive when the two pre-stellar concentrations have nearly equal gravitational attraction (i.e. mass); low-q systems follow Matthew 13:12 ('To them that have, more shall be given'). Pre-stellar cores at wider separations suffer less mutual interference with the result that lower-q systems can form at those separations. Dynamical evolution, due to encounters with others stars in the parent cluster or with higher-mass objects in the general field, then shape the evolution of the overall distribution. If such is the case, this may provide a guide as to how one should include binaries in computing the stellar initial mass function, as discussed in the following section.

16.5 The Stellar Mass Function: Present Day and Initial

Mass is the key parameter that defines the characteristics of individual stars; it is also the most difficult parameter to measure directly. We are only able to estimate the overall mass distribution of stars, the mass function, because stars spend most of their lifetime on the main-sequence, following a relatively tight correlation between mass and luminosity. As a result, one can use mass measurements derived from observations of individual stars in binary systems to determine an average mass-luminosity relation (MLR), and then apply that relation to observations to transform the stellar luminosity function to the stellar mass function.

In bolometric terms, the luminosity, L, varies with mass as $L \sim M^{4.0}$ for stellar masses exceeding $\sim 0.4 \, M_\odot$; at lower masses, the MLR transitions to $L \sim M^{2.3}$ (Smith 1983). It is worth noting that the change in slope in the MLR accompanies the loss of the radiative core and the transition to fully convective stars. Deriving accurate bolometric magnitudes generally requires observations across a wide wavelength range, so for practical purposes, MLRs are usually calibrated for wavelength-specific absolute magnitudes. Figure 16.9 illustrates recent derivations of the MLR at visual and near-infrared (K-band) wavelengths (Chabrier 2005). The calibrating stars are eclipsing binaries at high masses and predominantly resolved systems with astrometric orbit determinations at masses below $\sim 0.7 \, M_\odot$ (a handful of M-dwarf eclipsing binaries are known, including YY Gem, CM Dra and CU Cnc). The near-infrared data show significantly less dispersion than the visual data, partly reflecting the lower sensitivity to metallicity variations at those wavelengths.

Formally, the mass function is given by:

$$\Psi(M) = dM/dM_{\text{bol}} \times \Phi M_{\text{bol}}, \qquad (16.1)$$

where dM/dM_{bol} is the mass-luminosity relation. This quantity has been the subject of many extensive reviews, notably by Miller and Scalo (1979), Scalo (1986), Chabrier (2005) and, most recently, Bastian et al. (2010; see also Reid and Hawley 2005). In practice, the mass function is usually computed on a star-by-star basis for a particular stellar sample, estimating individual masses using a particular MLR and allowing for the appropriate statistical biases (see the Appendix to this chapter for further discussion). For stars, that gives the *present-day mass function*, the PDMF; two

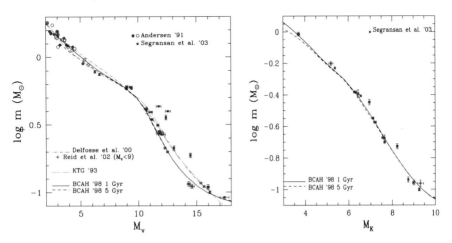

Fig. 16.9 Visual and near-infrared mass-luminosity relations defined by observations of binary stars; theoretical MLRs (from Baraffe et al. 1998) are also shown. Figure from Chabrier (2005)

issues need to be addressed regarding evolutionary phenomena at the high-mass and low-mass extremes, respectively, before deriving the *initial mass function*, the IMF.

Main-sequence lifetimes, τ_{ms}, decrease with increasing stellar mass. Clearly, only stars with ages $\tau < \tau_{ms}$ remain on the main-sequence at the present time, and contribute to measurements of the PDMF. Stars with masses greater than $\sim 1.3 \, M_\odot$ have $\tau_{ms} < 10 \, \text{Gyr}$, the approximate age of the Galactic disc; if we assume that the IMF has been largely invariant over the disc's lifetime, then the PDMF progressively underestimates the total contribution from high-mass stars. We can allow for this if we assume a star formation history for the disc, and adjust the observed densities to allow for stars that have evolved beyond the main-sequence. The available observations are generally consistent with a constant star formation rate, although the constraints are not strong (see Chap. 20), so the densities are typically scaled by a factor τ_{ms}/τ_{disc}.

At the low-mass limit, the main complication in deriving the IMF again centres on lifetimes. Substellar-mass brown dwarfs fail to ignite central fusion, and therefore lack any long-term energy source. As a result, they cool and fade on timescales that are rapid by astronomical standards, with the cooling rate scaling as $M^{0.8}$. As a further complication, the radii are set by electron degeneracy, and lie within 10 % that of Jupiter, regardless of mass. The net result is that brown dwarfs evolve along very similar tracks in the (L, T_{eff}) diagram, and it is not possible to either define a meaningful MLR or estimate masses for isolated objects unless their age is known. This issue can be addressed through observations of star clusters (e.g. the Pleiades) where age is a (relatively) known quantity and where the cluster is sufficiently young that dynamical evolution has not yet depleted the low-mass complement, but sufficiently old that theoretical models can offer at least moderately reliable mass estimates. The major additional caveat rests with the small number statistics that are usually presented by young star clusters.

The concept of the IMF originated with Edwin Salpeter, who derived the first observational estimate based on data for nearby stars (Salpeter 1955). That analysis was limited to stars more massive than the Sun, and Salpeter found that the overall distribution was well-represented by a single power-law, $\Psi(M) \propto M^{-2.3}$. Power-laws have since been used extensively to represent the mass function, and a power-law, slope -2.3, is known as a Salpeter mass function. The power-law index is often designated as α, and the mass function sometimes binned in units of $\log M$ (as in Fig. 16.10); in that case, a power-law, index α, becomes a power-law, index $\alpha + 1$.

The principal alternative mathematical representation of the IMF is the log-normal distribution:

$$\zeta(\log M) = A \exp\left[-\frac{(\log M - \log M_c)^2}{2\sigma^2} \right],\qquad(16.2)$$

where M_c is the characteristic mass, and σ the width of the distribution. This formulation was introduced by Miller and Scalo (1979) in their classic analysis of the stellar mass function. It is vitally important to recognise that, absent an underlying

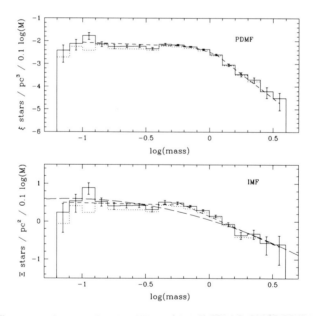

Fig. 16.10 The present-day mass function (PDMF; *upper panel*) and initial mass function (IMF; *lower panel*) derived for nearby stars; note the shallower slope at super-solar masses in the IMF. In both panels the *dotted line* shows the distribution if secondary components are ignored. The *bold dashed line* in both panels represents the broken power-law PDMF and IMF relations of Reid and Gizis (1997), whereas the *thin dashed line* in the *bottom panel* denotes the log-normal IMF relation of Chabrier (2005). Note that the data are binned logarithmically in mass, so a flat distribution corresponds to $\Psi(M) \propto M^{-1}$. Figure adapted from Reid and Hawley (2005)

physical mechanism, power-laws and log-normal distributions are merely mathematical representations of an observed distribution, with no intrinsic value in their own right.

Bearing that caveat in mind, Fig. 16.10 presents recent analyses of mass functions derived from nearby star samples. The contribution from secondary components in multiple systems is shown, but we should note that they have relatively little impact on the overall distribution. The most important characteristics to notice are, first, the shallower slope (i.e. slower decrease in numbers with increasing mass) at high masses as one goes from the PDMF to the IMF and allows for stellar evolution, and the flattening in the mass distribution below $\sim 1\,M_\odot$. Fitting the distribution with a power-law, the best fit to the PDMF at high masses is $\Psi(M) \propto M^{-4.3}$, changing to close to Salpeter, $\Psi(M) \propto M^{-2.9}$, in the IMF. We should note that the most massive star in the immediate Solar Neighbourhood is Sirius (at $\sim 2\,M_\odot$). Below $\sim 1\,M_\odot$, the mass function is significantly flatter, with a distribution closer to $\Psi(M) \propto M^{-1}$.

Other representations of the IMF are also possible: for example, Kroupa et al. (1993) proposed a 3-part power-law representation, with slopes of -2.7 for $M > 1\,M_\odot$, -2.2 for $0.5 \le M \le 1\,M_\odot$, and -1.3 for $0.08 < M < 0.5\,M_\odot$; Kroupa (2001) simplifies the representation to a two-part power-law, with Salpeter slope for $M > 0.5\,M_\odot$ and an index -1.3 at lower masses; Scalo (1986) coupled a log-normal distribution at lower masses with a Salpeter-like exponential at masses exceeding $\sim 1.5 - 2\,M_\odot$; and Chabrier (2005) has proposed a log-normal distribution, with a characteristic mass of $M_c = 0.25\,M_\odot$, and $\sigma = 0.55\,M_\odot$. All of these provide a reasonable representation of the data; none of them provides particular insight to the underlying physics.

In the substellar regime, field studies can only set constraints on the mass function. The typical approach is to adopt a mass function and a star formation history, generate the expected luminosity function in the substellar regime, and match against the observations. Figure 16.11 shows a typical example; further analyses have been completed by Reid et al. (1999), Burgasser et al. (2003), Allen et al. (2005), and Metchev et al. (2008). While the data remain sparse, there is broad concurrence among these investigations that the observed numbers of field brown dwarfs is consistent with a mass function that is no steeper than that measured for subsolar-mass stars; that is, matched against a power-law, the slope is no steeper than $\Psi(M) \propto M^{-1}$. This implies that the number density of brown dwarfs per decade in mass is no more than the number density of stars with masses between 0.1 and $1\,M_\odot$; that, in turn, implies that the total contribution made by brown dwarfs to the local mass density is no more than $\sim 10\,\%$ the stellar contribution. Thus, these results firmly and finally rule out any prospect of brown dwarfs making a significant contribution to dark matter.

For a more quantitative assessment of the substellar mass function, it is necessary to turn to star clusters of known age, particularly young clusters where brown dwarfs are significantly more luminous and dynamical evolution has not had sufficient time to deplete lower-mass members. These younger clusters also include stars that extend to significantly higher masses than in the Solar Neighbourhood. The results show some dispersion (some examples are shown in Fig. 16.12), which is not unexpected given the limited numbers in some clusters. There have been some recent suggestions

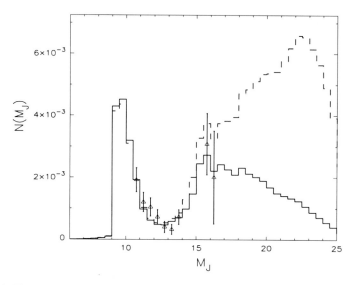

Fig. 16.11 The observed near-infrared luminosity function for L and T dwarfs (marked as *triangles*) matched against predictions for a constant star formation rate and a mass function index of -1 (*dashed line*) and 0 (*solid line*). Based on model calculations by Allen et al. (2005)

that there is evidence for a discontinuity at the hydrogen-burning limit, suggestive of a different star formation process at the lowest masses (e.g. Thies and Kroupa 2007). However, the evidence for systematic differences is very weak. Moreover, there seems little rationale for any connection between two such disparate processes as fragmentation and collapse in a molecular cloud and triggering nuclear fusion. When a mass of gas starts to collapse, how does it know that it cannot support central fusion and will form a brown dwarf, and therefore must fragment through a different process?

Overall, there is general agreement with the field in supporting a mass function at high masses that is consistent with a power-law with slope close to Salpeters original index of -2.3. In the brown dwarf regime, the distributions match power-laws that are no steeper than $\Psi(M) \propto M^{-1}$, and may even support a tendency for a turnover and a flatter distribution.

16.6 Is There a Universal IMF?

The IMF is a key quantity not only for understanding star formation, but also for understanding the overall evolution of galaxies. Perhaps illustrating the benefits of a classical education, Lynden-Bell (1977) has pointed out that the IMF can be divided into three parts: high-mass stars, with masses greater than $\sim 1.5 \, M_\odot$, that evolve relatively rapidly and dominate element production, enriching the ISM and future stellar

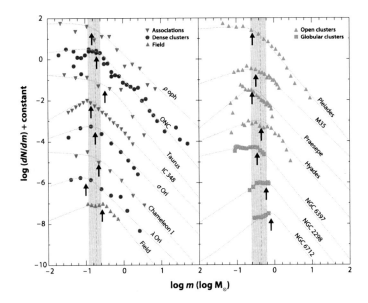

Fig. 16.12 The mass function measured for star clusters and stellar associations spanning a range of ages. Note mass segregation and preferential stripping of low-mass stars has a significant effect on the mass function of all clusters older than Praesepe and the Hyades. Figure from Bastian et al. (2010)

generations; intermediate-mass stars, with masses between $\sim 0.5\,M_\odot$ and $\sim 1.5\,M_\odot$, that are largely responsible for the light produced by old and intermediate-age stellar populations; and low-mass stars and brown dwarfs, below $\sim 0.5\,M_\odot$, that lock up a significant fraction of the total mass. Only a subset (sometimes a very limited subset) of this mass range is available to direct observation in most circumstances; consequently, our estimates of global parameters almost always rest on significant extrapolations. It is therefore crucial that we understand not only the constraints that observations set on possible variation in the IMF, but also gain insight into what physical mechanism(s) might drive the functional form and whether there might be significant variations in extreme star-forming environments.

The results summarised in the previous section are largely drawn from observations of stars in the local field or moderately populous clusters in the Galactic disc, and therefore do not span the extremes in terms of parameters such as density and metallicity. Overall, and perhaps not surprisingly, the IMF appears to follow the luminosity function in the Anne Elk (Miss) prescription. As Bastian et al. (2010) have shown, this type of IMF can be parameterised by 8 quantities (see Fig. 16.13). The observed mass function is less symmetric than the ideal, rising steeply at the high-mass extreme, flattens and peaks around $0.1–0.4\,M_\odot$, and exhibits a shallower decline towards lower masses. The upper mass limit lies around $\sim 150\,M_\odot$ (Zinnecker and Yorke 2007), and the lower mass limit is as-yet undefined, but likely lies at or below $\sim 0.01\,M_\odot$.

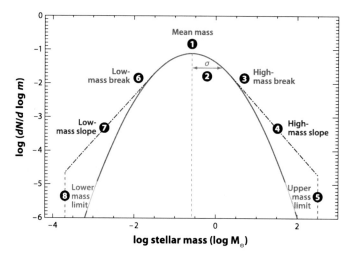

Fig. 16.13 Parameterising the IMF. Figure from Bastian et al. (2010)

Potential mechanisms for generating an IMF of this form are discussed earlier in this volume (see Clarke Chap. 3). Those mechanisms might be constrained if we were able to identify correlations between changes in the shape of the IMF (or any of the 8 parameters in Fig. 16.13) and changes in physical characteristics, such as metallicity, density of the star-forming environment, or location within the parent galaxy. Bastian et al. (2010) have undertaken a thorough analysis of the available observations, and find no evidence for any significant variations from observation of resolved stars within either the Milky Way or neighbouring galaxies. There are, however, a handful of cases that provide a hint of deviations from the standard form.

Microlensing

The first case centres on analyses of microlensing events within the Milky Way. Microlensing events occur when a foreground object passes very close to the direct line-of-sight to a more distant object. As the foreground source moves through the field-of-view, the gravitational potential focuses and amplifies light from a background source. The potential application to astronomical observations was first highlighted by Refsdal (1964) and identified as a means of probing dark constituents of the Milky Way by Paczynski (1986). Following Paczynski's work, several large-scale surveys were initiated, notably EROS, MACHO and OGLE, in a search for dark matter in the form of massive compact halo objects (hence MACHO): brown dwarfs, white dwarfs, black holes (Gates et al. 1996). These surveys require many background sources, and therefore targeted the Magellanic Clouds and the Galactic Bulge. In the event, the results rule out MACHOs as significant dark matter constituents, but offer other interesting possibilities.

Microlensing lightcurves are characteristic in form, being both symmetric and achromatic. In brief, the amplitude of the magnificent depends on how close the lens passes to the direct line-of-sight to the background source and the mass of the lens; the duration of the event depends inversely on the relative velocity of the lens and source, and directly on the mass of the lens. If we assume that the lenses within a given region of the sky are all drawn from the same parent population (e.g. the Disc) with the same kinematics, then we can associate an average transverse velocity with the observations. In that case, the mass distribution of the lenses can be inferred from the distribution of lensing timescales. Sumi et al. (2011) recently analysed data from the MOA survey and found a significant excess of short-timescale events, suggesting an upturn in the IMF at Jovian masses over a simple extrapolation of the brown dwarf mass function. This result has only moderate statistical significance, and might stem from detection of exoplanet companions at wide separations, rather than planetary-mass brown dwarfs.

Star formation on the edge

In Chap. 15 we noted that galactic discs tend to show a sharp edge in their radial density distribution. Stars and star formation, however, exist beyond that edge, albeit at a relatively low level. Observations with the GALEX satellite have been particularly effective in UV flux at large radii, while HI observations have traced neutral gas within these extended discs (e.g. M83; Thilker et al. 2005). The presence of flux at near-UV wavelengths ($>2000\,\text{Å}$) clearly indicates the presence of on-going star formation. However, were the IMF to follow a Salpeter-like distribution at high masses, as in young Milky Way star clusters, one would also expect significant Hα radiation which is not present. Similarly, Meurer et al. (2009) find reduced a Hα/UV ratio in galaxies with low star formation rates.

The ionising radiation that is responsible for the observed Hα emission is generated by short-lived ($\sim$10–20 Myr) massive stars ($>$10 $M_\odot$) which dominate the production of flux at far-UV wavelengths ($<$1000 Å); the near-UV is largely produced by less massive ($\sim$2–4 $M_\odot$), longer-lived ($\sim$100–200 Myr) stars. Thus, the low-levels of Hα emission indicate a scarcity of high-mass stars, implying either that star-forming regions are all sufficiently old that those stars have evolved off the main-sequence or, more probably, a high-mass truncation in the IMF in environments with low star formation rates.

Discussions on the origin of this discrepancy have centred on whether the probability for forming a high-mass star of a given mass scales with the mass of the star-forming cluster in other words, are massive star clusters capable of producing higher-mass stars than lower-mass clusters, or is the process purely stochastic? The resolution of this debate has interesting consequences for the parameter known as the Integrated Galaxy Initial Mass Function (IGIMF; Weidner and Kroupa 2006), measuring the total star production by a given galaxy. If the upper mass-limit (parameter 5 in Fig. 16.13) scales with cluster size, as favoured by Weidner and Kroupa (2006), then the IGIMF will have a steeper slope at high masses than the IMF for a given cluster, since only the highest-mass star-forming regions are able to contribute the highest mass stars. Alternatively, a stochastic star formation scenario, where lower-mass star-forming regions simply have a lower probability of forming very high-mass stars, as supported by Goddard et al. (2010), leaves the IGIMF identical with an individual IMF.

Star formation in elliptical galaxies

The stellar population(s?) in elliptical galaxies are characterised as old (several Gyr) and evolved, with little evidence for extensive on-going star formation at the current epoch. The more massive systems, such as M87 in the Virgo cluster or NGC 4874 and NGC 4889 in Coma, have stellar masses exceeding 10^{12} $M_\odot$, substantially more massive than disc galaxies, and may have formed through extensive mergers of smaller, gas-rich disc galaxies, triggering extremely vigorous star formation at early epochs. The star-forming environment may have been more extreme than the range currently on view in present-day galaxies like the Milky Way or the Magellanic Clouds. Recent spectroscopic observations by van Dokkum and Conroy (2011) and Conroy and van Dokkum (2012) have led to suggestions that those systems have an IMF that is skewed to substantially steeper slopes at the low-mass end, potentially even steeper than Salpeter.

The observational basis for this hypothesis rests on observations of line-strengths in the far-red spectral regions, where elliptical galaxies in Virgo and Coma show moderately broad features that have been interpreted as due to the presence of molecular bands due to iron hydride. Metal hydrides, including CaH, MgH and TiH as well as FeH, form in the high-gravity atmospheres of K and M dwarf stars, but are generally weak or absent in the lower-gravity K giants and subgiants that would be expected to dominate the flux from the old stellar populations in elliptical galaxies. K dwarfs are at least 8–10 magnitudes fainter than K giants; as a consequence, dwarf stars need to outnumber giants by 4–5 orders of magnitude if they are to make a significant contribution to the integrated spectrum. This leads to the requirement for the steep IMF at low masses.

As discussed earlier, observational analyses of the low-mass IMF within the Milky Way and its immediate neighbours are broadly consistent in favouring an index of $\alpha \sim -1$ rather than the -2.5 to -3 required in Virgo and Coma. The van Dokkum/Conroy result implies a radical change in the star formation process, prompting one to echo a question raised by Donald Lynden-Bell 'How does a mass of gas in a very massive galaxy know the mass of the galaxy it is in?' (Lynden-Bell 1977). What radical change in the star-forming environment (UV flux levels? gas density? systemic shocks?) in these systems produced such a dramatic change in the underlying IMF? Are there any possibilities for probing intermediate systems where some transition in properties might be observed?

Conclusions

These three cases all present interesting observational conundra that deserve to be resolved. However, it is important to recognise that none of the *data* analysed actually include visually-resolved images; wherever we can construct the IMF on an individual, star-by-star basis, the results are consistent with the Anne Elk function represented schematically in Fig. 16.13. This broad consistency must give one pause when considering the suggested possible deviations, and demands that we set an appropriately high bar before giving wide acceptance to those results.

16.7 Summary

Reviewing the results presented in this chapter, the Sun lies within a relatively quiet neighbourhood within the Galactic disc. The local stars have been studied extensively, and provide valuable information on stellar multiplicity and the overall mass distribution, as characterised by the IMF. The binary characteristics of the field stars are well-defined, and show a clear trend in properties with decreasing primary mass, both in the overall fraction and (likely correlated) the maximum separation. Binary mass ratios tend to be near unity for systems at small separation, trending towards a broader distribution at wider separations. The stellar IMF rises steeply from high masses to low masses, flattening near $\sim0.3\,M_\odot$, and at substellar masses.

A number of interesting questions remain, of course. The full connection between stellar multiplicity and the IMF remains to be established: How do you count stars and brown dwarfs in binary and multiple systems to give a distribution function that is directly relevant to star formation theory? For the IMF, we may have a good picture of what it is, but we do not really understand how it got there: What physics underlies the form of the IMF? And is the IMF really invariant, or are these tantalising indications of significant departures in other stellar systems pointing to key physical parameters that govern its properties?

Appendix: Some Useful Techniques

Introduction

Many quantities that play a key role in understanding the star formation process are derived from statistical analysis of ensembles of data. Those analyses are subject to a variety of uncertainties, including circumstances where random errors can combine to give rise to systematic bias. This appendix briefly outlines some of the more significant potential sources of error in this type of analysis, and summarises some methods for characterising distributions and refining samples.

Systematic bias is introduced through two main avenues: systematic errors in the calibration of properties of individual stars (e.g. temperature, absolute magnitude, metallicity); or statistical bias in the way that individual stars are compiled to form a final reference sample. We refer to the former as individual errors, and the latter as statistical bias. Individual errors in a given parameter can only be corrected through the determination of a more reliable calibration; the individual measurements for a given star represent the current best estimate of those quantities for that object.

In contrast, corrections for statistical bias are applied to the parameters adopted for a given star when combined within a particular dataset, and those individual corrections can change depending on the properties of the underlying parent dataset. Thus, individual errors are specific to a given object; statistical bias is tied to how a sample was constructed.

Fitting Bias

Measuring accurate intrinsic properties for individual stars is time-consuming and often involve complicated analyses. The stellar sample available for direct measurement is often limited in size, either because the observations are highly time-consuming (e.g. high-resolution spectroscopy), or because the number of stars that are accessible to observation is restricted by physical requirements (e.g. accurate parallax measurements require proximity to the Sun). There are therefore numerous instances in astronomy where a small number of objects are used to define mean relations that match an easily measurable quantity (colour, bandstrength) against an intrinsic parameter (luminosity, metallicity); that mean relation can then be used to infer intrinsic parameters for a much larger sample of objects. Clearly, any systematic inaccuracy in the mean calibration transfers directly to give systematic bias in the aggregate properties inferred for the broader sample. We consider two examples.

The Strömgren *uvby* photometric system defines magnitudes in bands of intermediate width ($\sim$300 Å) that sample the UV, blue and visual regions of the spectrum. Colour indices (flux ratios) constructed from those magnitudes can measure line blanketing as a function of temperature. The classic calibration of those indices against metallicity was established by Schuster and Nissen (1989; hereafter SN89) using stars with high-resolution spectroscopic abundance estimates as standards and fixing the zeropoint against data for the metal-rich ([Fe/H] = +0.14 dex) Hyades cluster. This calibration is widely used, including in the 14,000 star Geneva-Copenhagen catalogue (Nordström et al. 2004). However, Haywood (2002) has pointed out that there are systematic differences between the photometric reference data used by SN89 to set their Hyades reference (taken from Crawford 1975) and direct observations of Hyades stars. The net result is that applying the SN89 calibration to a stellar sample introduces significant colour-dependent bias in the inferred metallicity distribution. Those systematic biases affect the ages and masses inferred for stars, distorting estimates of the age-metallicity relation for disc stars.

The second example centres on dealing with fine structure in the stellar main-sequence, as in the M_V, $V-I$ colour-magnitude diagram. Until the advent of *Gaia*, reliable distance determinations are limited to nearby ($<$100 pc) stars observed by *Hipparcos* and Solar Neighborhood objects that are accessible to ground-based observations ($\lesssim$20 pc). Those parallax stars serve as key calibrators, defining mean relations in a variety of colour-magnitude planes that can then be applied to estimating photometric parallaxes for more distant stars.

Fig. 16.14 M_V, $V-I$
colour-magnitude diagram
for stars with accurate
trigonometric parallax
measurements. Systematic
bias can be introduced by
failing to fit small-scale
structure in the observed
main-sequence. *Green
crosses* mark disc stars,
purple circles denote L
dwarfs, *cyan circles*
represent moderately
metal-poor subdwarfs and
the *blue triangles* correspond
to extremely metal-poor
subdwarfs. The *dashed line*
shows a solar metallicity
5 Gyr theoretical model
isochrone from Baraffe et al.
(1998)

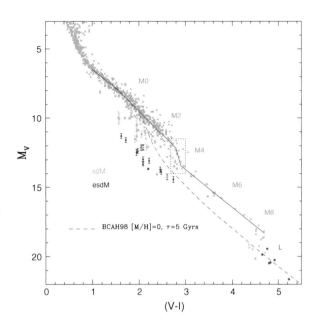

In deriving average colour-magnitude relations, the tendency is to opt for relatively smooth relations as exemplified by low-order polynomial fits to the data. This minimises the potential for small-scale distortions caused by individual datapoints that are outliers to the main distribution. There are circumstances, however, when that approach is not valid, as illustrated in Fig. 16.14. Over most of its range, the stellar main-sequence follows a relatively smooth relation. However, there is a clear discontinuity at $V-I \sim 3.0$ mag (spectral type $\sim$M3, $T_{\text{eff}} \sim 3000$ K), with the stars spanning a range of ~ 1.5 magnitudes ($12 < M_V < 13.5$ mag). The origins of this feature remain unclear, although one should note that the Baraffe et al. (1998) theoretical models predict a similar morphology, but with the relation steepening at $V-I \sim 2.2$ mag or effective temperatures ~ 250 K hotter than observed. Baraffe et al. (1998) ascribe this feature to the onset of molecular hydrogen formation near the photosphere in these cooler atmospheres, lowering the adiabatic gradient and favouring convection within the envelope (Copeland et al. 1970). Might the discrepancy reflect a ~ 250 K mismatch between the surface temperatures predicted by the models and the observations?

The origin of the discontinuity, however, is less important for present considerations than the consequences of fitting a mean colour-magnitude relation that does not take the feature adequately into account. A smooth fit through all the data leads to systematically underestimating absolute magnitudes (i.e. estimated values are too faint) for stars slightly bluer than $V-I = 3.0$ mag, and systematically overestimating absolute magnitudes (i.e. estimated values too bright) for stars slightly redder than $V-I = 3.0$ mag. If one then applies the incorrect calibration to wide-field photometric survey data, and bin the results to estimate number densities, the net result is that one overestimates the number of stars near $M_V \sim 12.5$ mag at the expense of brighter and fainter bins, leading to a spuriously strong peak in the derived luminosity function (Reid 1991).

In brief, these two examples show that it is very important to ensure that one is measuring what one thinks one is measuring.

Malmquist Bias

Malmquist bias is a statistical correction applied to the average luminosity/absolute magnitude derived for a magnitude-limited sample. The Swedish astronomer Gunnar Malmquist originally devised this correction in his analysis of a sample of A stars (Malmquist 1922). The bias essentially stems from the fact that the volume of a spherical shell increases with increasing radius.

Malmquist's original formulation considers observations of stars of a particular spectral class that have a uniform density distribution within a galaxy with no intervening absorbing material. Suppose that those stars have a luminosity function that is invariant with distance, with mean absolute magnitude, M_o, and intrinsic dispersion, σ. Consider observations of a magnitude-limited sample of stars, i.e. a set of stars with magnitudes $m < m_{\lim}$. That magnitude limit corresponds to a limiting distance modulus, $m_{\lim} - M_o$, for stars of mean absolute magnitude. The stellar sample, however, has an intrinsic dispersion in the individual absolute magnitudes: stars brighter than the mean absolute magnitude, M_o, have a larger effective distance limit for inclusion in the sample; stars fainter than M_o have a smaller effective distance limit. Stars brighter than M_o are drawn from a larger sampling volume, and the final sample therefore includes a greater proportion of those stars. As a result, the mean magnitude of the observed sample is brighter than M_o, the true mean absolute magnitude for stars of that spectral type.

Malmquist demonstrated that, given those assumptions, the correction to the mean absolute magnitude is given by:

$$\Delta M = -(\sigma^2/\log_{10}e) * (d\log A_S(m)/dm), \qquad (16.3)$$

where $A_S(m)$ represents the number-magnitude counts for stars of spectral type S, and $d\log A_S(m)$ represents the logarithmic gradient in those counts. For a uniform density distribution the number counts increase as $10^{0.6m}$, so the mean absolute magnitude of the sample is given by:

$$M(m) = M_o - 1.38\sigma^2. \qquad (16.4)$$

In most cases, the stellar density distribution is not uniform. For example, when one samples stars towards to Galactic Pole, the disc density law leads to a shallower slope in the stellar number counts, closer to $10^{0.4m}$. As a result, the appropriate correction for density law calculations is:

$$M(m) = M_\circ - 0.92\sigma^2. \qquad (16.5)$$

It is also more common to use photometric colours to determine absolute magnitude than spectral types. As Fig. 16.14 shows, there is an intrinsic dispersion in absolute magnitude of $\sim$0.25 $-$ 0.3 mag at a given colour, with the dispersion tied partly to metallicity variations. In addition, one must take account of uncertainties in measured for individual stars. The simplest means of doing so is to adjust the value of σ used in the Malmquist calculation to allow for the colour uncertainty. Stobie et al. (1989) provide a more rigorous analysis of how to deal with a continuous, rather than discrete, selecting variable.

Malmquist bias has wide implications across many astronomical disciplines. For example, it is instructive to consult the heated discussions on this topic that were prompted by the dueling teams of Sandage and Tamman and de Vaucouleurs at the height of the $H_\circ$ wars in the late 1970s and 1980s. As an aside, Malmquist was also a strong proponent of adopting the siriometer (the distance to Sirius) as the standard unit of distance for astronomical measurements (Malmquist 1925). The consequences of that effort have been less far-reaching.

Lutz-Kelker Corrections

Lutz-Kelker bias is a statistical correction that is applied to the average luminosity/absolute magnitude derived for a parallax-limited sample. The corrections were originally devised by Tom Lutz, an astronomer, and David Kelker, a statistician, to deal with analyses of stellar samples that are defined using a limiting parallax (Lutz and Kelker 1973). As with Malmquist corrections, the bias stems from the fact that the volume of a spherical shell increases with increasing radius, so there are always more stars with parallaxes $\pi_{in} < \pi_{out}$ than $\pi_{in} > \pi_{out}$.

For any given star, the measured trigonometric parallax provides the best estimate of its distance, and hence its absolute magnitude. However, each parallax measurement has an uncertainty, and those uncertainties combine to give a systematic bias when one considers that star as the member of a larger sample.

Consider stars drawn from a uniform space distribution. The number of stars in a radial shell, r to $r + dr$, is given by:

$$N(r) = 2\,V r^2 dr. \qquad (16.6)$$

Thus, since $r = 1/\pi$, the number of stars with parallax in the range $\pi_\circ$ to $\pi_\circ + d\pi$ is given by:

$$N(\pi_0)\,d\pi = 2\,V d\pi/\pi_0^4, \qquad (16.7)$$

where $V = 2\pi$ (and, in this case, $\pi = 3.14159\ldots$, not parallax).

Suppose we use parallax measurements to define a sample of stars within a given distance i.e. with $\pi > \pi_{\text{lim}}$. Each star has a measured parallax, π_i, and an associated uncertainty, usually taken symmetric, $\pm\sigma_\pi$. When we combine stars to give a parallax-limited sample, we need to take the measuring uncertainty and estimate the most probable value of the parallax, π_i'. The number of stars increases with decreasing parallax, so, given symmetric measuring uncertainties, it is more likely that more distant stars (with smaller parallax) are scattered into the sample than nearer stars (with larger parallax) are scattered out of the sample. Consequently, for statistical analysis, the $\pi_i' < \pi_i$, and the most likely estimate for the star's absolute magnitude is correspondingly brighter.

The degree of the correction depends on the accuracy of the parallax measurement (see Fig. 16.15), usually expressed as σ_π/π (a star-specific parameter) and on the underlying density distribution of the parent sample (a contextual parameter). Lutz and Kelker (1973) calculated absolute magnitude corrections for a range of parallax accuracy and a uniform density distribution; note that the ΔM_{LK} corrections become increasingly large as σ_π/π increases beyond 10%. Hanson (1979) extended those calculations to derive a general relation for a population with a power-law distribution, $N(\pi) \propto \pi^n$. As one might expect, the corrections decline with a shallower density distribution.

We should emphasis that context is important: there is not a unique Lutz-Kelker correction for a given star; while σ_π/π is invariant, the underlying density distribution depends on how the sample was selected. Applying colour selection, or spectral type selection, to a magnitude-limited sample can change the density distribution of the selected sample; as a result, a given star might have a significantly different Lutz-Kelker correction when considered as a member of a magnitude-limited sample rather than a colour-selected sample.

Probability Plots

The normal probability plot is a graphical means of assessing whether data in a particular dataset are well matched to a normal distribution (Daniel and Wood 1971). These are also known as Q-Q plots. As with the Lutz-Kelker corrections, Tom Lutz and Bob Hanson play a role in highlighting the potential application of this tool to studying astronomical distributions (Lutz and Hanson 1992).

Consider observations of a particular parameter, for example space velocity. If that distribution is normal, then plotting the cumulative probability distribution gives a smooth curve. If $F(i)$ is the cumulative probability up to the ith point from a sample of N points, then $F(i) = (i - 1/2)/N$. We could compare the rank-ordered distribution of observed points against $F(i)$; however, it is simpler to take the inverse of $F(i)$, $F^{-1}(i)$, since that linearises the distribution.

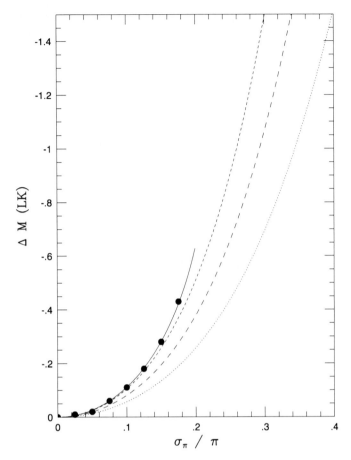

Fig. 16.15 Lutz-Kelker corrections. The *solid points* denote corrections for a uniform density distribution, corresponding to $N(\pi) \propto \pi^4$; the *dotted, dashed* and *solid lines* show the corrections for power-law distributions $n = 3, 2$ and 1 respectively. Figure from Reid (1997)

Lutz and Hanson (1992) describe how specialised 'probability plot' graph paper is available to test for normal distributions. One can also use simple computational methods. First, we rank-order a distribution (e.g. velocities), approximating $F(i)$. Then we compute the mean and standard deviation of the sample, and use those values to calculate the normalised offset of each datapoint from the mean in units of the calculated standard deviation, σ. Plotting velocity against the normalised offset gives a straight line, slope σ, if the distribution is exactly normal. In practice, the distribution often is near linear in the inner regions, with a shallower slope (see e.g. Figs. 16.16 and 17.5 in Chap. 17). This is consistent with either the presence of a few outliers, or the superposition of two or more near-normal distributions.

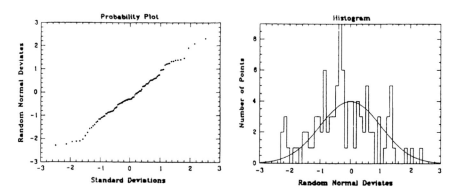

Fig. 16.16 *Left panel*: Probability plot. *Right panel*: Histogram of the data. Figure adapted from Lutz and Hanson (1992)

Identifying Members of Star Clusters

Any investigation that uses star clusters as a means of probing statistical quantities, such as the mass function, stellar multiplicity or metallicity variations, must start with a reliable catalogue of cluster members. This is relatively straightforward for young clusters, whether still embedded in the parent molecular cloud or recently emerged into the field (although see Chap. 20 for caveats). As the cluster ages, however, the members disperse and the cluster boundaries become less distinguishable. In such cases, a variety of methods need to be applied to segregate cluster and field stars. These methods form a series of 'AND' conditions, progressively winnowing an initial optimistic sample of candidate cluster members to give a final set of confirmed cluster members.

Classical wide-field cluster surveys start with proper motion analysis (we should note that *Gaia* will open the door to numerous future investigations along these lines). The proper motion of a star is given by $\mu = v_t/(4.74r)$, where r is the distance in parsecs, v_t is the velocity transverse to the line-of-sight and μ is in arcsec yr^{-1}. Stars within a star cluster are born with near-identical space motions. Clusters are three-dimensional entities, so our perspective leads to a progression in proper motion from front-to-back and from left-to-right in the cluster, and the motions appear to converge on a particular point on the sky.

Once the location of the convergent point is known, we can search for candidate cluster members within a larger proper motion catalogue. The proper motions for each star need to be transformed into motions towards, μ_u, and perpendicular to, μ_t, the direction of convergent point; the convergent motions, μ_u, are also adjusted to give the equivalent motion at the cluster centre, correcting for the longitudinal variation (see Hanson 1975 and Reid 1992 for further details of the transformation process).

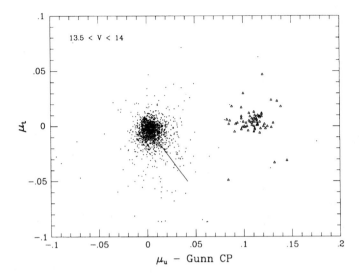

Fig. 16.17 The proper motion distribution for bright stars towards the Hyades cluster. The proper motions have been transformed to the convergent point system and cluster members form a well-defined sequence, distinct from the field stars. Figure from Reid (1992)

The resulting proper motion diagram generally shows a strong central concentration of points, representing stars in the general field, with cluster members forming a separate linear sequence, as in Fig. 16.17 for the Hyades cluster. The length of the cluster sequence reflects the line-of-sight depth of the cluster; since proper motion varies inversely with distance, the nearer cluster members have larger motions. Turning this around, measuring μ_u gives the distance to each star in the cluster.

The effectiveness of this technique can be judged from the wide-field analysis of the Hyades cluster that I carried out in the early 1990s. The original dataset was drawn from photographic photometry and astrometry of UK Schmidt plates of 4 fields covering an area of ~112 sq. deg. Coupled with plates from the original Palomar Sky Survey, the measurements spanned a baseline of 33–37 years and, scanned by the COSMOS measuring machine, gave motions accurate to 6–12 mas yr^{-1}. The initial sample included 450,000 stars brighter than 19th magnitude. Applying the convergent point method reduced that sample to 393 candidate cluster members.

Proper motion selection represents only the first step, however. Cluster members must also satisfy other criteria, including:

- Radial velocities consistent with membership, although care must be taken to allow for velocity variations introduced by binary systems.
- Photometric properties consistent with cluster membership.
- Levels of chromospheric and coronal activity consistent with membership.
- Metallicity consistent with the cluster chemical abundance.

Each step eliminates a few more candidate cluster members.

The first step in this process, applying proper motion selection, is usually the most onerous. To date, our inability to push wide-field astrometry below the milliarcsecond level has limited studies to the nearest clusters. *Gaia* will be transformative, and we should expect widespread application of these techniques to much more distant clusters in the coming years.

References

Allen, P. R., Koerner, D. W., Reid, I. N., & Trilling, D. E. 2005, ApJ, 625, 385
Baraffe, I., Chabrier, G., Allard, F., & Hauschildt, P. H. 1998, A&A, 337, 403
Bastian, N., Covey, K. R., & Meyer, M. R. 2010, ARA&A, 48, 339
Bouy, H., Martín, E., Brandner, W., & Bouvier, J. 2005, Astronomische Nachrichten, 326, 969
Burgasser, A. J., Kirkpatrick, J. D., Reid, I. N., et al. 2003, ApJ, 586, 512
Burgasser, A. J., Reid, I. N., Siegler, N., et al. 2007, Protostars and Planets V, ed. B. Reipurth, D. Jewitt & K. Keil, University of Arizona Press, 427
Chabrier, G. 2005, in Astrophysics and Space Science Library, Vol. 327, The Initial Mass Function 50 Years Later, ed. E. Corbelli, F. Palla, & H. Zinnecker, Springer, 41
Chauvin, G., Lagrange, A.-M., Dumas, C., et al. 2004, A&A, 425, L29
Close, L. M., Siegler, N., Freed, M., & Biller, B. 2003, ApJ, 587, 407
Conroy, C. & van Dokkum, P. G. 2012, ApJ, 760, 71
Copeland, H., Jensen, J. O., & Jorgensen, H. E. 1970, A&A, 5, 12
Cox, D. P. & Reynolds, R. J. 1987, ARA&A, 25, 303
Crawford, D. L. 1975, AJ, 80, 955
Daniel, C. & Wood, F. S. 1971, Fitting Equations to Data: Computer Analysis of Multifactor Data for Scientists and Engineers, ed. Daniel, C. and Wood, F. S., New York: Wiley
Duquennoy, A. & Mayor, M. 1991, A&A, 248, 485
Fischer, D. A. & Marcy, G. W. 1992, ApJ, 396, 178
Frisch, P. C. 2007, Space Sci. Rev., 130, 355
Gates, E. I., Gyuk, G., & Turner, M. S. 1996, Phys. Rev. D, 53, 4138
Goddard, Q. E., Kennicutt, R. C., & Ryan-Weber, E. V. 2010, MNRAS, 405, 2791
Hanson, R. B. 1975, AJ, 80, 379
Hanson, R. B. 1979, MNRAS, 186, 875
Hawley, S. L., Gizis, J. E., & Reid, I. N. 1996, AJ, 112, 2799
Haywood, M. 2002, MNRAS, 337, 151
Henbest, N. & Couper, H. 1994, The Guide to the Galaxy, ed. Henbest, N. and Couper, H., Cambridge University Press
Henry, T. J. & McCarthy, Jr., D. W. 1993, AJ, 106, 773
Jahreiß, H. & Gliese, W. 1993, in IAU Symposium, Vol. 156, Developments in Astrometry and their Impact on Astrophysics and Geodynamics, ed. I. I. Mueller & B. Kolaczek, Cambridge University Press, 107
Jahreiß, H. & Wielen, R. 1997, in Astronomische Gesellschaft Abstract Series, Vol. 13, Astronomische Gesellschaft Abstract Series, ed. R. E. Schielicke, 43
Kroupa, P., Tout, C. A., & Gilmore, G. 1993, MNRAS, 262, 545
Kroupa, P. 2001, MNRAS, 322, 231
Luhman, K. L. 2004, ApJ, 617, 1216
Luhman, K. L. 2013, ApJ, 767, L1
Luhman, K. L. 2014, ApJ, 786, L18
Lutz, T. E. & Hanson, R. B. 1992, in Astronomical Society of the Pacific Conference Series, Vol. 25, Astronomical Data Analysis Software and Systems I, ed. D. M. Worrall, C. Biemesderfer, & J. Barnes, 257

Lutz, T. E. & Kelker, D. H. 1973, PASP, 85, 573

Lynden-Bell, D. 1977, in IAU Symposium, Vol. 75, Star Formation, ed. T. de Jong & A. Maeder, Reidel Publishing Company, Dordrecht, 291

Malmquist, K. G. 1922, Jour. Medd. Lund Astron. Obs. Ser. II, 32, 64

Malmquist, K. G. 1925, The Observatory, 48, 142

Marcy, G. W. & Butler, R. P. 2000, PASP, 112, 137

Metchev, S. A., Kirkpatrick, J. D., Berriman, G. B., & Looper, D. 2008, ApJ, 676, 1281

Meurer, G. R., Wong, O. I., Kim, J. H., et al. 2009, ApJ, 695, 765

Miller, G. E. & Scalo, J. M. 1979, ApJS, 41, 513

Nordström, B., Mayor, M., Andersen, J., et al. 2004, A&A, 418, 989

Paczynski, B. 1986, ApJ, 304, 1

Raghavan, D., McAlister, H. A., Henry, T. J., et al. 2010, ApJS, 190, 1

Refsdal, S. 1964, MNRAS, 128, 307

Reggiani, M. M. & Meyer, M. R. 2011, ApJ, 738, 60

Reid, I. N. & Gizis, J. E. 1997, AJ, 113, 2246

Reid, I. N. & Hawley, S. L. 2005, New Light on Dark Stars: Red Dwarfs, Low-Mass Stars, Brown Dwarfs, ed. Reid, I. N. and Hawley, S. L., Springer-Praxis

Reid, I. N., Cruz, K. L., Kirkpatrick, J. D., et al. 2008, AJ, 136, 1290

Reid, I. N., Gizis, J. E., & Hawley, S. L. 2002, AJ, 124, 2721

Reid, I. N., Hawley, S. L., & Gizis, J. E. 1995, AJ, 110, 1838

Reid, I. N., Kirkpatrick, J. D., Liebert, J., et al. 1999, ApJ, 521, 613

Reid, N. 1991, AJ, 102, 1428

Reid, N. 1992, MNRAS, 257, 257

Reid, I. N. 1997, AJ, 114, 161

Riedel, A. R., Murphy, S. J., Henry, T. J., et al. 2011, AJ, 142, 104

Salpeter, E. E. 1955, ApJ, 121, 161

Scalo, J. M. 1986, Fund. Cosmic Phys., 11, 1

Schuster, W. J. & Nissen, P. E. 1989, A&A, 222, 69

Smith, R. C. 1983, The Observatory, 103, 29

Stobie, R. S., Ishida, K., & Peacock, J. A. 1989, MNRAS, 238, 709

Sumi, T., Kamiya, K., Bennett, D. P., et al. 2011, Nature, 473, 349

Thies, I. & Kroupa, P. 2007, ApJ, 671, 767

Thilker, D. A., Bianchi, L., Boissier, S., et al. 2005, ApJ, 619, L79

Trimble, V. 2008, The Observatory, 128, 286

van de Kamp, P. & Lippincott, S. L. 1975, Vistas in Astronomy, 19, 225

van de Kamp, P. 1930, Popular Astronomy, 38, 17

van Dokkum, P. G. & Conroy, C. 2011, ApJ, 735, L13

Weidner, C. & Kroupa, P. 2006, MNRAS, 365, 1333

Zinnecker, H. & Yorke, H. W. 2007, ARA&A, 45, 481

Zuckerman, B. & Song, I. 2004, ARA&A, 42, 685

Chapter 17
Stellar Kinematics and the Dynamical Evolution of the Disc

I. Neill Reid

17.1 Introduction

The discipline of stellar kinematics examines the velocity distributions of stars in the Milky Way and uses those results to investigate the origins of stars, their subsequent dynamical evolution within the Galaxy and, potentially, processes that govern the large-scale formation and evolution of spiral systems. In a nutshell, stars form within clusters and associations where all members have near-identical space motions; those clusters disperse through gravitational encounters with other objects, whether individual stars or molecular clouds; once in the Galactic field, stellar motions can change through encounters with large-scale phenomena, such as spiral structure or central bars. This chapter summarises early developments in this field and goes on to discuss the kinematics of stars in the Galactic disc and some of the processes that are believed to have played a significant role in shaping the overall distribution. For a more extensive discussion of the historical background, the interested reader is referred to the 1985 IAU Symposium 106, celebrating the centenary of Kapteyn's laboratory at Groningen, and the excellent text on the development of statistical astronomy in the late 19th and early 20th century by Paul (1993).

17.2 Building Stellar Kinematics as a Discipline

Statistical astronomy had its practical origins with William Herschel's star-gaging in the late 18th century. Herschel's observations depended on the greater light-grasp provided by his large-aperture, high-quality telescopes. The stellar distance scale started to come into focus in the 1830s with the first measurements of stellar parallax

I.N. Reid (✉)
Space Telescope Science Institute, Baltimore, MD, USA
e-mail: inr@stsci.edu

© Springer-Verlag Berlin Heidelberg 2015
C.P.M. Bell et al. (eds.), *Dynamics of Young Star Clusters and Associations*,
Saas-Fee Advanced Course 42, DOI 10.1007/978-3-662-47290-3_17

by Bessel, Henderson and Struve with precision transit instruments. Extensive analyses of stellar motions became possible in the latter part of that century, largely through the efforts of a Scotsman working in South Africa and a Dutch professor in charge of an observatory without any telescopes. Again, advances in technology played an appropriate role with the application of photography to astronomical imaging.

The Scotsman in this unexpected pairing was David Gill. The son of an Aberdonian clock maker, Gill had spent two years at Marischal College, Aberdeen (including taking some lectures from James Clerk Maxwell) before taking up work in his father's business. His long-term interests lay elsewhere, however. In the course of investigating how to establish a standard time service for Aberdeen, he became acquainted with astronomers at the Royal Observatory Edinburgh (then at Calton Hill), and started making his own observations. Shortly thereafter, he was recruited by James Ludovic Lindsay, Lord Lindsay, to help equip the private observatory that was being built at Dun Echt, about thirty miles from Aberdeen. In 1872, Gill became Director of the observatory, which by that time was second only to the Royal Greenwich Observatory in terms of telescopes and equipment. Some of Dun Echt observatory is still on view: Lord Lindsay became the 26th Earl of Crawford and in 1892, well after Gill left, he transferred his observatory equipment, including the characteristic cylindrical green domes, to the fledgling Blackford Hill site of the Royal Observatory, Edinburgh.

Gill was a first-rate observer, able to coax the most out of the available instrumentation, and he participated in two major astronomical expeditions, to Mauritius to observe the 1874 transit of Venus, and to Ascension Island in 1877 to use observations of Mars at opposition to refine the astronomical unit. While at Ascension, he learned that he had been appointed (by the First Lord of the Admiralty) as Her Majesty's Astronomer at the Cape, succeeding Edward James Stone. Gill took up the position in 1879 and revitalised the observatory, completely modernising the instrumentation and adding several new telescopes and heliometers.

Gill's major advance, however, centred on astronomical photography. He had experimented with astrophotography while at Dun Echt, mainly targeting the Sun and the Moon and using wet photographic processes which were messy and inefficient. Shortly after he took over at the Cape, he agreed to allow a local photographer, Mr. E.H. Allis, to attach his portrait camera to the side of the 6-inch Grubb equatorial telescope and take a time exposure of the comet (Gill 1882, see Fig. 17.1). The resulting image, taken using dry emulsion, not only gave a dramatic view of the comet and its tail, but also revealed numerous background stars. Gill immediately recognised the potential for stellar astronomy, and telegraphed a report to the Paris Academy of Sciences. He also applied for, and received, a grant from the Royal Society to undertake the first southern photographic sky survey, the Cape Photographic Durchmusterung (CPD).

Gill's communication with Paris had excited attention and inspired Admiral Amédée Mouchez, the Director of the Paris Observatory, to propose a photographic survey of the entire sky, the Carte du Ciel (CdC). This program was envisaged as being shared among many observatories, each of whom would contribute photographic plates covering a specific region of the sky. The final plan included 20 observatories, ranging from Greenwich, Catania, Bordeaux and Toulouse, to Hyderabad,

Fig. 17.1 A photograph of
the 1882 Great Comet
obtained by D. Gill and
E.H. Allis (see Gill 1882)

1882 Nov 7ᵈ.

Cordoba and Perth (Australia). As one might expect from such a complex, multi-national enterprise, progress was slow, with the project initiated in 1887 and the first plate not taken until 1891. Moreover, the inception of the CdC had repercussions for Gill, since the Royal Society refused to fund his CPD on the grounds that the work would duplicate the CdC observations. This decision was no doubt completely unrelated to the fact that the then Astronomer Royal, Sir William Christie, a Cambridge Wrangler, had been beaten out by the unlettered Gill for the position at the Cape. Undaunted, Gill, a supporter of the CdC, funded much of the CPD from his own salary, and the project was completed by 1890.

Once the plates were taken, they needed to be carefully scanned and processed to provide accurate astrometry for the stars detected. This is where Gill built a strong partnership with Jacobus Kapteyn, the recently appointed Professor of Astronomy and Theoretical Mechanics at the University of Groningen. Kapteyn was eight years younger than Gill, and had followed a much more traditional route through academia. His father was a mathematician and Kapteyn studied mathematics and physics at Utrecht University, moving to Leiden Observatory for three years after he completed his thesis. In 1878, he took up his position at Groningen, where he remained until his death in 1921.

Kapteyn had hoped to establish an observatory at Groningen but circumstances decreed otherwise, and he found himself spending his vacation getting occasional observations at Leiden Observatory. The development of astronomical photography offered Kapteyn another avenue for exploration, and he established the Astronomical Laboratory at Groningen for the express purpose of measuring photographic plates taken at working observatories. In particular, in 1885 Kapteyn wrote to Gill to ask for a few sample plates from the upcoming CPD, offering to devote the next 6 or 7 years of the Laboratory's work to measuring the full catalogue. Gill took up the offer, and, demonstrating the usual accuracy of project timeline predictions, Kapteyn spent the next 12–13 years processing the survey. The final catalogue included some 454,875 stars brighter that ~11th magnitude between declination −18° and the South Pole, and while Kapteyn enlisted some help (including the occasional convict from the local prison), he carried out the bulk of the measurements himself.

The CPD provided a clear indication of the important contribution that photography could make to astronomy in general and astrometry in particular, but single images provide only a static picture of the sky. Clearly, differencing photographic images taken at separate epochs offered the potential of adding time resolution, expanding proper motion determinations from painstaking one-star-at-a-time telescopic measurements to wide-field measurements of multiple stars within a given field. Kapteyn recognised the potential of the medium and the prospects for probing larger questions. In 1906 he proposed his plan of Selected Areas, which involved setting up a grid of 206 fields distributed across the sky to be targeted for concentrated photographic and photometric observations. The ultimate aim was to use the resulting star counts and motions to map the large-scale structure of the Galaxy. Kapteyn developed close links with Mt. Wilson Observatory, which contributed many observations for the selected area program.

In laying out the plan, Kapteyn already had an interesting conundrum in mind based on his prior analyses of proper motion data. As described in more detail by Paul (1985), Kapteyn anticipated a relatively uniform distribution of stellar motion. Instead, the observations appeared to reveal the presence of systematic trends that could be interpreted as motions along two preferred directions, towards Sagittarius and Orion, respectively. These results were first presented in his lecture to the Congress of Arts and Sciences held at the 1904 St. Louis Universal Exposition, followed by a 1905 lecture to the British Association for the Advancement of Science. Kapteyn suggested that the Sun sat in the midst of two independent streams of stars with significantly different mean motions.

Kapteyn did not venture to advance an hypothesis for the origin of the two star streams, but focussed on compiling additional observations to better describe the stellar motions and refine calculations of the parameters associated with each stream (see e.g. Kapteyn and Weersma 1912). In the meantime, other astronomers turned their attention to stellar kinematics, including Sir Arthur Eddington (1906, 1912) and Karl Schwarzschild (1907, 1908). Schwarzschild, in particular, developed the velocity ellipsoid model as a means of explaining the phenomenon. In this model, a star's motion is characterised by its velocities along three orthogonal axes (U, V, W); considering the population of local stars, the velocity dispersions along each axis are described by Gaussians $(\sigma_U, \sigma_V, \sigma_W)$; if the velocity ellipsoid is asymmetric (i.e. $\sigma_U \neq \sigma_V \neq \sigma_W$), then projecting the motions onto the plane of the sky leads to preferred directions of motion, i.e. star streams.

As the Selected Areas program progressed and data became available, Kapteyn turned his attentions to developing his model of the Milky Way. As noted in Chap. 15, elucidating the nature of the Milky Way was still hampered by a lack of understanding of the importance of interstellar absorption, particularly in the Galactic plane. Kapteyn recognised that absorption could lead to severe modifications to his analysis, but, at that time, he lacked any means of gauging its importance and allowing for its effects. The model of the Galaxy he developed was a highly flattened local system, perhaps ~1 kpc in diameter along the long axis and less than one-third that perpendicular to the Plane. The system was centred on the Sun and lay at the centre of a much larger Galaxy that extended to ~15 kpc distance. This model, the Kapteyn Universe, underlay Curtis' presentation at the Great Debate. Kapteyn summarised his life's work in 1922 as his *'First attempt at a theory of the arrangement and motion of the sidereal system'*. Trumpler revealed the importance of interstellar absorption within the decade.

The nature of Kapteyn's two star streams also came into better focus in the years immediately following his death. As more observations, particularly radial velocities and spectral types, were accumulated, it became possible to examine the motions of different types of stars. Gustaf Strömberg demonstrated that there was a clear asymmetry in the mean motions; grouping stars by their velocity relative to the Sun, the mean motion showed an offset that increased from lower to higher velocity stars. The velocity centroids lay along an axis directed towards galactic longitude 250°. This characteristic is known as the Strömberg asymmetric drift. It is important to remember that at that time the direction towards the Galactic Centre ($l = 0°$ in today's co-ordinates) was given as $l = 323°$, so the motions line up along $l \sim 280°$. [The present co-ordinate system was adopted by the IAU in 1958.]

The Swedish astronomer, Bertil Lindblad, recognised that Strömberg's results might well give a clue to the large-scale nature of the Galactic system. In a series of papers, he developed the concept of the Sun and the local stars forming part of a larger system that was rotating about an axis perpendicular to the Galactic Plane and about a centre that lay at considerable distance in the direction of Sagittarius, i.e. the region we now know as the Galactic Centre (Lindblad 1925). Jan Oort, Kapteyn's former student, expanded on Lindblad's work to develop a general description of the motions in a rotation system, including defining the Oort A and B constants (Oort 1927). Lindblad further hypothesised that the stars in this system formed several sub-systems, and that Strömberg's asymmetric drift reflected equipartition of energy within a differentially rotating system, with slower rotating sub-systems having higher random motions (i.e. higher pressure support). As to Kapteyn's star streams, Lindblad suggested that those arose as a natural course within a differentially rotating system, due to systematic differences in the mean motions of stars with orbits of different Galactic radii (Lindblad 1927).

There are two interesting lessons to carry forward from this discussion. First, Lindblad and Oort demonstrated that even though analyses of stellar kinematics were based on stars relatively near the Sun, the characteristic motions of those stars were indicative of a much larger-scale phenomenon, specifically differential rotation within the Galactic disc. Second, while Kaptyen's two star streams were laid to rest by this Galactic model, the concept of star streams persisted for many decades, particularly in the guise of moving groups associated with dispersing clusters and stellar associations. Both of these lessons return later in this chapter.

17.3 Kinematics of Local Stars

In order to determine the three-dimensional space motion of an individual star, we require measurements of the proper motion, the radial velocity and the distance. Proper motions are usually measured in the equatorial (α, δ) frame, and the transverse velocities given by $v = 4.74 \, \mu r$, where r is the distance in parsecs, μ is the proper motion in arcsec yr^{-1} and v is in km s^{-1}. In that case, the space motions, relative to the Sun, are given by the velocity triad $(v_\alpha, v_\delta, v_r)$. It is more meaningful, however, to transform those measurements to Galactic co-ordinates, (U, V, W), where U is positive towards the Galactic Centre, V is positive in the direction of Galactic rotation $(l = 90°)$, and W is positive towards the North Galactic Pole.

Given the prerequisite for accurate astrometry, particularly distance measurements, the catalogue produced by the *Hipparcos* satellite has played a major role in recent analyses of the kinematics of nearby stars. That dataset served as the focus for a series of papers by James Binney and collaborators (Dehnen and Binney 1998; Binney et al. 2000; Aumer and Binney 2009; Binney 2010; Schönrich et al. 2010), and the results presented in those papers form the basis for much of the discussion in this section. [As a personal aside, James Binney was one of the three lecturers, with Simon White and John Kormendy, in the Saas Fee school that I attended 30 years ago.] *Hipparcos* was limited to observations of stars brighter than 13th magnitude, and was complete only to 9th magnitude, so that dataset includes relatively few late-K and M dwarfs. We have therefore supplemented the *Hipparcos* analyses with results from the PMSU M-dwarf surveys (Reid et al. 1995; Hawley et al. 1996; Reid et al. 2002).

As a starting point, the *Hipparcos* sample is limited to single stars with parallax uncertainties $\sigma_\pi/\pi < 10\%$, corresponding to Lutz-Kelker corrections less than ~ 0.11 mag (see Appendix to Chap. 16). The sample is further refined to exclude most evolved stars (a handful of subgiants may be included) and stars with *Hipparcos* photometry discordant with the standard main-sequence (see Fig. 17.2). For the initial Dehnen and Binney (1998) analysis, the sample was further limited to stars with spectral types from the Michigan catalogues, giving a total of 11,865 stars; Aumer and Binney (2009) took advantage of the revised *Hipparcos* catalogue (van Leeuwen 2007) to expand the sample to some 15,113 stars, including 6918 stars with radial velocity and metallicity measurements from the Geneva-Copenhagen Survey (GCS; Nordström et al. 2004; Holmberg et al. 2009). Qualitatively, the results from these two analyses are very similar; the addition of the GCS data allows estimates of the age-metallicity relation, although lingering doubts remain over the accuracy of the metallicities and ages given for lower-mass stars in the GCS (see the Appendix to Chap. 16).

With typical astrometric uncertainties of ~ 0.8–1 mas, the *Hipparcos* sample extends to ~ 130 pc from the Sun, although most stars lie within 100 pc. This still falls short of the nearest star-forming regions, so, as Fig. 17.2 shows, the sample includes relatively few stars with spectral type A and only a handful of B stars.

Fig. 17.2 The
colour-magnitude limits
employed to select a nearby
star reference sample from
the *Hipparcos* catalogue; the
sample includes few stars
later than spectral type K.
Figure from Aumer and
Binney (2009)

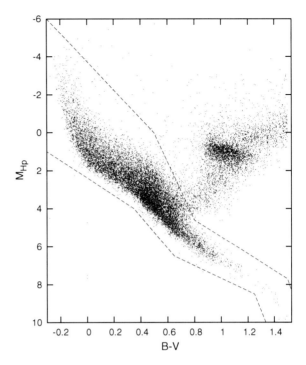

The overall sample is sufficiently large to allow sub-division, and Fig. 17.3 shows
how the mean motion relative to the Sun, the solar motion, and the velocity dispersion
vary with $B-V$ colour. The mean motions in U and W are relatively insensitive to
colour, but V, the rotational component, and the velocity dispersion show a sharp
increase between $B-V$ colours of 0.4 and 0.6 mag. This feature was originally highlighted in the 1950s by the Soviet astronomer, Pavel Parenago, and is known as
Parenago's discontinuity. Its origin and form lie in the dependence of evolutionary
lifetime with stellar mass. As one moves down through spectral types B, A, F to G
and K, the main-sequence lifetimes vary as $\tau_{MS} \sim M^{-2.5}$, running from ~ 10 Myr
for mid-type B stars to ~ 10 Gyr for a star like the Sun. Thus, for late-A/early-F type
stars, the local sample only includes stars that formed within the past ~ 100 Myr, by
$B-V \sim 0.4$ mag ($M \sim 2.5\,M_{\odot}$) $\tau_{MS} \sim 500$ Myr, while G-type (and later) stars have
lifetimes exceeding 8–10 Gyr, and therefore encompass stars from the full lifetime
of the Galactic disc.

Figure 17.3 strongly suggests that stellar velocities evolve with time, with the
overall velocity dispersion of a population increasing as it ages. Stars form in clusters, gravitationally bound entities with low velocity dispersions, <0.5 km s^{-1}. The
obvious mechanism for increasing the velocity dispersion is gravitational scattering, but star-star interactions only produce significant scattering at small separations, and are therefore ineffective in increasing the velocity dispersions of groups
of stars. Spitzer and Schwarzschild (1951) first proposed an alternative mechanism,

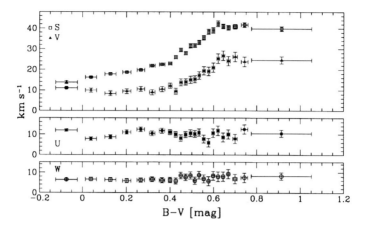

Fig. 17.3 Parenago's discontinuity (Parenago 1950). The sharp increase in the velocity dispersion *S* and *V* velocity at $B-V \sim 0.5$ mag reflects the increasing evolutionary lifetime of main-sequence stars, and the consequent larger span in age covered by G and K dwarfs. Figure from Dehnen and Binney (1998)

encounters between stars and molecular clouds. For two-body encounters, the relaxation time T, defined as the time when the velocity differences achieve the same order-of-magnitude as the original velocities, is given by:

$$T \propto V^3/M_c^2 n_c, \tag{17.1}$$

where M_c is the average mass of the clouds and n_c the number of clouds per unit volume. For large cloud complexes, $M_c \sim 10^5\,M_\odot$, the relaxation time is $\sim 10^9$ yr, as compared with $\sim 10^{14}$ yr for star-star interactions. Spitzer and Schwarzschild (1951) noted that this timescale is sufficient to account for the higher velocity dispersions of later-type dwarfs and red giants as compared with earlier-type stars. Gravitational interactions with molecular clouds remain a strongly favoured mechanism for accounting for the increase in velocity dispersion with age.

A couple of asides: first, the final section of Spitzer and Schwarzschild (1951) provides one of the first qualitative models for the formation of the Milky Way galaxy, suggesting that Population II stars formed within a highly turbulent interstellar medium that subsequently settled to the present lower-velocity configuration. We return to this in the following chapter. Second, interactions with massive objects in the Galaxy came up in a different context in the 1980s, when Bahcall et al. (1985) used the existence of wide, weakly-bound stellar binaries to set upper limits of $\sim 2\,M_\odot$ on the mass of objects that might contribute dark matter in the disc. We now recognise that there is no strong requirement for disc dark matter.

Figure 17.3 offers a path to probing the timescale for velocity evolution. This question was tackled by Wielen (1974, 1977), who took the nearby star sample from Gliese's 1957 catalogue, sub-divided them by colour and assigned likely ages based

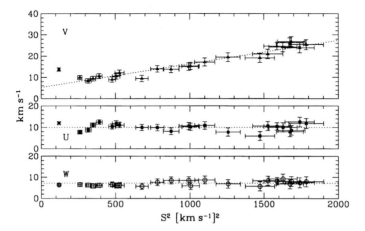

Fig. 17.4 Asymmetric drift within the *Hipparcos* sample. The data are binned by colour, as in Fig. 17.3, with U, V, W the mean motions and S^2 representing the square of the velocity dispersion. The *dotted lines* mark a linear to the trend in V, and show the values of U, W defined by stars bluer than $B-V = 0.0$ mag. Figure from Dehnen and Binney (1998)

on stellar evolutionary models. The results were broadly consistent with a velocity variation $\sigma \propto \tau^{1/3}$. Wielen (1974, 1977) analysed those data in terms of the Fokker-Planck diffusion equation, and noted that the acceleration process appears to 'top out' at large σ. This is not unexpected for a process that involves interactions with high-mass objects that are largely confined within the Plane of the Milky Way, like molecular clouds, since the scattering process can change the velocity in all three co-ordinates, and therefore ensures that higher velocity stars spend less time close to the Plane, therefore lowering the chance of further scattering.

As stars undergo gravitational interactions, they acquire random impulses that increase the velocity dispersion, usually at the expense of ordered rotational energy. This leads to the Strömberg asymmetric drift, the correlation between velocity dispersion and the mean motion in V relative to the Sun; Fig. 17.4 shows the results outlined by the colour-selected sub-samples from Dehnen and Binney (1998). There are no systematic trends with U and W. The correlation with V indicates lower Galactic rotational velocities in sub-systems with higher velocity dispersions, representing energy equipartition between systemic rotation and pressure (velocity dispersion).

The youngest stars in the Solar Neighbourhood would be expected to largely retain the motions of their parent clouds, which, in turn, are expected to be closely matched to the velocity of an object on a closed circular orbit at the Solar Radius. Those velocities are characterised as the Local Standard of Rest (LSR) and the Sun's motion with respect to the LSR is the solar motion. This quantity can be deduced from the *Hipparcos* data presented in Fig. 17.2; Dehnen and Binney (1998) deduced a solar motion of $[(U, V, W) = (10.0, 5.25, 7.17)$ km s$^{-1}]$. Binney (2010) and Schönrich et al. (2010) have since revised that result to $[(U, V, W) = (11.1, 12.2, 7.25)$ km s$^{-1}]$. Thus, relative to the LSR, the Sun has a mean motion that is directed towards the Galactic Centre, leads the LSR in Galactic rotation, and is moving from south-to-north across

Table 17.1 Velocity distribution of M dwarfs

	U	V	W	σ_U	σ_V	σ_W
All M	−9	−22	−7	43	31	25
Hα	−13	−14	−7	27	23	21

the Galactic plane. As a reminder, the Sun currently lies ∼20 pc above the mid-Plane. The circular velocity associated with the LSR is generally estimated as ∼240 km s^{-1} (Schönrich 2012).

The change in kinematic characteristics with age is also illustrated by the velocity distribution of local M dwarfs. Hα emission is a characteristic feature of chromospheric activity in these low-mass stars, and is known, from observations of M dwarfs in star clusters, to decline with time, with the timescale increasing in later-type dwarfs. Segregating field M dwarfs with prominent Hα emission therefore gives a sample of younger M dwarfs in the Solar Neighbourhood. Table 17.1 shows the results of such an analysis (from PMSU1); active M dwarfs have lower velocity dispersions and a higher mean rotational velocity than the full sample.

The M dwarf velocity distribution shows interesting structure. Figure 17.5 shows the W velocity probability plot for all 514 stars in the PMSU sample. This type of plot

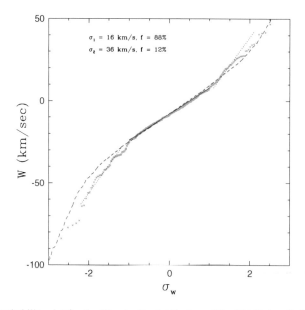

Fig. 17.5 A probability plot for the W velocity distribution of the 514 M dwarfs from the PMSU sample. The *green crosses* mark the individual observed velocities, rank-ordered against the overall rms dispersion of the sample. The *central solid line* represents the expected distribution for the stars with a Gaussian dispersion of 16 km s^{-1}, whereas the *outer dotted lines* denote those with a dispersion of 36 km s^{-1}. The *dashed line* illustrates the velocity distribution derived by combining the two populations with those velocity dispersions and the same average velocity (with a relative normalisation of 4:1 in favour of the lower dispersion component)

linearises a normal distribution, with the slope of the best-fit line corresponding to the dispersion (see Chap. 16). The W velocities follow a nearly linear distribution in the central (low-velocity) region with a slope corresponding to $\sigma_W = 16$–$20\,\mathrm{km\,s^{-1}}$. However, there are significant deviations at both the high- and low-velocity extremes that could be construed as due to a stellar component with a higher velocity dispersion. The overall distribution is well-represented by a combination of two stellar samples, with approximately 90 % of the stars in the lower dispersion component The U velocity distribution is also well-matched by two components with similar proportions, although the V velocity distribution is more complex. Analysis of data for a sample of 532 long-lived FGK dwarfs within 25 pc of the Sun gives very similar results (Reid et al. 2002).

These velocity distributions clearly suggest that, while there may be a single dominant population within the Solar Neighbourhood, there is a significant contribution from a higher velocity component. We return to this matter in Chap. 19.

17.4 Star Streams and Moving Groups

Kapteyn's star streams were envisaged as large-scale features that embraced all stars within the Sun's immediate vicinity. The realisation that the velocity trends that gave rise to this hypothesis could be explained naturally in the context of a differentially rotating Galaxy removed the rationale for this hypothesis. Nonetheless, the notion of star streams persisted, albeit in the context of the dispersing remnants of extensive, recent star-forming regions. Thus, Bart Bok refers to the Ursa Major and Taurus star streams in his 'Report on the Progress of Astronomy' to the Royal Astronomical Society (Bok 1946), with the Hyades and Ursa Major clusters representing the nuclei of the respective streams. Those two moving groups were first hypothesised over 60 years previously by R.A. Proctor (1869), who noted the alignment in motion of five stars in the plough and the presence of field stars near the Hyades with Hyades-like motion. Assuming a velocity dispersion of $\sim 0.5\,\mathrm{km\,s^{-1}}$, Bok envisages the cluster stars expanding into an ever larger volume and essentially becoming indistinguishable from the field within ~ 1 Gyr of their formation.

Bok's report also highlights one of the major problems faced by searches for stellar streams. The identification usually rests on matching space motions for stars distributed over a large spatial volume; spurious star streams, or moving groups, can be concocted by associating random stars that happen to have similar space motions. Bok comments that a dozen or so star streams had been identified among early-type stars in the Scorpius-Centaurus, Perseus and Orion associations, but few withstood more detailed scrutiny. Spurious association is a particular issue for stars that lack accurate radial velocities and/or distances, and therefore have poorly-determined space motions.

That being said, the concept of star streams, or moving groups, persisted, with the most enthusiastic proponent being the American astronomer, Olin Eggen. Eggen's career took him from a Ph.D. at Wisconsin (after serving in World War II in the OSS) to Cerro Tololo Inter-American Observatory via Lick Observatory, the Royal Green-wich Observatory, Caltech and Mt. Wilson Observatory, and Mount Stromlo Observa-tory. As the institutions might suggest, Eggen was one of the most active observational astronomers of the 20th century, a prolific publisher of colour-colour diagrams, colour-magnitude relations and tables of stellar photometry and, it seems, an avid collector of historical documents. Philosophically, he viewed astronomy as an obser-vationally driven science: theoretical astrophysicists often remark that they are 'get-ting their hands dirty' when they choose to work with observational data; Eggen once commented that, as an observer, when he had to deal with theory, he thought of himself as sitting back in his armchair and 'getting his mind dirty'.

Eggen's work on moving groups spanned most his career and took as its starting point the two star streams highlighted by Bok, the Hyades moving group, Eggen's Stream I and the Ursa Major moving group, Stream II (Eggen 1958, 1996). Eggen identified numerous sub-components of these streams, including:

- The Hyades and Praesepe clusters within Stream I
- The Pleiades group, including NGC 2516 and α Persei
- The Sirius moving group (67 stars)
- The ζ Herculis group (22 stars)
- The ϵ Indi group (14 stars)
- The 61 Cygni group (45 stars)
- The σ Puppis group (9 stars)
- The η Cephei group (5 stars)
- The Wolf 630 (Gl 644) group ($\sim$140 stars, including the low mass star, van Bies-broeck 8)
- The HR 1614 group (34 stars)
- The Ursa Major moving group, in Stream II (378 stars)
- The Arcturus moving group (53 stars), named for the brightest star in Bootes and associated with mildly metal-poor stars ([Fe/H] $= -0.6$ dex).
- The Groombridge 1830 (HD 103095) group of halo stars (5 stars, including RR Lyrae).

In several cases, the proposed members of these groups span a substantial range in age; for example, the components within the Pleiades group range in age from $\sim$100 Myr for α Persei to 120–130 Myr for the Pleiades itself. The average age of the Hyades group is pegged at the cluster age, $\sim$650 Myr, and the Ursa Major cluster is usually estimated as having a similar age, between 300 and 500 Myr. Note that both the Arcturus group and the Groombridge 1830 group are ostensibly made up of metal-poor stars, either from the earliest stages of star formation in the disc or Population II halo stars. Maintaining a dynamical association over this timescale is clearly statistically very unlikely, particularly for the sparse Groombridge 1830 group. On the other hand, Eggen was able to point to apparent similarities in chemical abundance between the members of some groups that were identified kinematically; for example, the stars in the HR 1614 group appear to show photometric properties consistent with higher metallicity than the average in the field (Eggen 1992).

Eggen is not the only astronomer to have shown a certain untoward exuberance in identifying moving groups. As an example, several groups have claimed to identify field L and T dwarfs associated with Eggen's two star streams (Bannister and Jameson 2007; Jameson et al. 2008) on the basis of proper motions and crude distance estimates. Those investigations fail to take into account several factors. First, the parent sample for these studies is the catalogue of L and T dwarfs within 20–30 pc of the Sun; if 10–15 % of those brown dwarfs are members of the Hyades or Ursa Major moving groups, then what are the implications for the total population summed over the full volume occupied by the hypothetical star stream, given that the clusters themselves lie at distances 40 to 60 pc? Second, L and T dwarfs associated with either moving group would have ages of $\sim$400 Myr and masses, $\sim$0.04–0.05 $M_\odot$; detecting so many such objects would clearly imply a much steeper initial mass function than observed in any cluster. Finally, as youthful, low-mass brown dwarfs, one would expect the spectra to show features characteristic of low gravity; none have been observed among the proffered candidate members.

All of Eggen's analyses were based on ground-based astrometry, generally of only modest accuracy. The completion of the *Hipparcos* catalogue offered an opportunity to take a new look at these issues, using a much larger sample with accurate, uniform astrometry. Such an investigation was undertaken by Dehnen (1998), who examined the velocity distributions of the kinematically-unbiased *Hipparcos* dataset. This sample includes $\sim$10,800 stars at distances up to $\sim$100 pc from the Sun.

Dehnen (1998) sub-divided the stars by colour, and the resultant distributions show evidence for sub-structure in velocity space (see Fig. 17.6). Moreover, some concentrations line up with velocities traditionally associated with moving groups: the concentration of stars at $[(U, V) = (-50, -30)\,\text{km s}^{-1}]$ in Fig. 17.6d lies close to the Hyades moving group; the knot of stars at $[(U, V) = (0, -30)\,\text{km s}^{-1}]$ in Fig. 17.6b marks the locus of Pleiades; others lie near the loci for the HR 1614 and Ursa Major moving groups. Subsequent analyses reveal lower-level features even near some of Eggen's sparser moving groups; for example, the Hercules star stream identified by Gardner and Flynn (2010) at $[(U, V) = (-50, -50)\,\text{km s}^{-1}]$ matches Eggens ζ Herculis group.

Do these kinematic features actually represent the remnants of dispersed clusters? If the stars in these kinematic concentrations really are products of coeval star-forming events, then one would expect them to share other characteristics, notably age and chemical abundance. We already noted the unusually large age spread in the so-called Pleiades group. The RAVE program targeted kinematic members of the Arcturus group for spectroscopy, and their observations confirm the existence of a group of stars with consistent motions at this location in velocity space (Williams et al. 2009). However, spectroscopy also shows that these stars span a significant range in metallicity, concentrated at the near-solar values typical of average disc stars. Similarly, Bovy and Hogg (2010) have shown that there are few indications of self-consistent chemical abundances among stars drawn from other concentrations in velocity space, strongly arguing against these representing moving groups in the Bok/Eggen tradition.

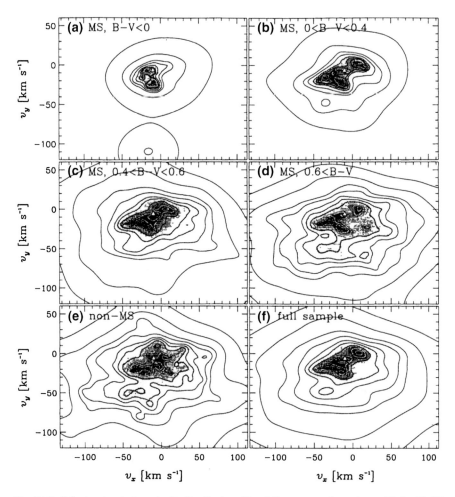

Fig. 17.6 Sub-structure in the velocity distribution of local disc stars, as shown here with the (U, V) distributions for colour-selected sub-samples of *Hipparcos* stars. Figure adapted from Dehnen (1998)

If we rule out coeval formation ('nature') as the explanation for the apparent moving groups in Fig. 17.6, then the likeliest alternative is that the observed sub-structure must arise through processes that affect the dynamical evolution of the stars that now reside in the Solar Neighbourhood (i.e. 'nurture'). Dehnen (1998) highlighted this possibility in his original discussion of the *Hipparcos* dataset. In particular, he noted that the possibility for resonances in local motions due to '... forcing by regular non-axisymmetric components of the gravitational potential, like the central bar, a long-living spiral pattern, a triaxial halo, or the Magellanic Clouds'. As with Kapteyn's original star streams, moving groups would then be a feature of the large-scale structure of the Galaxy, rather than the product of individual small-scale star formation events. We return to this possibility towards the conclusion of this chapter.

17.5 Stellar Migration, Radial Mixing and the Galactic Bar

How far can stars move within the Galaxy? Can global Galactic effects lead to significant radial redistribution of the stars within the disc? These questions were raised originally by Wielen et al. (1996) prompted by two observations: the metallicity of the Sun as compared with the local ISM; and the broad distribution shown by local stars in the age-metallicity diagram. Both seem to run counter to the well-defined, low-dispersion radial abundance gradient evident from measurements of star-forming regions in the Milky Way. We will return to the former issue in the final chapter in this series, and will discuss the age-metallicity relation in Chap. 19. Here, we consider the general question of what mechanisms might lead to radial stellar migration within the disc.

Stars are born within giant molecular clouds that generally have space motions within a few km s^{-1} of the circular velocity for a Galactic orbit (i.e. the Local Standard of Rest) at their radial location in the disc. As stars acquire larger peculiar motions through successive gravitational encounters, their orbits start to diverge, and they show significantly larger excursions from the location of their birth. It is relatively simple to estimate the extent of those excursions from:

$$\Delta \sim (2)^{0.5} \sigma_U / \kappa, \tag{17.2}$$

where κ is the epicyclic frequency,

$$\kappa = 2(B^2 - AB)^{0.5}, \tag{17.3}$$

in which A and B are the Oort constants. Taking A as 14.8 km s^{-1} kpc^{-1} and B as -12.4 km s^{-1} kpc^{-1} (Feast and Whitelock 1997), then $\kappa \sim 37.6$ km s^{-1} kpc^{-1}, so:

$$\Delta \sim 1.0 \text{ kpc} \quad \text{for} \quad \sigma_U = 27 \text{ km s}^{-1} \quad \text{(dMe stars)}, \tag{17.4}$$

$$\Delta \sim 1.6 \text{ kpc} \quad \text{for} \quad \sigma_U = 43 \text{ km s}^{-1} \quad \text{(all M dwarfs)}. \tag{17.5}$$

Taken by themselves, these excursions are not sufficient to provide sufficient mixing to explain the breadth of the age-metallicity relation, while the Sun's low velocity with respect to the local Standard of Rest argues that it has undergone even smaller radial excursions through this mechanism.

More radical excursions are possible. Spiral discs include larger-scale phenomena than even molecular clouds, notably the spiral density wave that drives the overall structure. Prompted by the same issues that intrigued Wielen et al. (1996), Sellwood and Binney (2002) explored the potential for spiral waves producing substantial changes in the orbits, and angular momenta, of individual stars without necessarily affecting the angular momentum of the system as a whole. They hypothesised that stars whose orbits were in resonance with a large-scale feature, such as a spiral wave, could transfer radial kinetic energy, leading to orbit migration.

This is indeed the result to emerge from N-body simulations that develop a single spiral wave. The spiral wave 'churns the disc'. Stars near the co-rotation radius can move onto closed horseshoe orbits, where they experience successive gain and loss of angular momentum, L, as they advance and then fall behind the spiral wave. Stars on circular orbits can experience large changes in L without significant change in eccentricity, that is, the orbit expands, but the overall velocity dispersion of the population is essentially unchanged. Indeed, gas clouds can also participate in the radial migration, potentially broadening the range of initial conditions for star formation at different Galactic radii. Figure 17.7 shows the degree of radial excursion possible within the simulation, suggesting that radial translations up to 3–4 kpc, either inward or outward, are possible.

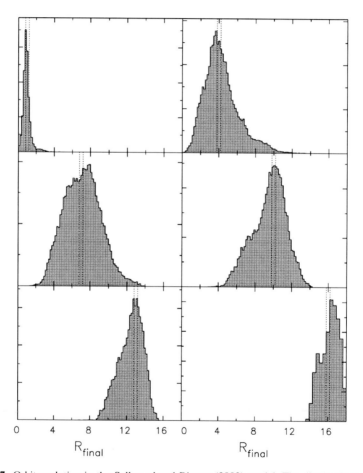

Fig. 17.7 Orbit evolution in the Sellwood and Binney (2002) model. The distributions plot the final radial distances for six sets of particles that originated within the radial region marked by the *dashed lines* in each sub-panel; the model shows the potential for radial migration over distances of 3–5 kpc. Figure from Sellwood and Binney (2002)

The calculations of Sellwood and Binney (2002) have since been expanded in a number of ways. Solway et al. (2012) examined whether the extent of the vertical motion exhibited by individual stars affects radial orbit migration. This is relevant to whether stars in the thick disc (Chap. 19) are liable to the same radial migration as thin disc stars, and, to first order, the answer appears to be that they are. In another vein, Roškar et al. (2008, 2012) undertook a more extensive series of N-body simulations, and they find that a single spiral pattern is unlikely to persist within a disc. Rather, discs are likely to be subject to a series of transient spiral features, following multiple patterns. As a result, stars and gas will move in and out of co-rotation orbits over the lifetime of the Galaxy, leading to multiple opportunities for radial migration and mixing within the disc.

Resonances with spiral arms may lead to migration within the Galactic disc; turning back to moving groups, resonances with the Galactic Bar could account for some kinematic sub-structure in the Solar Neighbourhood. The Bar's existence was initially hinted at by measurements of gas velocities in the inner Galaxy (e.g. de Vaucouleurs 1964; Binney et al. 1991). Infrared photometry, both star counts and surface brightness measurements, confirmed its presence and substantiated its nature (e.g. Dwek et al. 1995). In particular, data taken by the *Spitzer* space telescope for the GLIMPSE Galactic plane survey (Benjamin et al. 2005) shows a N/S asymmetry at longitude, $l < 30°$ and a pronounced excess at $10° < l < 30°$ that can be accounted for by a bar-like feature with half-length ~ 4.4 kpc, tilted at an angle $\Phi \sim 44°$ with respect to the direct line-of-sight towards the Galactic centre. Similar conclusions follow from recent analyses of the distribution of old population tracers, such as red clump giants (e.g. Cabrera-Lavers 2008).

The Bar is a relatively massive structure. As such, it has the potential to drive structure at large Galactocentric radii. As with stellar migration, the key resonances are at co-rotation, and at the inner and outer Lindblad radii, given by:

$$\Omega_b = \omega_\varphi \pm \omega_R, \tag{17.6}$$

where Ω_b is the bar pattern speed, and ω_φ and ω_R the azimuthal and radial orbital frequencies. Stars and gas at the inner and outer Lindblad radii 'see' the Bar at the same orientation on each passage, and therefore receive a consistent set of gravitational impulses. This can drive both components onto preferential orbits, leading to limited-lifetime concentrations within the velocity plane.

The outer Lindblad radius of the Galactic Bar likely lies at a galactocentric radius between 6 and 9 kpc, straddling the Solar Neighbourhood. A series of dynamical models computed by Dehnen (2000), Gardner and Flynn (2010), Bovy (2010) and others strongly suggest that the Bar's outer Lindblad radius resonance can produce significant structure in the local velocity plane, particularly developing the feature associated with the Hercules stream and ζ Herculis moving group. Hahn et al. (2011) suggest that the Hyades 'moving group' may have originated through an inner Lindblad radius resonance, while Minchev et al. (2010) argue that the Pleiades and Hyades moving groups have a common origin, and the number of stars in the Pleiades moving group suggests that the Galactic Bar formed ~ 2 Gyr ago.

17.6 Summary

As a discipline, stellar kinematics has developed considerably since its initiation with the collaboration between David Gill and Jacobus Kapteyn. Building on ground-based photometric and astrometric surveys and, particularly, the catalogue produced by the *Hipparcos* astrometric satellite, we now have a clear picture of the kinematics of the stars in the Solar Neighbourhood and a qualitative understanding of how those distributions may have arisen. That structure is fairly complex, with the suggestion of extended wings to the overall velocity distribution, suggestive of the superposition of two stellar populations, while the distributions in the (U, V), (U, W), and (V, W) planes show clear sub-structure. These knots and concentrations are coincident with features that have been identified as moving groups, dispersing cluster remnants.

The classical picture of the evolution of stellar velocities envisages stars forming at velocities matched to the LSR, with random motions increasing with time due to stochastic encounters with massive objects close to the Galactic Plane. Those encounters preserve angular momentum, but lead to increased orbital eccentricity. More recent calculations show that other processes can occur, such as interactions with spiral arms that lead to angular momentum exchange, transporting stars and gas over radial distances up to ~4 kpc without affecting orbital eccentricity. This complicates the process of back-tracking the orbit of any given star (e.g. the Sun) to its place of origin. In addition, large-scale features like the Galactic Bar can set up resonances that lead to small-scale transient structure within the local velocity plane, collecting diverse stellar constituents that share orbital parameters, but formed over a wide variety of times and conditions.

Observations of stars in moving groups have generally aimed at determining that nature of the star formation region that might have spawned the group. These new results turn that analysis on its head. Rather than probing local star formation, sub-structure in the velocity distribution may provide a means of investigating large-scale Galactic phenomena. This is a particularly interesting idea to bear in mind with *Gaia* coming on line, offering the potential for detailed photometric and kinematic information for stars at distances more than 10 kpc from the Sun.

References

Aumer, M. & Binney, J. J. 2009, MNRAS, 397, 1286
Bahcall, J. N., Hut, P., & Tremaine, S. 1985, ApJ, 290, 15
Bannister, N. P. & Jameson, R. F. 2007, MNRAS, 378, L24
Benjamin, R. A., Churchwell, E., Babler, B. L., et al. 2005, ApJ, 630, L149
Binney, J. 2010, MNRAS, 401, 2318
Binney, J., Gerhard, O. E., Stark, A. A., Bally, J., & Uchida, K. I. 1991, MNRAS, 252, 210
Binney, J., Dehnen, W., & Bertelli, G. 2000, MNRAS, 318, 658
Bok, B. J. 1946, MNRAS, 106, 61
Bovy, J. 2010, ApJ, 725, 1676
Bovy, J. & Hogg, D. W. 2010, ApJ, 717, 617

Cabrera-Lavers, A. et al, 2008, A&A, 491, 781
de Vaucouleurs, G. 1964, in IAU Symposium, Vol. 20, The Galaxy and the Magellanic Clouds, ed. F. J. Kerr, Australian Academy of Science, Canberra, 195
Dehnen, W. 1998, AJ, 115, 2384
Dehnen, W. 2000, AJ, 119, 800
Dehnen, W. & Binney, J. J. 1998, MNRAS, 298, 387
Dwek, E., Arendt, R. G., Hauser, M. G., et al. 1995, ApJ, 445, 716
Eddington, A. S. 1906, MNRAS, 67, 34
Eddington, A. S. 1912, MNRAS, 72, 368
Eggen, O. J. 1958, MNRAS, 118, 65
Eggen, O. J. 1992, AJ, 104, 1906
Eggen, O. J. 1996, AJ, 111, 1615
Feast, M. & Whitelock, P. 1997, MNRAS, 291, 683
Gardner, E. & Flynn, C. 2010, MNRAS, 405, 545
Gill, D. 1882, MNRAS, 43, 53
Hahn, C. H., Sellwood, J. A., & Pryor, C. 2011, MNRAS, 418, 2459
Hawley, S. L., Gizis, J. E., & Reid, I. N. 1996, AJ, 112, 2799
Holmberg, J., Nordström, B., & Andersen, J. 2009, A&A, 501, 941
Jameson, R. F., Casewell, S. L., Bannister, N. P., et al. 2008, MNRAS, 384, 1399
Kapteyn, J. C. & Weersma, H. A. 1912, MNRAS, 72, 743
Lindblad, B. 1925, ApJ, 62, 191
Lindblad, B. 1927, MNRAS, 87, 553
Minchev, I., Boily, C., Siebert, A., & Bienayme, O. 2010, MNRAS, 407, 2122
Nordström, B., Mayor, M., Andersen, J., et al. 2004, A&A, 418, 989
Oort, J. H. 1927, Bull. Astron. Inst. Netherlands, 3, 275
Parenago, P. P. 1950, AZh, 27, 41
Paul, E. R. 1985, in IAU Symposium, Vol. 106, The Milky Way Galaxy, ed. H. van Woerden, R. J. Allen, & W. B. Burton, Reidel Publishing Company, Dordrecht, 25
Paul, E. R. 1993, The Milky Way Galaxy and Statistical Cosmology, 1890–1924, ed. Paul, E. R., Cambridge University Press
Proctor, R. A. 1869, Royal Society of London Proceedings Series I, 18, 169
Reid, I. N., Hawley, S. L., & Gizis, J. E. 1995, AJ, 110, 1838
Reid, I. N., Gizis, J. E., & Hawley, S. L. 2002, AJ, 124, 2721
Roškar, R., Debattista, V. P., Quinn, T. R., Stinson, G. S., & Wadsley, J. 2008, ApJ, 684, L79
Roškar, R., Debattista, V. P., Quinn, T. R., & Wadsley, J. 2012, MNRAS, 426, 2089
Schönrich, R. 2012, MNRAS, 427, 274
Schönrich, R., Binney, J., & Dehnen, W. 2010, MNRAS, 403, 1829
Schwarzschild, K. G. 1907, Akad. Wissenschaft Göttingen Nach., 614
Schwarzschild, K. G. 1908, Akad. Wissenschaft Göttingen Nach., 191
Sellwood, J. A. & Binney, J. J. 2002, MNRAS, 336, 785
Solway, M., Sellwood, J. A., & Schönrich, R. 2012, MNRAS, 422, 1363
Spitzer, Jr., L. & Schwarzschild, M. 1951, ApJ, 114, 385
van Leeuwen, F., ed. 2007, Astrophysics and Space Science Library, Vol. 350, Hipparcos, the New Reduction of the Raw Data
Wielen, R. 1974, Highlights of Astronomy, 3, 395
Wielen, R. 1977, A&A, 60, 263
Wielen, R., Fuchs, B., & Dettbarn, C. 1996, A&A, 314, 438
Williams, M. E. K., Freeman, K. C., Helmi, A., & RAVE Collaboration. 2009, in IAU Symposium, Vol. 254, The Galaxy Disk in Cosmological Context, ed. J. Andersen, B. Nordström, & J. Bland-Hawthorn, Cambridge University Press, 139

Chapter 18
Clusters and the Galactic Halo

I. Neill Reid

18.1 Introduction

Devoting a full chapter to the halo, the oldest stars in the Milky Way, may seem a
shade contrarian in a conference that is nominally aimed at discussing young, massive star clusters. But the halo was young once, and the globular clusters that are
its most prominent constituents must have rivalled and even exceeded present-day
structures like NGC 3603 or 30 Doradus. More to the point, our understanding of
the nature of globular clusters has undergone a paradigm shift in the last half-decade
with the realisation that many (most?) are not simple, single-starburst entities, but
harbour evidence of a complex star-forming history and multiple constituent populations. Equally to the point, 2012 represents an important anniversary for one of
the most influential papers of 20th century astrophysics, the first serious attempt 'to
reconstruct the galactic past', by Eggen et al. (1962, hereafter ELS). This chapter
gives a brief introduction to globular cluster properties and places the ELS paper in
its contemporary context. We conclude with discussion of results from more recent
investigations of globular clusters, particularly the discovery of multiple stellar populations, and consider the implications for cluster formation and the early life of the
Milky Way.

18.2 Globular Clusters and the Galactic Halo

The members of the Galactic halo have been recognised as likely to be the oldest
constituents of the Milky Way since Baade's identification of Population I and II in
the 1940s. The field halo stars in the Solar Neighbourhood, the 'high velocity stars',

I.N. Reid (✉)
Space Telescope Science Institute, Baltimore, MD, USA
e-mail: inr@stsci.edu

© Springer-Verlag Berlin Heidelberg 2015

C.P.M. Bell et al. (eds.), *Dynamics of Young Star Clusters and Associations*,
Saas-Fee Advanced Course 42, DOI 10.1007/978-3-662-47290-3_18

had already played a role in the 1920s in Oorts expansion of Lindblad's differential rotation system into a (static) model of the Milky Way (Oort 1928). Globular clusters famously played a key role in the construction of Harlow Shapley's 'Large Galaxy' model for the Milky Way, in large part because of Solon Bailey's discovery of RR Lyrae variables in many such systems.

One hundred and fifty seven globular clusters are currently known (see Fig. 18.1), and the total population is estimated as around 160 systems (Harris 2010). Most are prominent stellar concentrations, easily discernible against the stellar background and, as a result, part and parcel of nebular catalogues since before the Herschels scanned the skies. Twenty-nine of the most prominent northern systems were included in Charles Messier's famous 'not-comets' catalogues published between 1771 and 1787. Other systems are more obscure. In particular, one of the products of the first Palomar Sky Survey was a list of 15 globulars that had previously escaped detection (Abell 1955). Those clusters, identified prosaically as Palomar 1–15, are either highly-obscured systems towards the Galactic Centre or less massive, more diffuse and much more distant than the classical globulars. A further three obscured Galactic Centre systems were identified from plates taken by the ESO Schmidt telescope, while the 2MASS infrared survey has added two additional highly-obscured systems. Those surveys have also been very effective at identifying dwarf galaxy satellites of the Milky Way, with SDSS extending detections to ultra-faint systems (Brown et al. 2012) that are comparable in mass to the larger globulars, an issue we return to towards the end of this chapter.

Parameterising the known systems, clusters range in mass from $\sim 10^6$ to $\sim 10^4$ M$_\odot$, with ω Cen the largest and Palomar 13 the smallest systems currently known. In comparison, ultra-faint dwarf galaxies identified by SDSS have masses of $\sim 10^4$ M$_\odot$, although most of that mass is likely dark matter while there is no evidence for significant dark matter associated with globulars. The majority lie within 5–10 kpc

Fig. 18.1 Extreme globular clusters: ω Cen (*left*, figure taken from http://www.spitzer.caltech. edu/images/1908-ssc2008-07a-Globular-Cluster-Omega-Centauri-Looks-Radiant-in-Infrared), the largest Galactic cluster, and Palomar 13 (*right*, figure taken from http://www.galaxyzooforum. org/index.php?topic=272726.0), among the smallest currently known

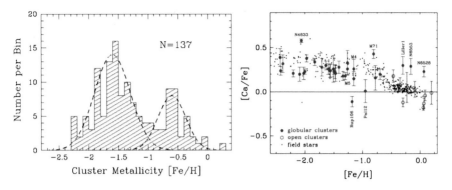

Fig. 18.2 *Left panel*: The metallicity distribution [Fe/H] of globular clusters (from Harris 2001). *Right panel*: The α/Fe abundance ratio as characterised by measurements of calcium abundance in a range of stellar systems (adapted from Gratton et al. 2004)

of the Galactic centre, but the outer clusters lie at galactocentric radii extending beyond 50 kpc, with the most distant system, AM 1, at RGC ∼125 kpc. Overall, the radial distribution is consistent with a power-law distribution, $\rho \propto r^{-3.5}$ (Harris 2001). Kinematically, the clusters show substantial velocity dispersion with no evidence for net rotation, indicating a pressure-supported, rather than rotation-supported, system. Determining proper motions (and hence three-dimensional space motions) can be problematic given the distances, and the overall kinematics can still be characterised as consistent with the determination of Frenk and White (1980), $[(U, V, W), (\sigma_U, \sigma_V, \sigma_W) \sim (0, -160, 0), (120, 120, 120)$ km s$^{-1}]$. Many clusters are on nearly radial orbits that lead to periodic passages through the inner Galactic disc.

Metallicities for globular clusters can be determined either by matching the colour-magnitude diagram (primarily the red giant branch) against theoretical models or through direct analysis of spectroscopic observations of main-sequence and evolved stars. The results show that they are metal-poor systems, with abundances extending to below [Fe/H] = −2.5 dex or less than 1/300th that of the Sun (see Fig. 18.2). There are some hints of a bimodal distribution (see left panel of Fig. 18.2). In addition, detailed analysis of elemental abundances shows that globular cluster stars, and halo stars in general, have enhanced abundances of α elements (Ca, Ti, Mg, etc.) when compared with disc stars. As discussed in the opening chapter, this indicates that those stars formed early in the star formation history of the Milky Way, before Type Ia supernovae could drive up iron production.

Globular clusters have well-defined colour-magnitude diagrams that readily lend themselves to age estimation by matching against theoretical isochrones for the appropriate abundance distribution (see Fig. 18.3). The crucial step is determining reliable cluster distances, since these systems lie well beyond the reach of (current) direct trigonometric parallax measurements. A variety of distance estimators can be used, including horizontal branch stars, RR Lyraes or local main-sequence halo stars (subdwarfs). In each case, the process involves taking nearby stars whose distances

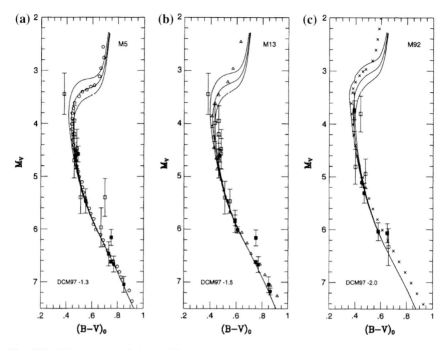

Fig. 18.3 Main-sequence fitting to fiducial colour-magnitude diagrams for globular clusters. The isochrones are drawn from the theoretical models by D'Antona et al. (1997) and are for ages of 10, 12 and 14 Gyr and with abundances of [Fe/H] = −1.3 (M5), −1.5 (M13) and −2.0 dex (M92). Figure from Reid (1997)

can be determined through some method (direct trigonometric parallax, statistical parallax, Baade-Wesselink analysis), and using those stars as templates to match against similar stars in the cluster. Of the local templates, FGK subdwarfs offer the best option since they have sufficiently high space density that reasonable numbers (approximately a dozen) are near enough the Sun for direct parallax measurements.

In main-sequence fitting, the fiducial cluster sequence (apparent magnitude, colour) is corrected for line-of-sight reddening and then matched against the (absolute magnitude, colour) sequence mapped out by the local subdwarfs. The latter stars span a range of abundance, so appropriate corrections need to be applied to the absolute magnitude or colour before matching against a specific cluster. Once the zeropoint is determined, the cluster age follows by matching against theoretical isochrones. Prior to *Hipparcos*, this technique led to ages generally estimated as between 14 and 17 Gyr, which posed something of a problem, given than the age of the Universe was generally pinned at 13–14 Gyr. *Hipparcos* targeted many of the local FGK subdwarfs, determining more accurate parallaxes and expanding the sample with reliable astrometry. The results show a systematic reduction in the average parallax of the calibrators by ∼3 mas. This leads to larger distance estimates to individual clusters, brighter turn-off magnitudes and younger ages (Gratton et al. 1997; Reid 1997).

Typical age estimates for globular clusters now lie in the range 11–13 Gyr, entirely compatible with the age of 13.772 ± 0.059 Gyr derived from analysis of the WMAP measurements of the microwave background (Bennett et al. 2013). There remain a few residual doubts over the new cluster distance scale (e.g. Harris 2001), but *Gaia* will clearly have a big impact here, not only by expanding coverage to many more local subdwarfs, but also by providing direct parallax measurements for the nearest globular clusters.

As the prior discussion suggests, field halo stars are rare. Locally, disc main-sequence stars outnumber halo subdwarfs by $\sim$200 to 1—there are only 4 FGK subdwarfs within 25 pc. Nonetheless, there are sufficient stars to trace the general properties of the population, which are fairly consistent with the cluster system. The radial density distribution is consistent with a power-law, $\rho \propto r^{-3}$. Field halo stars are also α-enhanced, but the metallicity distribution shows a much larger tail extending to significantly lower abundances, [Fe/H] < -4 dex, with the most metal-poor star currently known, HE 1327-2326, having [Fe/H] ~ -5.5 dex. The kinematics indicate a non-rotating, pressure-supported system, like the clusters, with perhaps some indications of triaxiality $[(U, V, W), (\sigma_U, \sigma_V, \sigma_W) \sim (-20, -190, -3), (152, 104, 95)$ km s$^{-1}]$ (Carney et al. 1994). There have also been suggestions that there is duality in the field halo star populations, with inner and outer components (Sommer-Larsen and Zhen 1990; Beers et al. 2012).

Local subdwarfs are sparsely distributed, but there are sufficient numbers to set constraints on the mass function for halo stars. The results are broadly consistent with those derived for the mass function derived for solar-metallicity disc stars (Gizis and Reid 1999). Globular clusters provide further insight into the halo mass function, although those analyses are tempered by the fact that those systems undergo significant dynamical evolution. Internal relaxation leads to mass segregation that concentrates higher mass stars (and binary systems) towards the centre, and external interactions, primarily as the cluster passes through the disc, preferentially strip less tightly-bound, lower-mass stars from the system. Nonetheless, analyses of globular cluster show little evidence for significant departures from the disc prescription outlined in Chap. 16, despite the substantial differences in metallicity (Paust et al. 2010).

18.3 Setting the Context

The previous section sketches out much of what we know about the halo now. Much has changed over the last half century. Before discussing the ELS model and how that paper impacted the field, it is useful to spend a little time considering the broader state of astrophysics at that time.

Starting on the largest scales, Hubble had identified the expanding nature of the universe in the late 1920s, but the underlying model still remained a matter of debate. Ralph Alpher and George Gamow laid the basis for the 'hot start' theory in the famous alphabetical article (Alpher et al. 1948), proposing that the universe had an origin at a fixed point of time when all matter was concentrated at extremely high density and high temperature, synthesising the chemical elements. The discovery of the microwave background, the redshifted echo of those high temperatures, still lay several years in the future, so there was on-going debate with the Cambridge-centred steady-state theory favoured by Fred Hoyle, Hermann Bondi and Thomas Gold. Driven partly by the philosophical concept of the Perfect Cosmological Principle (the Universe looks the same at all times and places), steady-state theory envisioned an expanding universe of infinite age and duration, with matter created to maintain uniformity (Hoyle 1948). Competition between the two theories was fairly intense. Indeed, the name 'Big Bang' now widely associated with Alpher and Gamow's work was coined by Fred Hoyle, perhaps as a dismissive label, in one of a series of BBC radio lectures given in March 1949.

Wide-field photographic surveys such as the 10-year program undertaken by C.D. Shane and C.A. Wirtanen, and George Abell's analysis of the Palomar Sky Survey (POSS I) plates revealed that galaxies were clustered. Quasars were on the point of discovery. In the early 1960s, Allan Sandage, Thomas Matthews and collaborators at Carnegie Observatories and Caltech identified several very blue objects that appeared to be the source of very strong radio emission. Cyril Hazard and John Bolton localised the position of the radio source, 3C 273, by timing its occultation by the moon. Maarten Schmidt obtained a spectrum with the Palomar 200-inch telescope in August 1962, and recognised that those features could be explained if the source were at a redshift of $z = 0.157$.

Looking within galaxies, Walter Baade's wartime observations with the Mt. Wilson 100-inch resolved stars in of M32 and the bulge of the Andromeda galaxy, and led to the development of the concept of stellar populations. Baade drew a link between those extragalactic stellar populations and the Galactic globulars, dominated by red giants, and contrasted that with the bright blue stars, and on-going star formation, with the Galactic disc. Allan Sandage and Halton Arp pursued more detailed studies of Galactic clusters, pushing colour-magnitude diagrams to fainter magnitudes to reveal the main-sequence turn-off in globular clusters, and adding observations of extensive numbers of Galactic open clusters. Those observational results, summarised in the composite diagram reproduced in the right panel of Fig. 18.4, provided observational incentive for theoretical work on stellar evolution, particularly the transition onto the red giant branch, pursued by Martin Schwarzschild, Roger Tayler and Fred Hoyle, among others.

The range of stellar chemical abundances was also starting to come into focus. In 1951, Joseph Chamberlain and Lawrence Aller undertook a detailed analysis of the spectrum of two so-called A-type subdwarfs, HD 19445 and HD 140283, comparing their analysis against the main-sequence A star, 95 Leonis, as a reference. The results were surprising, indicating that the subdwarfs had temperatures more consistent with spectral type F and metallicities significantly lower than the solar value (Chamberlain and Aller 1951). Indeed, Aller subsequently admitted that their formally derived metallicity was significantly lower than the $[Fe/H] = -1$ dex quoted in the paper, but they were deliberately cautious because the result was so surprising. Nancy Roman subsequently demonstrated that many high-velocity stars shared these characteristics,

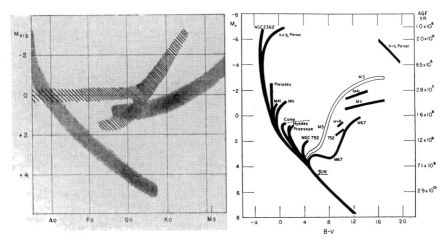

Fig. 18.4 *Left panel*: Schematic representation of Population I (*shaded*) and Population II stars (*hatched*) in M32 (from Baade 1944). *Right panel*: Composite cluster colour-magnitude diagram (from Sandage 1956)

with F-type spectra but a substantial UV-excess due to the reduced line blanketing in the metal-poor atmospheres. The orbits of those stars are highly eccentric, passing through the Galactic bulge, and Roman drew a direct analogy with main-sequence stars in globular clusters (Roman 1954).

At the same time, significant progress was being made on identifying the likely origins of the heavy elements. Gamow had postulated that all elements could be generated by neutron addition to the lightweight elements generated in the Big Bang. However, it became clear that there was no element with a stable isotope of mass 8, setting an insurmountable roadblock to this hypothesis. Partly prompted, no doubt, by his cosmological views, Fred Hoyle had developed the notion of nucleosynthesis, formation of heavy elements by fusion within stars. In the early 1950s, Edwin Salpeter suggested that carbon might be formed by fusing three helium atoms, and Hoyle demonstrated that this process was feasible in red giants. Building on that foundation, Hoyle worked with Geoffrey Burbidge, Margaret Burbidge and Willy Fowler at Caltech to develop a more detailed theory, published in Reviews of Modern Physics in 1957 as the famous B^2FH paper (Burbidge et al. 1957). Their work showed that stars could change the overall composition of the Milky Way (and other galaxies) by transforming hydrogen and helium into increasingly heavier elements. Willy Fowler received the Nobel prize for physics in 1983 '... for his theoretical and experimental studies of the nuclear reactions of importance in the formation of the chemical elements in the universe'.

These separate threads were drawn together almost fifty-five years ago, in May 1957 at the Vatican conference on Stellar Populations. Bringing together most of the major astronomers of the time (including Baade, Blaauw, Fowler, Hoyle, Morgan, Oort, Salpeter, and Sandage), that conference took Baade's simple concept of stellar populations, and applied it to observations of stars in the Milky Way. As described by Blaauw (1995), Baade's original separation of the M31 stars into Populations I and II was based on observational properties, specifically the colour-magnitude diagram; stellar evolution was not part of the initial picture. By the Vatican conference, however, the theoretical work by Hoyle, Tayler, Schwarzschild and Fowler, among others, had turned the focus on colour-magnitude diagrams from 'what?' to 'why?'. At the same time, the extensive measurements of stellar motions were being combined with the new radio observations of the gas content of the Milky Way to probe its overall dynamics. In the final presentation of the conference, Oort wove these threads into a complex picture of the Milky Way that included six stellar populations, ranging from Extreme Population I, characterised by OB stars, supergiants and Cepheids within the Galaxy's spiral arms, to the Halo Population, characterised by globular clusters and the high-velocity stars. Crucially, each population had an associated age estimate, heavy element abundance and spatial distribution. The potential for probing the past history of the Galaxy became apparent.

The most important outcome of the Vatican conference was a schematic outline of the current structure of the Milky Way, with clear hints as to the historical progression of star formation within the different sub-systems. Oort's discussion touched on the likely implications for Galactic evolution, and Hoyle discussed how star formation may proceed on smaller scales, but no model emerged providing an overall architecture for the formation of the Milky Way. That important step was first taken by Eggen, Lynden-Bell and Sandage in 1962.

18.4 The ELS Model

The observational foundation of the ELS model lay in Sandage and Eggen's collaborative work on the properties of high-velocity stars. The two astronomers had been close colleagues for some years, but this collaboration developed while Sandage was an invited visitor to the Royal Greenwich Observatory, Herstmonceux Castle, where Eggen was the chief assistant to the Astronomer Royal, Sir Richard Woolley. Eggen had compiled extensive photometric observations of numerous stars in the relatively recently-established Johnson UBV system, including observations of high-velocity stars. A subset of those stars had trigonometric parallaxes, and clearly lay below the main-sequence in the standard M_V, $B-V$ diagram. Indeed, this property led Kuiper (1939) to identify these stars as 'subdwarfs', a term previously applied more broadly to stars with unusual spectral properties.

Spectroscopic observations by Nancy Roman and others had shown that subdwarfs have weak spectral lines, leading to their being assigned earlier spectral types than might be appropriate. The weaker lines also lead to changes in the photometric properties, with lower line blanketing leading to more emission at blue and UV wavelengths. Sandage and Eggen (1959) developed simple models to estimate the effect on the UBV colour indices of reduced line blanketing. Decreasing metal content moves stars to bluer $B-V$ and $U-B$ colours, outlining blanketing lines in the

$U-B$, $B-V$ two-colour diagram. Taking field subdwarfs with trigonometric parallax measurements, they showed that a mean relation could be applied to adjust those stars onto the disc main-sequence. Thus, to first order, the offset of a star from the main-sequence in the M_V, $B-V$ colour-magnitude diagram could be tied to the level of line blanketing, and hence the metallicity.

Sandage and Eggen's 1959 paper defined UV-excess, $\delta U-B$, the vertical offset in $U-B$ between the observed location of a star in the UBV diagram and the location of the Hyades sequence at that $B-V$ colour. The UV-excess not only measures the metallicity of the field subdwarf, but also enables an estimate of its subluminosity, and hence its absolute magnitude. Given that estimate, photometric parallaxes (and distances) could be estimated for high-velocity stars that lacked direct trigonometric parallax measurements, and space motions derived from the radial velocity and proper motions. This was crucial in enabling ELS, since even now, after *Hipparcos*, only a few tens of subdwarfs have reliable trigonometric parallax measurements.

Buoyed by the initial results, Sandage and Eggen acquired further photometric and radial velocity measurements of high-velocity stars. By 1962, Sandage had returned to Mt. Wilson and Eggen had taken up a professorship at Caltech, following a rather public disagreement with Woolley on a policy issue. They were joined by Donald Lynden-Bell, on leave from Clare College, Cambridge, on a postdoctoral appointment with Sandage. Together, the three researchers provided a prime example of the whole exceeding the sum of the parts, combining Eggen's detailed and compendious observations with Lynden-Bell's theoretical knowledge of dynamics and Sandage's broad perspective on large-scale issues.

Observationally, ELS focuses on the analysis of data for only 221 F, G and K stars, comprising 108 nearby stars from Eggen's Solar Neighborhood sample together with 113 high-velocity stars. Besides UBV photometry, all stars had proper motion and radial velocity measurements as well as distance estimates, either from trigonometric parallax or UV-excess-corrected photometric values. As a result, three dimensional (U, V, W) space motions could be derived for each star. Placing those motions in a model potential enabled an estimate of the Galactic orbits.

Lynden-Bell's Galactic potential was derived from an axisymmetric model, set to match the rotation curve derived from radio observations of HI gas clouds in the disc. Stellar orbits were characterised using the angular momentum, h; eccentricity, $e = (R_{max} - R_{min}/R_{max} + R_{min})$, where R_{max} and R_{max} are the apogalacticon and perigalacticon of the orbit; and $|W|$, the velocity perpendicular to the Plane. Integrating the motions over time within a static potential (note that all these calculations were made by hand, not by computer) showed that the orbits were generally unclosed, but maintained the same values of e, h and $|W|$. Similar circumstances hold within a slowly varying potential (a potential that changes on a time-scale that is long relative to the orbital period). However, if the potential varies rapidly, then, on average, the eccentricity increases significantly while conserving angular momentum. The net result in the last case is that test particles will evolve onto highly eccentric, radial orbits.

Working within this theoretical framework, with the knowledge that the sample stars lie within a few hundred parsecs of the Sun, one can calculate the orbital

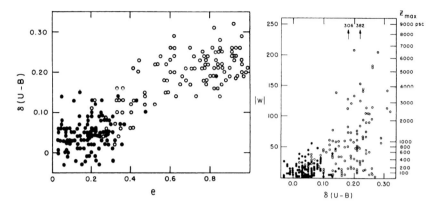

Fig. 18.5 *Left panel*: The correlation between orbital eccentricity and UV-excess (metallicity). *Right panel*: The correlation between the W velocity perpendicular to the Galactic plane and UV-excess. The *filled circles* represent stars from the nearby star catalogue, while the *open circles* denote high-velocity stars. Figure adapted from Eggen et al. (1962)

energy and angular momentum, and estimate eccentricities and vertical velocities for each object. The results are shown in Fig. 18.5, which includes the two most influential figures from ELS. The left panel plots orbital eccentricity against UV-excess, showing that the nearby stars tend to have relatively circular orbits while stars with the strongest UV-excess (i.e. the lowest metallicities) have the most eccentric orbits. The right panel maps out the vertical velocities, and the consequent maximum height from the Galactic Plane; it is clear that lower abundance stars have a spatial distribution that extends to significantly larger distances from the Plane.

It is important to note that there is the potential for two systematic selection effects in these figures. Metal-poor stars on circular orbits would not necessarily appear in this sample, since their velocities relative to the Sun would be low, leading to their non-inclusion in the high-velocity star sample, but their local density might preclude inclusion in the nearby star sample. In the same vein, stars with modest UV-excess and intermediate eccentricities might be under-represented in such a relatively small sample.

These concerns aside, ELS highlighted the parallels between the local subdwarf sample and main-sequence stars in globular clusters, known to be amongst the oldest objects in the Milky Way with ages of $\sim 10^{10}$ yr. On that basis, they identified the correlation with UV-excess as a correlation with age, and deduced that the large Z reached by subdwarfs was tied to the likely location of the parent star-forming clouds. Coupled with the theoretical calculations, they concluded that the observations could be explained if the oldest stars in the Galaxy (the subdwarfs) formed during a phase when the galactic potential was undergoing rapid evolution (timescale of a few $\times 10^8$ yr) driven by the radial collapse of the protogalactic cloud, whose initial extent was perhaps ten times the size of the present Milky Way. The net angular momentum of the cloud slowed and stopped the collapse in R, but the collapse continued in Z, leading to the formation of a rotating, gas-rich disc, where continuing star formation led to the formation of the present-day disc population. Globular clusters and halo field subdwarfs formed during the collapse, leading to their having strongly radial orbits.

The ELS paper was extremely influential at the time, and continues to play an important role in galaxy formation.

18.5 Searle and Zinn and Galaxy Mergers

ELS established a framework for discussing potential formations scenarios for spiral galaxies like the Milky Way. Conceptually, their model envisages the Milky Way as the product of the monolithic collapse of the gaseous protogalaxy, implying an overall coherence in structure within the Galactic halo. As observations accumulated of larger samples of halo objects more complex circumstances emerged that cast some doubt on that relatively simple model. In particular, more extensive photometry revealed that clusters with the same overall metallicity could have radically different horizontal branch morphologies. For example, both NGC 7006 and M2 have average metal abundances [M/H] ~ -1.5 dex, or 1/30th solar; M2 has an extended blue horizontal branch, while NGC 7006 has a relatively short horizontal branch that does not extend far beyond the RR Lyrae instability strip (see Fig. 18.6). This dichotomy has become known as the second-parameter effect. Other examples of second-parameter cluster pairs are M13 and M3, and NGC 288 and NGC 362.

The ELS model envisages rapid, free-fall collapse of the initial protogalactic gas cloud. Under those circumstances, one would not expect significant abundance gradients to develop since the timescale for collapse is short compared with the evolutionary timescale for recycling stellar ejecta in the ISM. Following ELS, alternative models were developed that involved slower, pressure-supported collapse (e.g. Yoshii and Saio 1979); under those circumstances an abundance gradient might develop. The existence of moderately metal-rich globulars in the inner Galaxy was well established by 1978, but the abundance distribution in the outer halo remained uncertain.

Searle and Zinn (1978) were among the first to undertake a major analysis that took advantage of the new information on the halo. They focussed on nineteen globular clusters, the majority of which lay in the outer parts of the Galactic halo. Employing a customised narrowband photometric system, they used red giants to determine the cluster abundances, sampling the CH and CN molecular bands in the giant's spectrum. The results showed no evidence for a systematic radial abundance gradient, arguing strongly against a slow, pressure-supported monolithic initial collapse. However, the clusters in the outer halo also exhibited diverse properties, particularly with regard to the second-parameter effect. This is in contrast to the inner halo clusters, which show little dispersion in morphology at a given metallicity.

There are three factors that drive horizontal branch morphology: metallicity (specifically C, N, O abundances), helium abundance and age (Rood and Iben 1968). Ruling out CNO variations, Searle and Zinn (1978) argued that age differences were more plausible than invoking some unknown additional mechanism for changing helium abundance without changing the overall metallicity. The required age differ-

Fig. 18.6 The second-parameter effect in globular clusters. *Upper panel*: V, B–V colour-magnitude diagram for the intermediate metallicity cluster M3, which has a relatively short horizontal branch. *Lower panel*: V, B–V colour-magnitude diagram for M13, which has a similar metallicity to M3, however the horizontal branch extends to bluer, hotter stars. Figure from Rey et al. (2001)

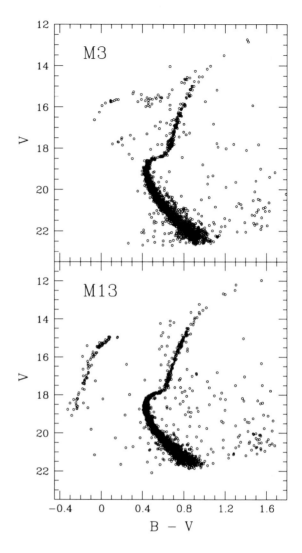

ences exceed 10^9 yr, and would therefore be incompatible with formation within an ELS-like initial collapse. Instead, Searle and Zinn (1978) argued for a more chaotic formation scenario, with the outer clusters accreted later in the Galaxys history; specifically, they hypothesised that those clusters '...originated in transient proto-galactic fragments that continued to fall into dynamical equilibrium with the Galaxy for some time after the collapse of its central regions had been completed'.

Following up on this work, Zinn (1980, 1985) extended observations to include 121 clusters, the majority of the Galactic population. Based on those data, he identified two subgroups: halo clusters, with typical Galactocentric distances exceeding 9 kpc, exhibiting a wide range in height above the Plane, predominantly metal-poor, with a high velocity dispersion and negligible net rotation; and disc clusters, confined within 5 kpc of the Plane with metallicities [Fe/H] > -0.8 dex, lower velocity dispersions and moderate rotation. Zinn's suggestion of a two-phase halo has been echoed by subsequent investigations, including analyses of the velocity distribution of nearby subdwarfs by Sommer-Larsen and Zhen (1990) and of the large-scale density distribution and kinematics of metal-poor stars in the SDSS (Beers et al. 2012). He also suggested that the disc clusters may have formed in a 'thick disc' phase of early Galactic evolution, while the halo clusters represent remnants of satellites accreted by the Milky Way after its initial collapse.

Stepping forward thirty years, satellite accretion is a key process in galaxy formation within the ΛCDM (cold dark matter) cosmological paradigm. Direct constraints on the power spectrum of the initial density functions from the cosmic microwave background, galaxy clustering and observations of the Lyman-α forest, suggest that structure formation is hierarchical (Bullock 2010). Dark matter simulations predict that the deep potential wells defined by the dark matter halos, the sites of future large galaxies, should be populated by hundreds of smaller dark-matter concentrations. Some will be accreted by the host galaxy; others may survive as ultra-faint dwarf galaxies (Brown et al. 2012), while still others may have masses that are too low to retain substantial baryonic material.

Satellite accretion and minor galaxy mergers, and their consequences, can be observed in many nearby galaxies. Indeed, the Milky Way itself is undergoing a merger at the present time. The dwarf galaxy in question lies towards the Galactic Bulge, and was only discovered by chance when a survey of the kinematics of the Bulge revealed an unusual feature—a significant number of stars, spanning a wide range of colour, at a specific velocity (Ibata et al. 1994, see Fig. 18.7). The stars were

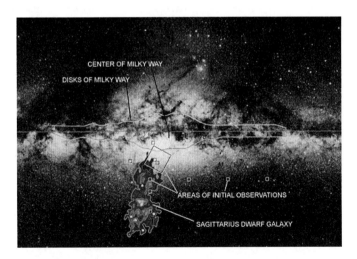

Fig. 18.7 The location of the Sagittarius dwarf galaxy. Figure from http://annesastronomynews. com/annes-picture-of-the-day-the-sagittarius-dwarf-elliptical-galaxy/. Image credit: R. Ibata (UBC), R. Wyse (JHU) and R. Sword (IoA)

identified as members of a dwarf galaxy lying beyond the Bulge, currently being torn apart by the Milky Way. Several globular clusters, including M54 and Terzan 7, are associated with the Sagittarius dwarf galaxy; analysis of data from the 2MASS survey has succeeded in tracing giant star members over more than 270° (Majewski et al. 2003); and theoretical models indicate that the system is likely to be in its third passage through the Milky Way's disc (Johnston et al. 2005).

Following Sagittarius' discovery, extensive searches were undertaken for evidence of past accretion events, and traces of other debris streams have been uncovered in analysis of SDSS data. In particular, several linear features have been uncovered in the so-called 'field of streams' (Belokurov et al. 2006). One such feature is the Monoceros ring (Casetti-Dinescu et al. 2006), which may have originated in the 'Canis Major dwarf', an over-density of stars in that region of the sky; an alternative possibility is that this feature has its origin in a (spatial) flare in the Galactic disc. Whatever the origin, these features are detectable because they lie at Galactocentric distances of ~ 17 kpc (Li et al. 2012), beyond the steep density decline that marks the edge of the disc.

What are the implications of these results for the origin and formation of the Milky Way? An important point to make is that both the Searle and Zinn model and its successors include an early monolithic collapse in the protogalaxy; that ELS component is responsible for forming the inner halo, and is supplemented by subsequent accretion of dwarf proto-galaxies to form some/most/all of the outer halo, with the accretion phase persisting even to the present day. The crucial question is how significant is the overall contribution to the halo from these accreted satellites?

The observed [α/Fe] ratios of stars in the present-day Milky Way dwarfs may offer some insight on that question. As Fig. 18.8 shows, those values are not consistent with observations of halo stars; the lower ratios imply that those stars formed in an ISM that had been enriched by Type I supernovae in addition to ejecta from short-lived, high-mass stars. Of course, present-day dwarf galaxies may not be representative of satellite galaxies that may have been absorbed earlier in the Galaxy's history; those dwarfs have existed as separate entities for a Hubble time, undergoing sustained periods of star formation. Moreover, while those galaxies appear to have a substantial dark matter component, there is no evidence of dark matter associated with any present-day globular cluster. Thus, if the outer globular clusters are the remnants of accreted dwarf galaxies, either the underlying dark matter was stripped through some mechanism and mixed within the Galactic halo, or the parent systems lacked such a component. In short, there is no question that accretion of satellite galaxies played a role in the formation of the Milky Way, but it may be that Sandage was correct in his characterisation of the Searle and Zinn model as merely ELS plus noise.

18.6 Globular Clusters Revisited

The three decades since Searle and Zinn's analysis have seen extensive observational and theoretical investigations of the properties of globular clusters. Many results from those studies are included in Bill Harris' 2001 Saas-Fee lectures on globulars. Most

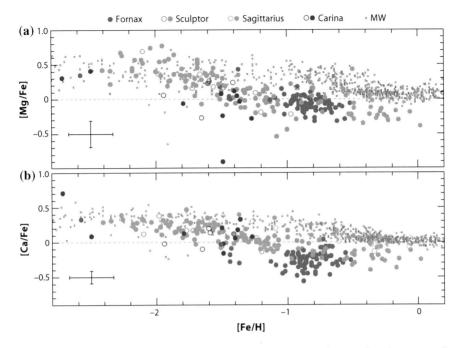

Fig. 18.8 α/Fe abundance ratio as characterised by measurements of magnesium (*upper panel*) and calcium abundance (*lower panel*) in today's dwarf spheroids compared with Milky Way stars. *Open circles* refer to single-slit spectroscopy measurements, while *filled circles* denote multi-object spectroscopy. Figure from Tolstoy et al. (2009)

observations contributing to that review were drawn from ground-based telescopes, but these are high star density systems, which can limit the potential for seeing-limited observations to probe key parameters. Space-based observations offer key advantages in depth and resolution. The initial imaging cameras on *Hubble*, Wide-Field Camera and Wide-Field Camera 2, had relatively small fields-of-view, but with the addition of the Advanced Camera for Surveys (ACS) in servicing Mission 3B (2004) and Wide-Fields Camera 3 (WFC3) in Serving Mission 4 (2009), HST has had the capability of providing observations of extraordinary sensitivity.

One of the most influential HST program for globular cluster studies involves an ACS two-colour (*V I*) imaging survey of 63 clusters, almost half the known sample (Sarajedini et al. 2007). Data from that program have been used to probe a wide variety of cluster properties, including the stellar mass function (Paust et al. 2010), the binary fraction (Milone et al. 2012), relative cluster ages and the second-parameter problem (Marín-Franch et al. 2009). The mass function results show a strong correlation in slope with cluster mass, underlining the importance of dynamical evolution; none of the results indicate a mass function steeper than that of the local disc stars. The binary fraction is anti-correlated with cluster luminosity, again indicating that mass segregation and tidal stripping have played a role, preferentially stripping single stars in the less massive clusters as the binaries sink towards the cluster centre. Finally, the consensus analysis points towards age as the second-parameter, with the relative age distribution indicating that most clusters formed almost coevally, consistent with an ELS-style monolithic collapse, with a subset that appear to have significantly younger ages, more consistent with the Searle and Zinn formation model (see Fig. 18.9).

Most significantly, HST observations have prompted a radical change in our understanding of the fundamental nature of (many) globular clusters. Classically, globulars were described as simple stellar population, the product of a single, rapid burst of star formation within an isolated gas cloud in the Milky Way's protohalo, with coeval stars of uniform metallicity and helium abundance. Stellar winds generated by either high-mass stars or supernovae explosions overcome the cluster self-gravity and sweep it clear of gas, eliminating the potential for subsequent star forming episodes. That simple picture has cracked in recent years.

ω Centauri, lying towards the Small Magellanic Cloud, represents the thin end of the wedge. The most massive globular cluster, this system has long been known to possess stars spanning a substantial range in metallicity. Red giant cluster members exhibit a significant range in CH and CN bandstrength (Dickens and Bell 1976), RR Lyraes show a mix of pulsational properties (Caputo and Castellani 1975) and main-sequence stars span metallicities spanning almost an order of magnitude, $-1.9 <$ [Fe/H] < -1 dex (Stanford et al. 2007). This diversity is clearly inconsistent with a single starburst, and suggests that the cluster is either the product of a merger, or that the parent entity was sufficiently massive that it could sustain multiple bursts of star formation. Indeed, as early as 1975, there were suggestions that ω Cen might be the remnant core of a dwarf galaxy (Freeman and Rodgers 1975).

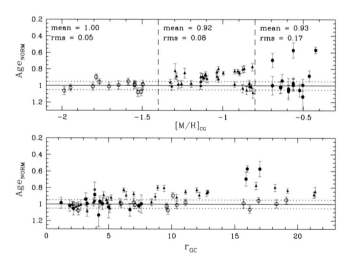

Fig. 18.9 Normalised globular cluster ages as a function of [M/H] (*upper panel*) and galactocentric distance (*lower panel*). *Open circles, filled triangles* and *filled circles* represent globular clusters within low-, intermediate- and high-metallicity groups respectively. Figure from Marín-Franch et al. (2009)

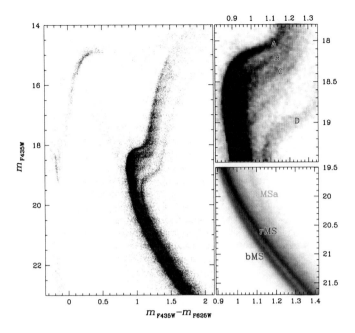

Fig. 18.10   The complex colour-magnitude diagram of ω Cen illustrating the main four sub-giant branches (*upper right panel*) and triple main-sequence (*lower right panel*). Figure from Bellini et al. (2010)

More detailed spectroscopic observations of red giants in other globulars were acquired through the 1970s and early 80s, and it became apparent that there was a dichotomy in CN and CH bandstrengths in a significant number of clusters (see Kraft 1994 for a review). In addition, evidence began to accumulate for an anti-correlation between oxygen and sodium abundances (Cohen 1978; Peterson 1980); Na-rich stars tend to also be CN-strong. These results sparked lively discussion on the origin, with the initial debate focused on either primordial variations in the protocluster gas (nature) or internal mixing of nucleosynthetic products within extended red giant atmospheres (nurture). The identification of carbon and nitrogen abundance variations among main-sequence stars (Suntzeff 1989), which are not expected to undergo extensive internal mixing, argued against the latter option. Similarly, mixing cannot account for the Na-O anti-correlation; that phenomenon can be explained if Na is synthesised from Ne within an environment deep within massive stars where oxygen is being depleted in the ON cycle (Gratton et al. 2004). That, in turn, requires that cluster stars form from material that has been polluted by ejecta from a previous stellar generation or generations.

HST provided the key observations that crystallised this debate. The initial hints came from analyses of WFPC2 observations of ω Cen (Bedin et al. 2004), which revealed that the main-sequence clearly became bimodal 1–2 mag below the turn-off. The full complexity of the cluster only became apparent with the application of refined photometric analysis techniques to the wider-field ACS and WFC3 observations. Figure 18.10 shows the extremely complex nature of this cluster with the high precision WFC3 data revealing as many as 5 distinct stellar sequences (Bellini et al. 2010), more than in many dwarf galaxies. There is a clear variation in age, with separation between the turn-offs, as well as abundance variations. Moreover, the bluer main-sequence has higher metallicity than the red, requiring that the former should have substantially enhance helium abundances, potentially as high as $Y = 0.38$ (Piotto et al. 2005). As with the Na-O anti-correlation, the additional helium can be generated in hot-H burning in massive stars; moreover, helium-rich stars evolve faster, leading to a lower turn-off mass for the same age and lower-mass, hotter (bluer) stars on the horizontal branch. The hypothesis that one could produce this complex morphology through external events, such as mergers of separate protoclusters or external pollution from nearby clusters, requires a sequence of exceedingly improbably events. The simplest explanation is that ω Cen underwent multiple star-forming episodes, and that it was sufficiently massive that it was able to retain a fraction of the original gas content, which was then polluted and enriched by stellar ejecta from the first generation stars before forming a second (and third and fourth) generation population.

Subsequent observations have shown that many globular clusters harbour multiple stellar populations (Piotto 2009). The initial focus was on massive clusters, such as NGC 2808, NGC 1851, NGC 6388 and NGC 6566 (M22), but subsequent observations have shown that anomalies are present in lower-mass clusters. Indeed, even the archetypical metal-poor globular NGC 6397 ([Fe/H] $= -1.9$ dex) shows evidence for main-sequence bimodality (Milone et al. 2012) and, to date, all globulars studied in sufficient detail exhibit the Na-O anticorrelation indicative of early pollution (Carretta et al. 2009).

How does this translate to a model for the origin of globular clusters within the earliest stages of formation of the Milky Way? Gratton et al. (2012) have proposed an interesting scenario. Taking a leaf from cosmological simulations, they envisage the Milky Way's halo forming by accretion of many smaller, gas-rich systems in the early universe. The interactions lead to substantial starbursts within the fragments, with the first stars that form producing a rapid increase in the metallicity of the system. The initial stars trigger a strong burst of star formation in the protoclusters. Massive stars and supernovae from that initial burst removes most of the residual gas, but the protoclusters are sufficiently massive that enough gas is retained to permit the formation of a second (and, for more massive systems, third) generation, whose stars are O-depleted, Na-rich and He-rich, thanks to the ISM pollution from the first stars.

What about the field halo? Approximately 2.5 % of field subdwarfs show evidence for Na-O anticorrelation, suggesting that those stars were formed as second (or later) generation cluster starbursts. Those stars entered the field through standard dynamical evolution processes, such as two-body encounters within the cluster and disc-shocking and encounters with massive objects external to the cluster. Looking at present-day clusters, second generation stars are estimated to contribute approximately two-thirds of the stellar complement, leaving the first generation stars as a minority constituent (Carretta et al. 2009). If we assume similar evaporation rates, the first generation cluster stars that were present when the second generation formed would contribute approximated 1 % of field halo. The cluster system itself contributes approximately 1 % of the mass of the halo ($\sim 10^7$ M$_\odot$); thus, ejected and retained globular cluster stars constitute at least 5 % of the mass of the current halo.

However, first generation cluster stars may make a larger contribution to the field. There are strong arguments that those stars must have been substantially more populous if they were to generate sufficient mass-loss to pollute the ISM to match the abundances observed in the second generation stars. Estimates are far from precise, but the case can be made the first generation of cluster stars had to be at least 5 to 10 times more populous than the second generation, with a correspondingly higher evaporation into the field. In that case, the original protoglobulars might well have constituted 25–50 % of the total halo mass. Indeed, given the uncertainties, the original globular clusters might be the parent star-forming sites for all of the stars in the Milky Way's present-day halo (see Gratton et al. 2012 for a more extensive discussion).

A major complication for this scenario is the absence of dark matter in present-day globulars. If one constitutes the initial formation phase as accretion of fragments, à la ΛCDM, one might expect those fragments to mimic current-day dwarf galaxies and have substantial dark matter content. One means of mitigating this issue would be to envisage collisional interactions either between the gas-rich fragments or between the accreted material and gas in the proto-Milky Way. Dark matter would not participate in these dissipational interactions, and would therefore part ways from the gas.

Alternatively, one might imagine protoglobulars as high-density regions within the proto-Galaxy itself, high-mass baryonic concentrations within the overall gravitational potential defined by the Milky Way's dark matter component. The initial star formation and subsequent evolution occurs as the overall system undergoes gravitational collapse, giving a cluster system whose members have very similar ages, consistent with the narrow relative age distribution for the majority of clusters (see Fig. 18.10). Younger clusters would be acquired through subsequent satellite accretion. This scenario thus envisages a lumpy version of ELS as the dominant factor in forming the Milky Way, with a more chaotic initial collapse, with a mild sprinkling of Searle and Zinn in later years. Overall, to echo Sandage, the Milky Way formation process is ELS plus noise.

18.7 Endword

The paper authored by Eggen, Lynden-Bell and Sandage has been extremely influential in shaping discussion over the years, and its results and conclusions continue to play an important role in studies of galaxy formation. As Lynden-Bell (2012) has pointed out, ELS remains the highest-cited paper for each of the three authors, a notable achievement given the competition in each case. As the first serious attempt to reconstruct the overall picture of the formation of the Milky Way, it set the scene for what has become known as galactic archaeology. The concepts of galaxy formation through large-scale collapse and rotational support within a disc still play a key role in the modern view of galaxy formation, albeit supplemented with the additional complexities of subsequent multiple mergers with lower-mass systems.

References

Abell, G. O. 1955, PASP, 67, 258

Alpher, R. A., Bethe, H., & Gamow, G. 1948, Phys. Rev., 73, 803

Baade, W. 1944, ApJ, 100, 137

Bedin, L. R., Piotto, G., Anderson, J., et al. 2004, ApJ, 605, L125

Beers, T. C., Carollo, D., Ivezić, Ž., et al. 2012, ApJ, 746, 34

Bellini, A., Bedin, L. R., Piotto, G., et al. 2010, AJ, 140, 631

Belokurov, V., Zucker, D. B., Evans, N. W., et al. 2006, ApJ, 642, L137

Bennett, C. L., Larson, D., Weiland, J. L., et al. 2013, ApJS, 208, 20

Blaauw, A. 1995, in IAU Symposium, Vol. 164, Stellar Populations, ed. P. C. van der Kruit & G. Gilmore, Kluwer Academic Publishers, 39

Brown, T. M., Tumlinson, J., Geha, M., et al. 2012, ApJ, 753, L21

Bullock, J. S. 2010, in Canary Islands Winter School of Astrophysics on Local Group Cosmology, ed. D. Martínez-Delgado, Cambridge University Press

Burbidge, E. M., Burbidge, G. R., Fowler, W. A., & Hoyle, F. 1957, Reviews of Modern Physics, 29, 547

Caputo, F. & Castellani, V. 1975, Ap&SS, 38, 39

Carney, B. W., Latham, D. W., Laird, J. B., & Aguilar, L. A. 1994, AJ, 107, 2240

Carretta, E., Bragaglia, A., Gratton, R. G., et al. 2009, A&A, 505, 117

Casetti-Dinescu, D. I., Majewski, S. R., Girard, T. M., et al. 2006, AJ, 132, 2082

Chamberlain, J. W. & Aller, L. H. 1951, ApJ, 114, 52

Cohen, J. G. 1978, ApJ, 223, 487

D'Antona, F., Caloi, V., & Mazzitelli, I. 1997, ApJ, 477, 519

Dickens, R. J. & Bell, R. A. 1976, ApJ, 207, 506

Eggen, O. J., Lynden-Bell, D., & Sandage, A. R. 1962, ApJ, 136, 748

Freeman, K. C. & Rodgers, A. W. 1975, ApJ, 201, L71

Frenk, C. S. & White, S. D. M. 1980, MNRAS, 193, 295

Gizis, J. E. & Reid, I. N. 1999, AJ, 117, 508

Gratton, R., Sneden, C., & Carretta, E. 2004, ARA&A, 42, 385

Gratton, R. G., Carretta, E., & Bragaglia, A. 2012, A&A Rev., 20, 50

Gratton, R. G., Fusi Pecci, F., Carretta, E., et al. 1997, ApJ, 491, 749

Harris, W. E. 2001, Star Clusters: Saas-Fee Advanced Course 28, ed. Labhardt, L. and Bingelli, B., Springer

Harris, W. E. 2010, arXiv:1012.3224. Available at http://physwww.physics.mcmaster.ca/~harris/mwgc.dat

Hoyle, F. 1948, MNRAS, 108, 372

Ibata, R. A., Gilmore, G., & Irwin, M. J. 1994, Nature, 370, 194

Johnston, K. V., Law, D. R., & Majewski, S. R. 2005, ApJ, 619, 800

Kraft, R. P. 1994, PASP, 106, 553

Kuiper, G. P. 1939, ApJ, 89, 548

Li, J., Newberg, H. J., Carlin, J. L., et al. 2012, ApJ, 757, 151

Lynden-Bell, D. 2012, Biogr. Mems Fell. R. Soc., 58, 245

Majewski, S. R., Skrutskie, M. F., Weinberg, M. D., & Ostheimer, J. C. 2003, ApJ, 599, 1082

Marín-Franch, A., Aparicio, A., Piotto, G., et al. 2009, ApJ, 694, 1498

Milone, A. P., Piotto, G., Bedin, L. R., et al. 2012, A&A, 540, 16

Oort, J. H. 1928, Bull. Astron. Inst. Netherlands, 4, 269

Paust, N. E. Q., Reid, I. N., Piotto, G., et al. 2010, AJ, 139, 476

Peterson, R. C. 1980, ApJ, 237, L87

Piotto, G. 2009, in IAU Symposium, Vol. 258, The Ages of Stars, ed. E. E. Mamajek, D. R. Soderblom, & R. F. G. Wyse, Cambridge University Press, 233

Piotto, G., Villanova, S., Bedin, L. R., et al. 2005, ApJ, 621, 777

Reid, I. N. 1997, AJ, 114, 161

Rey, S.-C., Yoon, S.-J., Lee, Y.-W., Chaboyer, B., & Sarajedini, A. 2001, AJ, 122, 3219

Roman, N. G. 1954, AJ, 59, 307

Rood, R. & Iben, Jr., I. 1968, ApJ, 154, 215

Sandage, A. 1956, PASP, 68, 498

Sandage, A. R. & Eggen, O. J. 1959, MNRAS, 119, 278

Sarajedini, A., Bedin, L. R., Chaboyer, B., et al. 2007, AJ, 133, 1658

Searle, L. & Zinn, R. 1978, ApJ, 225, 357

Sommer-Larsen, J. & Zhen, C. 1990, MNRAS, 242, 10

Stanford, L. M., Da Costa, G. S., Norris, J. E., & Cannon, R. D. 2007, ApJ, 667, 911

Suntzeff, N. B. 1989, in The Abundance Spread within Globular Clusters: Spectroscopy of Individual Stars, ed. G. Cayrel de Strobel, 71.

Tolstoy, E., Hill, V., & Tosi, M. 2009, ARA&A, 47, 371

Yoshii, Y. & Saio, H. 1979, PASJ, 31, 339

Zinn, R. 1980, ApJ, 241, 602

Zinn, R. 1985, ApJ, 293, 424

Chapter 19
Star Formation over Time

I. Neill Reid

19.1 Introduction

The bulk of the Galactic halo formed over a relatively short period between $\sim$12 and 13 Gyr ago. Since that time, the Galactic disc has been the main site of star formation in the Milky Way. The disc is the dominant stellar component in the Galaxy, perhaps fifty times more massive than the halo and accounting for $\sim$85 % of the baryonic matter. A substantial fraction of that matter remains in gaseous form, enabling the extensive star formation evident at the present time. This chapter gives a broad overview of our understanding of the star formation history of the disc. In particular, I consider the thick disc, and insights that may be gained regarding its formation. I also discuss the various techniques that can be used to estimate stellar ages, and the conclusions that can be drawn regarding the age distribution and the metallicity enrichment history of stars in the disc.

19.2 The Structure of the Disc

Chapter 15 included a summary of the large-scale properties of the disc. In particular, we noted the complex density structure exhibited perpendicular to the Plane. The basic result derived originally by Gilmore and Reid (1983), an excess stellar density at $z \gtrsim$ 1500 pc over a simple exponential, has been reproduced by many subsequent analyses, although, as previously noted, the interpretation remains ambiguous.

The most extensive investigation to date, by Jurić et al. (2008), uses multicolour data for $\sim$48 million stars from the Sloan Digital Sky Survey (SDSS) to derive structural parameters for the disc and halo. Characterising the disc as a two-component

I.N. Reid (✉)
Space Telescope Science Institute, Baltimore, MD, USA
e-mail: inr@stsci.edu

© Springer-Verlag Berlin Heidelberg 2015
C.P.M. Bell et al. (eds.), *Dynamics of Young Star Clusters and Associations*,
Saas-Fee Advanced Course 42, DOI 10.1007/978-3-662-47290-3_19

system, the starcounts are consistent with vertical scaleheights of 300 and 900 pc for the thin and thick disc, with a local normalisation thick/thin $\sim$12 %. Accepting this characterisation, the total mass of the thick disc is approximately one third that of the thin disc. Thus, the implication is that the thick disc contributes roughly 25 % of the total mass of the disc, or $\sim$1.25 $\times$ 10^{10} M$_\odot$. Indeed, whether one views the thin/thick disc as a continuum or a dichotomy, observations indicate that these stars are likely the product of a major event in the evolution of the Milky Way.

The intrinsic properties of the thick disc stars give important clues to their origin. From the outset, it is has been clear these stars represent an old population. Starcounts show that A and F stars are closely confined to the Plane; considered as a population, the colour-magnitude diagram for stars at $|z| > 1.5$ kpc has a turn-off at $B - V \sim$ 0.60 mag, corresponding to an age of $\sim$10 Gyr. This is broadly consistent with the age of the disc, and initial hypotheses tended to equate the thick disc with the oldest stars in the disc. The vertical density distribution implies a velocity dispersion of 30–40 km s^{-1} perpendicular to the Plane. This is extremely difficult to achieve through conventional scattering mechanisms (see Chap. 17), but could be the consequence of a major merger early in the history of the Milky Way.

The spatial distribution of the thick disc stars indicates that they are members of a rotationally support population. As a group, thick disc stars have been characterised as having a rotational lag (relative to the Sun) of $V \sim -30$ to -60 km s^{-1} and higher velocity dispersions than the thin disc; for example, Chiba and Beers (2000) derive $[\sigma_U, \sigma_V, \sigma_W, = 46, 50, 35$ km s$^{-1}]$ in their analysis of metal poor stars. The abundance distribution is skewed towards sub-solar metallicities, possibly extending to [Fe/H] ~ -1.5 dex. Deriving the overall characteristics, however, has proven difficult since those intrinsic properties overlap significantly with stars generally associated with the thin disc. That overlap in properties led to early suggestions that it may be more appropriate to represent the thick disc as the metal-poor tail of the thin disc, and this ambiguity in interpretation persists at some level to the present day.

The strongest observational evidence favouring the thick disc as a separate population lies in the detailed elemental abundance derived by several research groups, notably Fuhrmann (1998, 2004) and Bensby et al. (2003), for higher velocity stars in the Solar Neighbourhood. Figure 19.1 shows that the [α/Fe] distribution for those stars appears to exhibit two distinct sequences in the metallicity range $-0.6 <$ [Fe/H] < -0.2 dex. As described by Fuhrmann (2004), the less α-enhanced sequence is defined by stars with relatively low velocities relative to the Sun, and therefore likely represents the metal-poor tail of the thin disc; the overwhelming majority of stars in the higher α-enhanced sequence have space motions that are substantially lower than those expected for halo stars, placing them on the outskirts of the disc velocity distribution. The higher α-enhanced stars form a rotationally-supported population, and Fuhrmann and Bensby et al. identify these stars with the thick disc.

The strong α-enhancement suggests that the thick disc stars formed during a rapid burst of star formation early in the lifetime of the Milky Way. The disconnect between the thick disc and thin disc sequences also suggests a lull in star formation

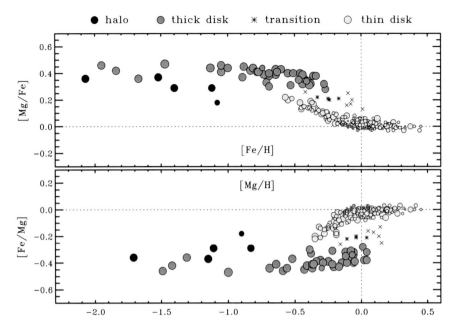

Fig. 19.1 *Upper panel*: Abundance ratio [Mg/Fe] versus [Fe/H] for nearby F and G stars. *Lower panel*: Same as upper panel but with Mg as the reference element. Circle diameters correspond to the age estimates of the stars, with the smallest diameters indicating the youngest stars. Figure from Fuhrmann (2004)

activity, allowing massive stars to evolve through the Type Ia supernova phase and enhance the iron abundance in the ISM before the formation of the earliest stars in the thin disc. These abundance-related constraints and the observed spatial and velocity distributions need to be taken into account in theoretical attempts to identify potential formation mechanisms.

Formation theories generally fall into two categories: formation of the thick disc in situ early on in the Milky Way's history, while material is still collapsing into the disc; and excitation due to an injection of energy from an external source, i.e. a merger. As an example of the latter approach, Abadi et al. (2003) envisage the thick disc being constituted of remnants of accreted satellites, merging with the Milky Way as part of the ΛCDM galaxy formation scenario. Brook et al. (2004) suggest that the thick disc forms in situ during the Milky Way's merger with of a gas-rich satellite. In contrast, Villalobos and Helmi (2009) postulate that the thick disc forms as a result of the accretion of one or more satellite galaxies; in this case, however, the present-day thick disc stars represent the disrupted members of the pre-merger thin disc. Villalobos and Helmi (2009) estimated that a satellite with mass 10–20 % that of the host galaxy heats the disc sufficiently to produce a thick disc. The second and third scenarios offer potential means of quenching on-going star formation and producing the two α-element sequence.

Schönrich and Binney (2009) have taken a different approach to examining the potential for in situ formation. They constructed a full-scale Galactic chemical evolution model, with star formation rates and metal production, coupled with dynamical interactions, including scattering by massive bodies (Spitzer-Schwarzschild) and resonance interactions (orbit migration through change in angular momentum). As discussed in Chap. 17, radial migration has the potential to bring stars from the innermost regions of the Galaxy to the Solar Neighbourhood. The higher surface mass densities in the inner Galaxy lead to higher velocities, and Schönrich and Binney (2009) suggest that the local thick disc component is contributed by the oldest inner disc stars that have migrated to the Sun's locality.

Taking this argument a step further, Bovy et al. (2012a, b) have recently analysed low-resolution spectroscopic data for 30,353 G stars from the SEGUE survey (Yanny et al. 2009). The spectroscopic data provide abundances and moderate-accuracy velocities, which the SDSS photometric data enable distances to be estimated from photometric parallax. The SEGUE sample is limited to stars at distances $z > 200$ pc above the Plane, and therefore does not include significant numbers of young disc stars. Bovy et al. (2012a, b) find that the data show a smooth progression in velocity dispersion with metallicity, with a broad correlation between [α/Fe] and [Fe/H]; subdividing by abundance gives sub-groups that are isothermal with height above the Plane, implying that the stars are well-mixed, i.e. relaxed and in equilibrium with the gravitational potential; the vertical velocity dispersion appears to increase with decreasing radius, with an e-folding scalelength of $\sim$7 kpc; and the radial scalelength of the density distribution decreases with decreasing abundance, implying that metal-poor disc stars are predominantly concentrated towards the Galactic Centre.

Synthesising these results, Bovy et al. (2012a, b) suggest that the disc shows a continuous distribution of properties, and the stars that we call the thick disc are simply the oldest and (usually) the most metal-poor stars from that continuum. Taking metallicity as a proxy for age, they argue that the oldest disc stars formed in a more centrally concentrated distribution, consistent with a general picture of disc galaxies forming from the inside out. As in the Schönrich and Binney (2009) scenario, the older disc stars have higher velocity distributions due to the higher mass density in the inner regions of the Galaxy, and they retain those dispersions as radial migration disperses them through the Milky Way.

Invoking radial migration coupled with a continuum of properties for the Galactic disc offers the advantage of a conceptually simple scenario, providing a potential mechanism of accounting for the hotter kinematics and spatial distribution of thick disc stars. However, the devil lies in the details, and it remains unclear whether the velocity distribution in the inner Galaxy can match the properties of the local thick disc. Moreover, the migration scenario fails to account for the distinct separation between thin disc and thick disc stars in the [α/Fe]/[Fe/H] plane (as in Fig. 19.1). That characteristic strongly suggests that these stars are drawn from separate populations, rather than from a spatially-dispersed continuous distribution. In summary, the events leading to the formation of the thick disc were clearly significant in the evolution of the Milky Way, but remain a matter of some considerable conjecture.

19.3 Measuring Stellar Ages

The formation of the thick disc represents an important discrete event in the history of the Milky Way, likely tied to stars within a relatively narrow range of ages. Identifying other significant star formation episodes over the past history of the Milky Way depends on our ability to estimate reliable ages for stars. That ability depends strongly on whether the star is still a member of its natal cluster, or whether it has escaped, alone, into the vasty deep. This is a very complex subject that can only be touched on in a single chapter; interested readers are referred to the recent IAU Symposium 258 and the thorough review article by my STScI colleague Dave Soderblom that go into much more detail on this important subject (Soderblom 2010).

19.3.1 Ages for Stars in Clusters

Determining ages for stars in intermediate-age or old clusters is a relatively straightforward process, as discussed in the context of globular clusters in the previous chapter. Given a sufficiently populous cluster, accurate photometry of members (and field stars) allows one to define the colour-magnitude diagram. Given a measure of the foreground reddening, the colour-magnitude diagram can be matched against theoretical isochrones to estimate the age, metallicity and distance (see Fig. 19.2). Further constraints can be drawn from inferred quantities, such as luminosity and temperature in the Hertzsprung-Russell (H-R) diagram, and the mass radius relation. Spectroscopic measurements can also be used to refine the metallicity estimate. The age estimate is generally set by matching the turn-off, and it is worth emphasising that the detailed physics included in the theoretical models (for example, how one treats convection, convective overshoot and other forms of internal mixing) can lead to systematic differences in the predictions different sets of models.

The situation is more complicated for young clusters, particularly those that are still partially embedded within the parental molecular cloud. In addition to the potential for small-scale variations in differential reddening across the cluster, individual members may be reddened to greater or lesser degrees by circumstellar material, and photometry can be affected by emission from accretion within the protoplanetary discs. Binary and multiple stars also increase the width, and hence the apparent dispersion of intrinsic properties, of the main-sequence; the same situation holds in older clusters, but it is usually easier to identify those stars as outliers in the cleaner CMDs in those systems. Star formation is unlikely to be instantaneous, and a significant (1–2 Myr) dispersion in ages will clearly have a greater proportional effect in younger clusters. All of these effects lead to greater uncertainties in both the intrinsic photometry for young stars and in the fundamental parameters, such as luminosity and temperature, inferred from observations. As a result, H-R diagrams inferred for young clusters tend to show substantial breadth (see e.g. Fig. 19.3). Finally, calculating pre-main-sequence isochrones from theoretical models is a process fraught

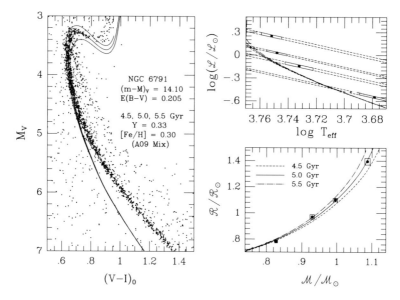

Fig. 19.2 *Left panel*: M_V, $(V-I)_0$ de-reddened colour-magnitude diagram for the old open cluster, NGC 6791, with theoretical model isochrones overplotted. *Upper right panel*: Hertzsprung-Russell diagram of the individual components of two member eclipsing binaries compared to the same theoretical models. *Lower right panel*: Same as upper right panel but showing the mass-radius relation of the binary components. The best-fit age is ~5 Gyr. Figure from Brogaard et al. (2011)

with significantly more concerns than generating isochrones for old or middle-aged star clusters, and cluster age estimates are less reliable and more likely to exhibit systematic differences between different model formulations.

19.3.2 Ages for Individual Stars

Determining ages for star clusters can be difficult; estimating ages for individual stars can be almost impossible. Stars evolve through different phases at different rates that are dependent on several factors, but primarily mass. In clusters, one is dealing with an ensemble of stars spanning a wide range of mass, increasing the likelihood that one or more stars will be found in relatively short-lived evolutionary states, making the transition from one phase to the next. It is those few stars that provide the strongest constraints on the age of the ensemble.

In general, field stars are unlikely to be caught in transition through a short-lived evolutionary phase. There are four techniques that can be employed to estimate ages: measuring the rotational properties of the star; measuring the level of stellar activity, either through X-ray observations of coronal emission or estimation of chromospheric activity through measurements of optical/UV emission lines and UV

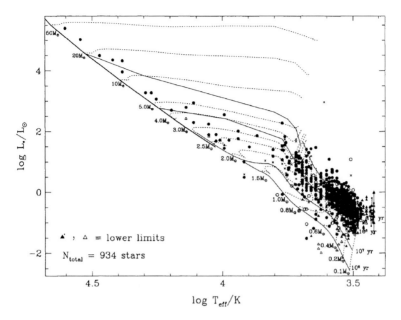

Fig. 19.3 The Hertzsprung-Russell diagram for the ∼1–3 Myr-old Orion Nebula Cluster. The *solid lines* represent model isochrones with ages of 0.1, 1, 10 and 100 Myr, while the *dotted lines* denote mass tracks in the range 0.1–50 M_⊙. Figure from Hillenbrand (1997)

continuum flux; asteroseismology, using the pulsational properties to set constraints on the mean molecular weight, and hence the helium content; and isochrone fitting, matching the luminosity, effective temperature and metallicity against theoretical models. Each of these techniques has its own strengths and weaknesses, and most become less effective the older the star.

Rotational age determination, or gyrochrononology, rests on the general principle that stars gradually lose angular momentum as they age, probably due to magnetic braking as the intrinsic magnetic field interacts with the stellar wind. The technique was introduced by Barnes (2007), who derived a conceptual calibration using observations of FGKM stars in clusters spanning ages from ∼30 Myr (IC 2391, IC 2602) to ∼ 600 Myr (Hyades, Coma). Computing the quantity $P/t^{1/2}$, where P is the rotational period and t the cluster age, and plotting that quantity against $B - V$ colour, Barnes (2007) found that stars tended to cluster around a broad sequence (see left panel of Fig. 19.4). The dispersion about that relation corresponds to age uncertainties between 13 and 20 %, depending on the colour, although there are obviously a significant number of outliers. Mamajek and Hillenbrand (2008) have refined the calibration using additional data and extended coverage to ages of several billion years (see right panel of Fig. 19.4). They estimate that, on average, their calibration can give ages accurate to ∼20 % for stars younger than the Sun.

There are clearly a number of drawbacks in this method. For a start, measuring rotational periods is not a simple task. If the star is moderately active, then it may be

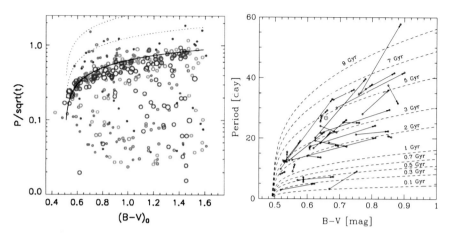

Fig. 19.4 Gyrochronology isochrones. *Left panel*: The correlation between the observed rotation periods of individual member stars in 10 clusters and $B - V$ colour (note the periods have been normalised by the square root of the cluster age). The *solid line* represents the calibrating relation (from Barnes 2007). *Right panel*: Predicted rotation periods for field binary stars. The dashed lines denote proposed period-$B - V$ gyrochrones (from Mamajek and Hillenbrand 2008)

possible to discern the rotational period by measuring periodic photometric variations as spots move in and out of the field of view. This generally requires millimagnitude-level photometry, and is less effective in older stars where activity is generally at lower levels. The alternative is measuring the rotational velocity by modelling spectral line widths: that approach require high signal-to-noise, high-resolution spectra, and only permits measurement of the line-of-sight component of the stellar rotation; one also requires an estimate of the stellar radius to derive the rotation period. Finally, there is obviously substantial intrinsic scatter in individual rotational periods that runs beyond the rms scatter estimated for the mean relations, which adds further uncertainty to deriving ages for individual field stars.

Stellar activity can enable measurement of rotation periods, but also serves as an age indicator in its own right. The activity is produced by non-thermal processes associated with stellar magnetic fields permeating the outer stellar atmosphere. In Sun-like stars (F to mid-M), the magnetic field is generated by an (α, Ω) dynamo, produced by shear at the boundary between the outer convective envelope and the inner rotational core (Babcock 1961). That feature does not exist in later-type stars, which are fully convective, and some 40 years ago the general expectation was that activity levels would be weak or absent in such stars. However, even at that juncture it was known that later-type M dwarfs were subject to spectacular flares (including the M8 dwarf, VB 10; Herbig 1956). Observations with the *Einstein* satellite demonstrated that there is no break in X-ray coronal activity as stars become fully convective (Vaiana 1980). It is now believed that magnetic fields, and stellar activity, in those stars is generated by shear between convective cells (Reid and Hawley 2005).

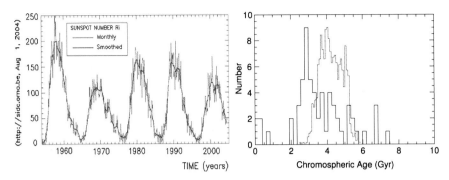

Fig. 19.5 *Left panel*: The variation in solar activity, as measured by sunspot counts (figure taken from http://sidc.oma.be). *Right panel*: The solid histogram shows the distribution of chromospheric activity in M67 G dwarfs, while the dotted histogram shows the range of activity corresponding to the solar cycle (from Giampapa et al. 2006)

Magnetic activity manifests itself in X-rays from stellar coronae, UV continuum flux and emission lines such as Ca II H & K, neutral alkaline lines, Balmer emission (notably Hα) and UV lines such as Mg II (2800 Å). The average level of activity generally declines with time, as one might expect if the driving force is rotational shear, and the rate of decline is faster in higher mass stars. Close binaries maintain rapid rotation through tidal locking and can mimic the activity levels of younger stars. As with rotation, calibrations rest on observations of stars in open clusters of known age, and mean relations have been derived for various parameterisations of Ca II H & K activity (e.g. Donahue 1993; Rocha-Pinto and Maciel 1998) and Hα emission (Gizis et al. 2002). Those different relations, however, can offer significantly different age estimates for the same star (see the discussion in Reid et al. 2007).

The principal caveat in using activity to determine age lies in the substantial intrinsic dispersion exhibited by stars of similar age. As with rotation, the average activity level is likely to vary from star to star; to further complicate the picture, we know from many years of detailed observation of the Sun that activity fluctuates on a timescale of years for individual stars. As with almost every other age parameter, these variations, global and individual, are likely to produce the highest uncertainties among the older stars. This is highlighted in Fig. 19.5, which shows the range of solar activity as measured by the sunspot cycle. Giampapa et al. (2006) have translated those cyclical variations to give the range of chromospheric (Ca II H & K) activity, and matched that distribution against measurements of Sun-like stars in the solar-aged open cluster, M67. Main-sequence fitting gives an age of ∼4.5 Gyr for that cluster, directly comparable with the Sun (see Chap. 21). In brief, applying standard age calibrations to the solar data would give age estimates between 2 and 6 Gyr, depending on the phase of the Solar cycle. Applying the same calibrations to the M67 stars gives individual ages between 1 and 7 Gyr.

Asteroseismology is the analysis of stellar pulsations (Gough 2003). As discussed further in Chap. 21, these pulsations are driven by acoustic waves, or p-mode oscillations. The frequencies depend on the internal sound speed which, in turn, depends on the mean molecular weight. The primary factor that determines the latter quantity is the relative abundance of hydrogen and helium in the stellar core which, in turn, depends on the how much helium has been produced by nucleosynthesis, and hence the age of the star.

Asteroseismology can provide very precise ages, but it also requires observations that are both accurate and precise. Stellar pulsations in solar-type stars have typical periods of minutes, so measuring those quantities demands high-accuracy, high time-cadence photometry, which limits observations to bright stars and requires observations with remarkable photometric stability. For the Sun and a handful of other stars, those criteria can be met with ground-based observations. Space offers greater photometric stability, and HST has been used to observe a few stars, including the $\sim$1 trillion photon observation of the transiting planet host, HD 17156 (Gilliland et al. 2011). However, dedicated satellite observatories such as MOST, COROT and, particularly, *Kepler* (Borucki et al. 2010) have proven most effective at providing the high-quality data required for asteroseismology (Huber et al. 2013). Given data with sufficient time resolution, this technique has the potential to measure stellar ages to an accuracy of a few percent.

Finally, the isochrone fitting process for single stars is similar to that for clusters, with the exception that one cannot fit for age and metallicity given a single point in the Hertzsprung-Russell or colour-magnitude diagram plane. Given an accurate estimate of the metallicity, however, one can match an individual star against a suite of theoretical isochrones and identify the best fit. As one might expect, the accuracy of this method is higher, in absolute terms, for higher mass stars with shorter evolutionary lifetimes. Moreover, any systematic errors in the abundance estimates will lead to systematic errors in the ages.

19.4 The Age-Metallicity Relation and the Age Distribution of Local Stars

The age distribution of local stars probes the past star formation activity in the Galactic disc, while the age-metallicity relation offers insight into the self-enrichment history. Given ages and metallicities for an ensemble of field stars, one can derive both of these quantities. The usual caveats apply: the mean relations are only as reliable as the input data; systematic errors in either age or metallicity will translate to systematic errors in the derived mean relations; and, most important, systematic bias in constructing the stellar reference sample will lead to systematic bias in the derived results.

19.4.1 Stellar Metallicities

Chapter 15 touched on the measurement of stellar metallicities. As we noted there, stellar abundances are usually measured relative to the Sun and various measurement techniques are available to the observer. Those techniques rest on either spectroscopic or photometric observations.

Spectroscopy provides the most accurate methods of deriving stellar abundances, providing direct access to absorption features produced by individual elements. Abundances can be derived using standard curve of growth techniques, combining equivalent width measurements of spectral lines of the same species with different oscillator strengths, or spectral synthesis, modelling the overall spectrum within a limited wavelength range. In general, the higher the spectral resolution of the observations, the more accurate, and detailed, the abundance estimates; but the higher the resolution, the more expensive the observation in terms of telescope aperture and exposure time, and the brighter the effective magnitude limit.

Photometric measurements sample broad regions of the spectrum, and are therefore easier to obtain for larger samples. Abundance estimates can be derived by choosing filters that sample spectral regions that include abundance-sensitive absorption features. One of the earliest examples is $\delta U - B$, discussed in the previous chapter in the context of the ELS model: the Johnson U-band samples the spectral region through the Balmer jump, which includes many metallic absorption features; reducing the metallicity leads to weaker features, and a corresponding increase in the U-band flux (see Fig. 19.6). The Johnson and Cousins photometric systems feature wide-bandpass ($\sim$800 Å) filters; intermediate-band systems, such as DDO or Strömgren $uvby$, use 200–400 Å filters that are tuned to better match particular absorption features. One can take this approach further, and use narrowband indices derived

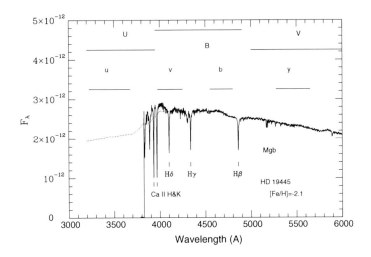

Fig. 19.6 The Johnson UBV and Strömgren $uvby$ bandpasses

from low-resolution spectrophotometric observations to measure the strengths of particular sets of features.

Photometric abundance indicators are calibrated using stars with spectroscopically-determined metallicities. These systems provide a broad measure of stellar metallicity, and are generally not capable of detecting element to element abundance differences, such as α-element enhancements. The accuracies achieved by this approach are typically ~ 0.2 dex, and the low-resolution measurements can be more susceptible to systematic error than spectroscopic analyses.

The Strömgren system is particularly useful for deriving metallicities for solar-type stars. The individual measurements are usually combined to give two indices, $m_1 = (v - b) - (b - y)$, measuring the line blanketing in blue spectral regions, and $c_1 = (u - v) - (v - b)$, which measures the Balmer discontinuity and the UV line blanketing. The indices have been calibrated against metallicity by using observations of Hyades stars to define a fiducial sequence, representing $[M/H] = +0.2$ dex (Schuster and Nissen 1989). That calibration has been widely applied, including calibration of the 14,000 FGK stars in the Geneva-Copenhagen catalogue (Nordström et al. 2004). As noted in the Appendix of Chap. 16, there are systematic offsets in this calibration. Haywood (2002) has recalibrated the Strömgren indices, deriving the metallicity distribution for local stars shown in Fig. 15.11 of Chap. 15, and Casagrande et al. (2011) have applied the revised calibration to the stars in the Geneva-Copenhagen catalogue.

19.4.2 The Age-Metallicity Relation

The Age-Metallicity relation, or AMR, traces the chemical enrichment history of the Galactic Disc. The general expectation is that metallicity should increase with time, as succeeding stellar generations pollute the ISM with nucleosynthetic products redistributed by stellar winds and supernovae. The classical analysis by Twarog (1980) is based on $uvby$-derived metallicities and isochrone-fitted ages for a sample of F stars. He derived a roughly linear decrease in [Fe/H] from $\sim +0.1$ dex at the present day to ~ -0.3 dex at $\tau \sim 8$ Gyr, with a steeper decline in metallicity at older ages.

A subsequent analysis using similar techniques by Edvardsson et al. (1993) found a somewhat steeper decline in metallicity within the last 1–2 Gyr, although that may well reflect the characteristics of the reference sample, which was biased towards metal-poor stars. The results show significant dispersion in metallicity, $\sim \pm 0.2$ dex, at a given age, which may include a significant contribution from age uncertainties. Overall, the trend is broadly consistent with the results of Twarog (1980). In sharp contrast, the analysis of the full Geneva-Copenhagen catalogue by Nordström et al. (2004) suggests that the AMR is flat, with a mean metallicity close to solar and a dispersion of $\sim \pm 0.2$ dex over the full lifetime of the Galactic Disc (see the Appendix of Chap. 16 for a discussion of some issues concerning the abundance calibration). That result persists in the recent analysis by Casagrande et al. (2011). Given that the metallicity of the ISM should increase with time, and that the Milky Way exhibits a radial metallicity gradient, a flat AMR near the Sun has been interpreted as suggestive of extensive radial mixing (see Chap. 17).

All these analyses use Strömgren photometry to derive metallicities and ages. An alternative approach is to focus on a smaller sample and derive the stellar parameters from spectroscopic measurements. One of the products of the Berkeley/Carnegie planet search program is a catalogue of metallicities and ages for ~1000 Sun-like nearby stars (Valenti and Fischer 2005). Beside Keck HIRES spectroscopy, the stars have *Hipparcos* astrometry including reliable parallax measurements. Metallicities are derived from spectral synthesis analysis of the high-resolution spectra used for the radial velocity analyses, with the ages, given as probability distributions, determined by isochrone fitting in the (L, T_{eff}) plane. The overall sample is not volume-complete; however, it includes ~50 % of F, G and early-K stars within 40 pc of the Sun. Limiting the sample to main-sequence stars, there is no evidence of significant selection bias (Reid et al. 2007).

The stars drawn from the Keck planet survey have absolute magnitudes in the range $4 < M_V < 6$ mag, corresponding to long-lived stars with masses within ~25 % of the Sun. Uncertainties in individual ages are significant, particularly for stars older than a few Gyr, as underlined by the presence of stars with formal ages of 14–15 Gyr, older than the Universe. The scatter at a given age is significant, ~±0.15 dex. Nonetheless, the overall trend indicates an average metallicity of ~+0.2 dex at the present day, declining by ~0.04 dex Gyr^{-1} for the last 10 Gyr. The mean relation matches the age and metallicity of the Sun and the Hyades cluster (see Fig. 19.7).

This is clearly a research area that requires further attention. Whether or not the Sun has an unusual metallicity for its age (or age for its metallicity) factors at some level into investigations of its birthplace and the need for radial migration (see Chaps. 17 and 21). On the one hand, the mean relation derived from the Berkeley/Carnegie Keck sample fits qualitative preconceptions and matches the age and metallicity of the Sun and the Hyades cluster; on the other, the flat AMR derived by Casagrande et al. (2011) can be accounted for through substantial mixing within the Disc (or by systematic errors in the age estimates). Ideally, one would like asteroseismology data for a substantial subset of the Berkeley/Carnegie Keck sample, providing reliable ages and metallicities for an unbiased sample of F and G dwarfs in the Solar Neighbourhood.

Fig. 19.7 The age-metallicity [Fe/H] relation derived from the analysis of nearby GK stars from the Keck planet survey. The *filled hexagon* represents the Sun's location. Figure from Reid et al. (2007)

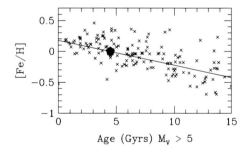

19.4.3 The Age Distribution in the Disc

Accepting that we have imperfect age estimators, we can still attempt to discern the age distribution of the nearby stars. Combining the individual probability distributions derived by Valenti and Fischer (2005) for the Berkeley/Carnegie Keck stars within 30 pc of the Sun gives a distribution that is relative flat from the present day to an age $\tau \sim 5$ Gyr, and declines thereafter by a factor of three by $\tau \sim 15$ Gyr. Analyses of the cosmic microwave background constrain the universe to $\tau \sim 13.4$ Gyr, so these age estimates clearly reflect the inherent uncertainties in the analysis. The Berkeley/Carnegie Keck results indicate that as many stars formed at $\tau > 5$ Gyr as at $\tau < 5$ Gyr, i.e. the star formation rate in the Galactic Disc has been relatively flat over the history of the Milky Way.

Similar conclusions can be drawn from the analysis by Rocha-Pinto et al. (2000) based on chromospheric age estimates for ~550 nearby late-type stars. The inferred star formation history of local stars is shown in Fig. 19.8. As with the Berkeley/Carnegie Keck analyses, the distribution includes a tail of stars with halo-like ages, extending beyond 12 Gyr to ~16 Gyr; this reinforces the limited accuracy possible in deriving ages for stars older than a few Gyr. As plotted, there appear to be several dips in the star formation rate, notably at $\tau \sim 1$–2 and ~5–7 Gyr, but it is not clear whether the data really have this fine a resolution. Again, the more significant conclusion that can be drawn is that the average star formation appears to have remained constant, within a factor of 2, over the history of the Galactic Disc.

The age distribution of local stars can be derived from their colour-magnitude distribution. Hernandez et al. (2000) have matched *Hipparcos* data for nearby stars against star formation models derived from theoretical isochrones, although their analysis is limited to stars brighter than $M_V \sim 3.15$ mag and therefore to ages less than ~3 Gyr. Over that period, variations in the star formation rate are limits to ~±50 %. Hernandez et al. (2000) find some evidence for oscillations with a period of ~0.5 Gyr superimposed on a constant star formation rate. They suggest

Fig. 19.8 The star formation history of the disc as deduced from the distribution of chromospheric activity in late-type dwarfs. Figure from Rocha-Pinto et al. (2000)

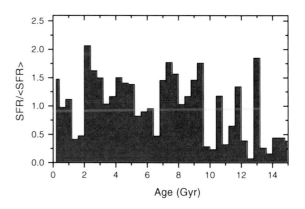

that the oscillatory component might arise from the spiral density wave pattern speed, dynamical interactions with the Galactic Bar or close encounters with the Magellanic Clouds.

Binney et al. (2000) also analysed *Hipparcos* astrometric and photometric data, using the Padova stellar isochrones to constrain the ages. They have not explicitly derived an age distribution or a star formation rate, but they estimate an age of $\sim 11.2 \pm 0.75$ Gyr for the oldest stars in the Disc; they note that an age limit as young as 9 Gyr is possible, but not likely. One can also use the luminosity function of disc white dwarfs to constrain this parameter; these represent the remnants of higher mass disc stars, and one can use theoretical models of cooling rates to estimate the age of the faintest white dwarfs in the vicinity of the Sun. Current estimates suggest that the cut-off in the luminosity function corresponds to a somewhat younger age of ~ 8–9 Gyr.

It is important to note that all the analyses discussed here are effectively based on distance-limited samples of disc stars; none attempt an explicit segregation of thin and thick disc stars. Given the discussion in the earlier part of this chapter, one might expect the thick disc to contribute many of the stars in the tail of the age distribution; indeed, one would expect the oldest stars in the Solar Neighbourhood to be thick disc, rather than thin disc.

19.5 Summary

This chapter, as with Chap. 18, has focussed less on star clusters per se than on how their products (i.e. stars) have been integrated into the larger fabric of the Milky Way. I have given a brief review of what we know about the thick disc, which continues to resist attempts to characterise its origins and relation to the conventional thin disc. The thick disc represents a substantial Galactic stellar component, likely contributing up to 25 % of the mass of the disc. Elucidating its origins, and how prevalent components like this may be in other disc galaxies, remains a matter of considerable interest. The thick disc also probably represents the oldest stars in the disc. The second part of this chapter reviews some of the methods available to estimate ages for stars, whether members of clusters or isolated in the field. Once one has age estimates for a well-defined sample of field stars, the individual estimates can be combined to map out the age distribution, and hence probe global parameters such as the age-metallicity relation and the star formation history of the disc. All of these investigations can shed light on how the Milky Way assembled itself, and therefore offer the potential to constrain the general properties of galaxy formation.

References

Abadi, M. G., Navarro, J. F., Steinmetz, M., & Eke, V. R. 2003, ApJ, 597, 21

Babcock, H. W. 1961, ApJ, 133, 572

Barnes, S. A. 2007, ApJ, 669, 1167

Bensby, T., Feltzing, S., & Lundström, I. 2003, A&A, 410, 527

Binney, J., Dehnen, W., & Bertelli, G. 2000, MNRAS, 318, 658

Borucki, W. J., Koch, D. G., Brown, T. M., et al. 2010, ApJ, 713, L126

Bovy, J., Rix, H.-W., & Hogg, D. W. 2012a, ApJ, 751, 131

Bovy, J., Rix, H.-W., Hogg, D. W., et al. 2012b, ApJ, 755, 115

Brogaard, K., Bruntt, H., Grundahl, F., et al. 2011, A&A, 525, 2

Brook, C. B., Kawata, D., Gibson, B. K., & Freeman, K. C. 2004, ApJ, 612, 894

Casagrande, L., Schönrich, R., Asplund, M., et al. 2011, A&A, 530, 138

Chiba, M. & Beers, T. C. 2000, AJ, 119, 2843

Donahue, R. A. 1993, PhD thesis, New Mexico State University, University Park

Edvardsson, B., Andersen, J., Gustafsson, B., et al. 1993, A&A, 275, 101

Fuhrmann, K. 1998, A&A, 338, 161

Fuhrmann, K. 2004, Astronomische Nachrichten, 325, 3

Giampapa, M. S., Hall, J. C., Radick, R. R., & Baliunas, S. L. 2006, ApJ, 651, 444

Gilliland, R. L., McCullough, P. R., Nelan, E. P., et al. 2011, ApJ, 726, 2

Gilmore, G. & Reid, N. 1983, MNRAS, 202, 1025

Gizis, J. E., Reid, I. N., & Hawley, S. L. 2002, AJ, 123, 3356

Gough, D. 2003, Ap&SS, 284, 165

Haywood, M. 2002, MNRAS, 337, 151

Herbig, G. H. 1956, PASP, 68, 531

Hernandez, X., Valls-Gabaud, D., & Gilmore, G. 2000, MNRAS, 316, 605

Hillenbrand, L. A. 1997, AJ, 113, 1733

Huber, D., Chaplin, W. J., Christensen-Dalsgaard, J., et al. 2013, ApJ, 767, 127

Jurić, M., Ivezić, Ž., Brooks, A., et al. 2008, ApJ, 673, 864

Mamajek, E. E. & Hillenbrand, L. A. 2008, ApJ, 687, 1264

Nordström, B., Mayor, M., Andersen, J., et al. 2004, A&A, 418, 989

Reid, I. N. & Hawley, S. L. 2005, New Light on Dark Stars: Red Dwarfs, Low-Mass Stars, Brown Dwarfs, ed. Reid, I. N. and Hawley, S. L., Springer-Praxis

Reid, I. N., Turner, E. L., Turnbull, M. C., Mountain, M., & Valenti, J. A. 2007, ApJ, 665, 767

Rocha-Pinto, H. J. & Maciel, W. J. 1998, MNRAS, 298, 332

Rocha-Pinto, H. J., Scalo, J., Maciel, W. J., & Flynn, C. 2000, A&A, 358, 869

Schönrich, R. & Binney, J. 2009, MNRAS, 399, 1145

Schuster, W. J. & Nissen, P. E. 1989, A&A, 222, 69

Soderblom, D. R. 2010, ARA&A, 48, 581

Twarog, B. A. 1980, ApJ, 242, 242

Vaiana, G. S. 1980, Highlights of Astronomy, 5, 419

Valenti, J. A. & Fischer, D. A. 2005, ApJS, 159, 141

Villalobos, Á. & Helmi, A. 2009, MNRAS, 399, 166

Yanny, B., Rockosi, C., Newberg, H. J., et al. 2009, AJ, 137, 4377

Chapter 20
Where Do Stars Form?

I. Neill Reid

20.1 Introduction

Stars form within a molecular cloud when over-dense regions reach critical density, form cores and collapse under self-gravity. Molecular clouds provide a wide variety of environments for the subsequent star formation. Those range from broadly dispersed groupings of high-mass stars, known as OB associations, through sparse star-forming regions with few stars more massive than the Sun, such as the Taurus-Auriga clouds, to compact, relatively low-mass clusters like IC 348 and λ Orionis, and dense, high-mass clusters like NGC 3603 and the Orion Nebula Cluster (ONC).

The properties of the local star-forming environment can have a strong influence on the distribution of properties of the final stellar systems that emerge into the general field. Stars in dense, populous clusters are not only liable to undergo more two-body interactions, but protoplanetary discs are subject to sculpting and truncation by photoionisation and interactions with stellar winds generated by massive stars; these effects are directly observable in the ONC. The relative proportion of stars forming within these different environments is therefore a matter of significant interest in estimating the likely properties of both the stars themselves and associated planetary systems. This chapter gives an overview of the statistical properties of different star-forming regions, and our current understanding of their likely contribution to the overall population in the disc. The detailed properties of star clusters are covered in greater detail by Bob Mathieu in his chapters, while Cathie Clarke focuses on the physics governing the formation of individual stars in clusters.

I.N. Reid (✉)
Space Telescope Science Institute, Baltimore, MD, USA
e-mail: inr@stsci.edu

© Springer-Verlag Berlin Heidelberg 2015
C.P.M. Bell et al. (eds.), *Dynamics of Young Star Clusters and Associations*,
Saas-Fee Advanced Course 42, DOI 10.1007/978-3-662-47290-3_20

317

20.2 OB Associations

The basic properties of OB associations are well summarised by Blaauw (1964). Their original identification dates back to the mid-teens of the 20th century, when Jacobus Kapteyn identified loose groups of 'helium stars'—that is, high-mass O and B stars whose spectra are characterised by absorption lines due to neutral and ionised helium. In the mid-1930s, Bok and Mineur demonstrated that the stellar density within these groups is sufficiently low that they are gravitationally unbound. In the late 1940s, Ambartsumian first designated these groups as 'OB associations', estimating ages of ~10 Myr years based on dynamical arguments that were later confirmed by Blaauw. This timescale was consistent with the evolutionary ages estimated previously for O and B stars by Unsold, and therefore suggested that these associations were entities that shared a common origin, rather than aggregates built on pre-existing, longer lived structures.

The spatial distribution of OB associations played a key role in illuminating the structure of Galaxy. Even with Kapteyn's early survey, it was clear that they were distributed neither randomly nor uniformly. As young, star-forming groups, OB associations lie close to the gas swept up by spiral density waves and their distribution traces out the Galactic spiral arms, as demonstrated by Morgan et al. (1952).

OB associations often combine to form extended structures, spanning linear scales exceeding 100 pc. The classic example is the Orion I association (see Fig. 20.1), lying

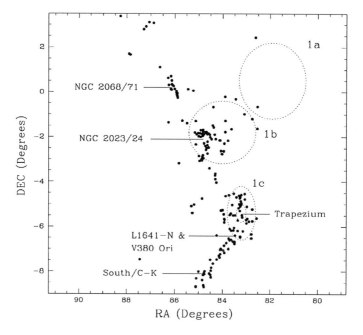

Fig. 20.1 The Orion I association in which the *solid points* mark individual OB stars. Figure from Meyer et al. (2008)

at a distance of between 400 and 500 pc and stretching over 10–15° on the sky or $\sim$100 pc. The nearer Scorpius-Centaurus (Sco-Cen) association extends over almost 200 pc. Most associations show evidence for significant sub-structure. In the case of Orion, millimetre-wave surveys of molecular gas (CO and CS) show that dense molecular gas (and dust) is present throughout an extended filamentary structure. That structure is punctuated by active star-forming clusters, notably the Orion Nebula Cluster and the NGC 2023/24 composite system. The total mass in the association, including gas, dust and stars, is generally estimated as $\sim 2 - 4 \times 10^5 \, M_\odot$ (e.g. Isobe 1987). The relatively low star density within associations makes it difficult to set boundaries between sub-regions and between the association and field; identifying lower-mass members can be particularly problematic.

Proper motions offer a possible avenue for expanding the OB association census to lower mass stars. However, there is little kinematic contrast with the field population and the dynamical state of OB associations still remains poorly understood. Ambartsumian originally suggested that the quasi-spherical distribution of stars in many of the sub-systems within associations indicated that they were undergoing expansion from their point of origin, reflecting the presence of random motions within a gravitationally unbound system. Testing this hypothesis requires highly accurate space motions. *Hipparcos* provided some initial insight, with milli-arcsecond astrometry of stars in several nearer associations, including Sco-Cen (de Zeeuw et al. 1997; see also Mathieu Chap. 8, Fig. 8.15). However, higher accuracies are required to adequately discriminate against field stars. *Gaia* will have a substantial impact in this area, with both increased precision and sensitivities that will extend coverage to sub-solar mass members of the association.

As with many OB associations, there is evidence for a dispersion of ages among the sub-groups within Sco-Cen and Orion I. In Orion, the Trapezium and the Orion Nebula Cluster have ages generally estimated as $1 - 3$ Myr; the Orion Ic and Ib system are approximately twice as old, while Orion Ia, with very few B stars, is likely as old as 10 Myr. Similarly, within Sco-Cen, Pecaut et al. (2012) have used isochrone fitting to estimate an age of $\sim$11 Myr for the Upper Scorpius region, while Upper Centaurus-Lupus is dated as $\sim$16 Myr and Lower Centaurus-Crux as $\sim$17 Myr. In both cases, there is an age progression along the filament, suggestive of a wave of star formation triggered by shock interactions.

20.3 The Taurus-Auriga Cloud

OB associations are characterised by the presence of massive stars. Such is not the case for the nearest star-forming region to the Sun, the Taurus-Auriga molecular cloud (see Fig. 20.2). Covering a total area of $\sim$250 deg^2 at a distance of $\sim$140 pc, this is the archetypical low-mass star-forming region. The region has been surveyed extensively at radio, millimetre, near-infrared and optical wavelengths, and Kenyon et al. (2008) provide an extensive, detailed review of the observational history and the derived properties. The Taurus cloud forms a linear system, extending over 10–15° or

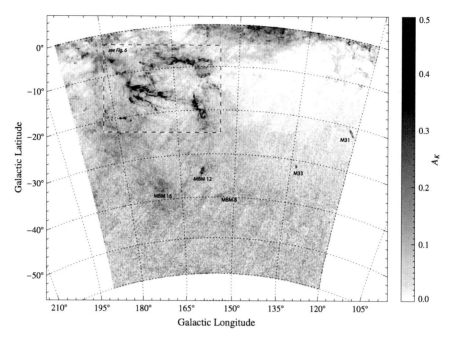

Fig. 20.2 The Taurus-Auriga and Perseus star-forming regions, based on extinction maps computed from the 2MASS near-infrared sky survey. Figure from Lombardi et al. (2010)

24 − 36 pc, but the total mass in the cloud, including stars, gas and dust, is generally estimated as only $\sim 10^4 \, M_\odot$. This is less than one-tenth that of the typical OB association and less than the total mass of the much more compact Orion Nebula Cluster.

Several hundred protostars are known within the Taurus-Auriga cloud. Most optically-identified members are T Tauri stars, including the eponymous prototype, with strong optical emission lines and infrared excess radiation indicating the continued presence of a gaseous protoplanetary disc. Typical ages are estimated as a few million years. Near-infrared and radio observations have revealed younger, highly obscured protostars, still embedded within the natal cloud. X-ray observations have led to the identification of weak-lined T Tauri stars, with lower levels of emission and negligible near-infrared excess. Those properties indicate that the circumstellar disc has dissipated. Statistical studies indicate that the fraction of stars with detectable protostellar discs drops to $\sim 40\,\%$ at ages of 3 Myr and $\sim 20\,\%$ at 5 Myr (Hernández et al. 2008).

Within the Taurus star-forming region, Luhman et al. (2010) calculated a disc fraction of $\sim 75\,\%$ for solar-type stars and $\sim 45\,\%$ for stars with masses $\lesssim 0.3\,M_\odot$, suggestive of an age of $\lesssim 3$ Myr. The negligible levels of near-infrared excess associated with the weak-lined T Tauri stars, however, implies an older age of ~ 10 Myr. Whilst there is a clear trend in disc fraction as a function of time, therefore providing a possible age diagnostic, it is only approximate at best due to several complicating factors including the scatter in disc fractions at a given age (possibly due to environment, stellar multiplicity, mass effects) and the uncertainty in the absolute ages of young regions ($\ll 10\,\mathrm{Myr}$). Hence, whilst rough ages can be inferred from the observed disc fraction, better precision is likely to come from other methods (e.g. model isochrones). The stellar density ranges from ~ 0.1 to $1\,\mathrm{star}\,\mathrm{pc}^{-3}$, with most stars concentrated in several small groups within the darkest clouds.

The stellar mass function in Taurus has long been a subject of interest. Unlike Orion or Sco-Cen, there are no OB stars; the most massive stars within the cloud are estimated at $\sim 2-3\,\mathrm{M}_\odot$ and are likely to evolve to become late-type A or early-type F stars on the main-sequence. There is also some evidence for a deficit at lower masses between Taurus and the other young clusters. The Taurus IMF appears to peak at $\sim 0.8\,\mathrm{M}_\odot$ and decline monotonically towards lower masses (Luhman 2004), whereas the mass function in the field (and denser star-forming regions) is significantly flatter through $\sim 0.3\,\mathrm{M}_\odot$ (as discussed in Chap. 16). Observations also suggest that the binary fraction among the lower-mass members of the Taurus-Auriga cloud tends to be higher than in the field (e.g. Konopacky et al. 2007).

In general, the Taurus-Auriga clouds provide a more quiescent star-forming environment than the typical OB association, and the differences in the mass function and multiplicity may be tied to the environmental differences. In particular, Kirk and Myers (2012) have shown that crowded stellar environments (which are relatively scarce in Taurus) tend to produce greater numbers of high-mass stars. One might also speculate that the scarcity of high-mass stars might lead to fewer shocks, reducing the number of low-mass cores that collapse to form stars. At the same time, the lower star density leads to fewer dynamical interactions, perhaps accounting for the higher frequency of low-mass binaries. In any event, star-forming regions like Taurus-Auriga appear to be rare, limiting their contribution to the overall field-star population.

20.4 Young Stars Near the Sun

The nearest active star-forming regions lie at distances of more than $100\,\mathrm{pc}$, with the low-density, sparely-populated Taurus-Auriga cloud at $\sim 140\,\mathrm{pc}$ and the Sco-Cen OB association at distances from 150 to $200\,\mathrm{pc}$. Nonetheless, relatively young stellar systems have been detected at much closer distances to the Sun (Zuckerman and Song 2004). Eight co-moving groups have been identified within the Local Bubble ($\lesssim 100\,\mathrm{pc}$; see e.g. Torres et al. 2008). Of these, four are associated with individual stars (TW Hydrae, β Pictoris, AB Doradus and η Cha), whereas the brightest members of the other four are α Pavonis (B2.5IV; Tucana-Horologium moving group), HR 1621 (B9Vann; Columba association), β Leonis (A3Va; Argus association) and the binary HD 83096 (F2V+G9V; Carina association).

The number of members in each group ranges from approximately 10 to 200, dispersed over a relatively wide region of space. The associations are based on the similarity of the individual space motions, together with evidence for youth, whether through the presence of strong optical emission indicating continued gas accretion from a protoplanetary disc, chromospheric activity and flares, enhanced X-ray fluxes and/or infrared excess consistent with thermal emission from a dusty debris disc.

The TW Hya association, for example, includes some $30-40$ low-mass members. There is only one early type star, the A0V primary TWA 11A (HR 4796A; mass $\sim 2\,M_\odot$), with all other members having spectral types later than $\sim$K3. The lowest mass members are the M8+mid/late-L binary pair, 2MASSW J1207334-393254; both are brown dwarfs, with the primary likely having a mass of $\sim 25\,M_{Jup}$ and the secondary $\sim 5\,M_{Jup}$ (Gizis 2002; Chauvin et al. 2004). TW Hydrae itself is a K8 dwarf with a likely mass of $\sim 0.6\,M_\odot$. Overall, the association includes 7 known binaries and 2 triple systems in addition to the quadruple system, HD 98800; this represents a higher multiplicity frequency than for field dwarfs in the same mass range, possibly reflecting the young age and limited time for dynamical interactions.

The observational properties of member stars suggest that these associations span a range of ages. TW Hydrae, for example, is a classical T Tauri star with emission lines throughout the optical and UV spectrum, lithium absorption at 6708 Å, strong X-ray emission and radio detection of HD and CO from its circumstellar disc. In contrast, the A6V star β Pictoris has a very prominent, edge-on dusty debris disc, extending to radii over $\sim$14 arcsec or $\sim$300 au. The original discovery, and the first image of any debris disc, came from ground-based coronagraphy following the detection of a strong infrared excess by the IRAS satellite (Aumann 1985). *Hubble* observations of the disc show complex structure (Golimowski et al. 2006) suggesting dynamical interactions with massive planetary companions, and ground-based observations with the ESO Very Large Telescope have succeeded in resolving a super-Jovian gas giant at a separation of $8-9$ au from the central star (Lagrange et al. 2009). α Pavonis, in the Tucana-Horologium association, is generally estimated as having a mass $\sim 6\,M_\odot$, implying an age less than $\sim$50 Myr; similar considerations apply to the B8V star, η Cha (mass $\sim 4\,M_\odot$), the prototype for its cluster. Finally, AB Doradus is a triple system lying at $\sim$15 pc that shows substantial chromospheric and coronal activity but no evidence for significant emission from a residual protoplanetary disc. These characteristics, in conjunction with positions of other AB Doradus moving group members in the colour-magnitude diagram, suggest it is roughly coeval with the Pleiades ($\sim 100-130$ Myr; Luhman et al. 2005).

Whilst it is possible to estimate the ages of these associations through the study of individual members, the best method is arguably to combine kinematic and spectroscopic data to establish a reliable list of member stars and then use photometric data to derive an age from either the Hertzsprung-Russell or colour-magnitude diagram. Despite the high level of model dependency using pre-main-sequence model isochrones to establish absolute ages (see e.g. Dahm 2005 who demonstrated a factor of 3 age difference depending on the models adopted), the relative ages are robust i.e. the η Cha cluster is roughly coeval with the TW Hya association, which is younger than the β Pic moving group, which in turn is younger than the Tucana-Horologium, Columba, Carina and Argus associations, which are all younger than the AB Doradus moving group.

Setting aside the AB Doradus moving group, there is a significant population of stars with ages less than $\sim$30 Myr within $\sim$50 pc of the Sun. It is clear from the absence of molecular material that they did not form at their present location. Their ages are broadly comparable with the nearest star-forming region, the

Sco-Cen OB association. Moreover, as one might infer from their names, the stars in these loose field aggregations are distributed across the southern hemisphere in the general direction of Sco-Cen. The space motions are also broadly consistent with the associations originating on the near edge of the association some $10 - 15$ Myr ago. Zuckerman and Song (2004) suggest that shocks from massive star formation triggered by the passage of a spiral arm or from the subsequent supernovae may have led to the formation of these small groups, whose peculiar motions of $6 - 10$ km s^{-1} were sufficient to bring them to the Solar Neighbourhood at the present time.

20.5 Clustered Star Formation: Open and Embedded Clusters

20.5.1 Open Clusters

Open star clusters are the most prominent and obvious sites of recent active star formation, at least at optical wavelengths. They are gravitationally bound entities that offer dense environments, with star densities that can exceed 10^4 stars pc^{-3}. Like globular clusters, open clusters undergo interactions with spiral arms, giant molecular clouds and other massive bodies in the Galaxy. As a result, they evolve dynamically leading to mass segregation and stellar evaporation, and eventually merge with the general field. We present only a brief summary of their properties here, since this topic is covered in detail in Bob Mathieu's chapters.

A census of the local systems offers a means of estimating both their lifetimes and their likely contribution to the field-star population. Initial calculations by Wielen (1971) rested on photographic catalogues of clusters within ~ 1 kpc of the Sun. Based on the age distribution, Wielen (1971) estimated that $\sim 50\%$ of the systems had dissipated within $\sim 2 \times 10^8$ yr of formation, while only 10 % survived for 5×10^8 yr and 2 % for $\sim 10^9$ yr. The longer-lived clusters tend to be more massive and have Galactic orbits that carry them relatively far from the Plane (e.g. M67, NGC 6791). Moreover, integrating the contribution made by (former) open cluster members to the general field population suggested that they make only a small fractional contribution.

Wielen's calculations were repeated three decades later by Piskunov et al. (2006) who derived very similar results from more extensive cluster catalogues. As shown in Fig. 20.3, the number of clusters as a function of age drops sharply beyond ~ 100 Myr, with a typical half-life estimated as ~ 250 Myr. Piskunov et al. (2006) estimate a cluster formation rate of ~ 0.3 clusters kpc^{-2} Myr^{-1}. Allowing for a typical cluster size of $\sim 1,000$ stars, this implies that $\sim 10\%$ of Galactic disc stars were born in open clusters.

A major limitation of these studies is that they rely on optical observations which are extremely ineffective at detecting even relatively massive clusters during their youth when they are shrouded by dust and gas within the parent molecular cloud. Observations at longer wavelengths, notably in the near- and mid-infrared, not only match the peak in the flux distribution of young stars, but are also subject

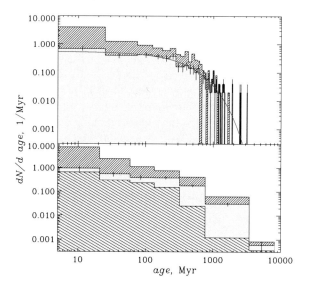

Fig. 20.3 The age distribution of open clusters. Figure from Piskunov et al. (2006)

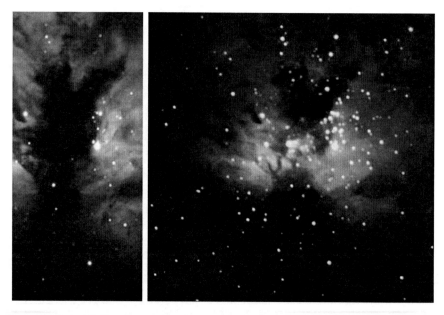

Fig. 20.4 Optical (*left*) and near-infrared (*right*) images of NGC 2024, the Flame Nebula, in Orion. The individual stars in the embedded young cluster are clearly evident in the near-infrared image (National Optical Astronomical Observatory: I. Gatley & R. Probst)

to substantially less absorption by intervening dust (see Fig. 20.4). Consequently, those observations have proven vital in gaining insight into the earliest stages of star formation.

20.5.2 *Infrared Astronomy and the Detection of Embedded Clusters*

Infrared astronomy dates back to William Herschel, but wide-scale application to stellar astronomy really began in the early 1960s with the availability of PbS and Si detectors that provided reasonable sensitivity to radiation beyond 1 μm. Neugebauer et al. (1965) carried out the first large-scale survey using a custom-built telescope with a 62-inch f/1 aluminised-plastic mirror to survey the 70 % of the sky accessible from Mt. Wilson. Their 'camera' consisted of 8 PbS detectors, mounted in pairs, that sampled the K-band (2.01 − 2.41 μm) and a single silicon detector with a 20 arcmin field-of-view that covered the Johnson I-band (0.68−0.92 μm). The resulting survey, the Two-Micron Sky Survey (TMSS), had a resolution of 10×10 arcmin and detected some 5700 sources, essentially providing the near-infrared equivalent of the Bright Star catalogue.

The first near-infrared survey of a star-forming region was carried out by Grasdalen et al. (1973), who used a single-channel detector to painstakingly map ~ 100 arcmin^2 of the ρ Ophiuchus dark cloud at a resolution of ~ 35 arcsec. The survey detected 41 sources brighter than 9th magnitude in K, with the brightest having JHK colours consistent with highly-reddened, luminous stars. The sources were clearly clustered, and Grasdalen et al. (1973) noted that there was '... much more compact clustering than is typical for OB associations'. Indeed, the authors hypothesised that '... the observed density suggests that we are observing a rich open cluster located at the centre of the ρ Ophiuchus cloud'. This was the first clear identification of an embedded star cluster.

The distribution of sources within ρ Oph suggested that a substantial majority are forming in a clustered environment. If ρ Oph were typical of star-forming regions, then this would appear to contradict the birthplace statistics deduced from the open cluster census. The two results might be reconciled if the average lifetime of embedded clusters were significantly shorter than the 250 Myr inferred for open clusters. Such circumstances are not implausible, given that gas and dust make a significant contribution to the total mass of an embedded cluster; as winds from high-mass stars and shocks from supernovae disperse the gas, the mass and binding energy may decrease to the point where the system becomes unbound before it becomes optically visible.

Understanding the evolution and dispersal of embedded clusters and their contribution to the field population demands reliable statistics for the distribution of protostars in star-forming clouds. Wide-field near-infrared imaging is essential for compiling a reliable stellar census within star-forming regions and calculating those statistics. Detailed surveys with the appropriate sensitivity and angular resolution only became possible in the late 1980s with the development of two-dimensional near-infrared arrays.

The first arrays pulled into service for astronomical observations marked a significant advance over single photodiodes, but surveying star-forming regions still required substantial effort. As an example, one of the earliest surveys covered part of the Lynds 1630 cloud in Orion Ib, including NGC 2068/71 and NGC 2024 (Lada et al. 1991); those observations, which reached 13th magnitude in K, used a 58×62 pixel InSb array that covered approximately 1×1 arcmin on the sky; the full survey required ~ 2800 separate pointings. Extinction within the cloud can exceed $A_V \sim 20$ mag, rendering the protostars completely invisible to even far-red optical imaging. The corresponding extinction at near-infrared wavelengths (JHK) is only $2 - 3$ mag and, as a result, near-infrared imaging reveals almost 1,000 sources, with the faintest corresponding to protostars with masses $\sim 0.6\,M_\odot$. The spatial distribution is highly non-uniform, although clustered sources do not necessarily equate to sources in bound clusters.

Observationally, these results suggest that embedded clusters, like ρ Oph or Taurus-Auriga, represent the evolutionary starting point of star clusters. After a few million years, stars emerge from the parent cloud, with examples ranging from massive clusters such as ONC and NGC 3603 to much lower-mass entities like λ Orionis and σ Orionis. Within $10 - 20$ Myr, all of the gas dissipates, revealing a young cluster like IC 2602. As the cluster matures and evolves, it relaxes dynamically and gradually disperses due to internal and external dynamical interactions. Only the most massive systems (Pleiades, Hyades, Praesepe) and those on highly favourable Galactic orbits (M67, NGC 6791) survive for more than a few hundred million years.

20.5.3 Statistics of Embedded Clusters

As technology advanced and infrared arrays increased in size, ground-based surveys became increasingly more efficient and capable of covering larger areas at higher sensitivity. Those efforts culminated in the DENIS and 2MASS surveys, covering the southern sky and the whole sky to fainter than 15th magnitude in the IJK and JHK_s passbands respectively. At the same time, space missions such as IRAS, ISO, *Spitzer* and, most recently, *Herschel* extended coverage to mid-infrared wavelengths. Together, these facilities have enabled a more reliable census of embedded star clusters.

Wide-field observations show that the typical molecular cloud complex includes multiple concentrations of sources, spanning a range of volume and density (see Fig. 20.5). Given this range of properties, it is important to set clear criteria on how one identifies an embedded star cluster. Lada and Lada (2003, hereafter LL03) developed a methodology based on dynamical evolution. The evaporation time for a cluster, τ_{ev}, can be quantified as:

$$\tau_{ev} \sim 100\tau_{relax} \sim 0.1N/\ln N \times \tau_{cross}, \tag{20.1}$$

where N is the number of stars and τ_{cross}, the cluster crossing time, is $\sim 10^6$ yr. Setting the requirement that $\tau_{ev} > 10^8$ yr sets a lower limit of $N = 35$ for a viable cluster.

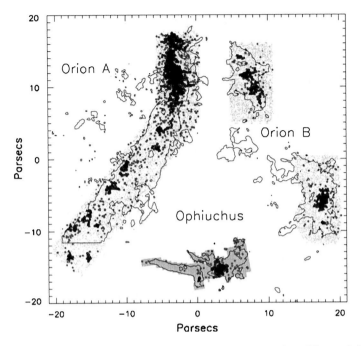

Fig. 20.5 Source maps for several molecular cloud complexes. Figure from Allen et al. (2007)

Using this criterion, LL03 compiled a census of local embedded clusters, analysing literature data for 76 clusters within 2 kpc of the Sun. While the sample is incomplete, perhaps by as much as a factor of 4, they argued that it is representative. Estimating the total mass within each cluster by using the Orion Nebula Cluster IMF as a template, they found that although massive clusters are rare, they likely account for ∼50 % of the stellar mass; moreover, the cluster mass function appears to turnover at low masses, with small clusters ($N < 50$) being relatively rare, although one might imagine that detection of those systems is most susceptible to observational selection effects. The overall star formation rate within the volume surveyed is estimated as $1 - 3 \times 10^{-9}\,M_\odot\,\mathrm{yr}^{-1}\,\mathrm{pc}^{-2}$ or $0.5 - 1.1\,M_\odot\,\mathrm{yr}^{-1}$ integrated over the Galactic disc ($R \sim 12\,\mathrm{kpc}$). This is comparable with other estimates of the global star formation rate (see Chap. 15).

Similar conclusions were drawn by Porras et al. (2003), who compiled and analysed literature data for 73 star-forming regions within 1 kpc of the Sun. Within that sample, the median cluster size was $N = 28$, while the mean cluster size was $N = 100$; 80 % of the stars are in clusters with $N > 100$. Both of these analyses suggest that most stars form in clustered environment. However, the birthrate of $2 - 4$ clusters $\mathrm{Myr}^{-1}\,\mathrm{kpc}^{-2}$, deduced by LL03 on the assumption that the embedded clusters have ages of $1 - 2\,\mathrm{Myr}$, demands rapid attrition and dissipation of even the more massive stellar concentrations in the star-forming clouds.

Subsequent analysis, incorporating *Spitzer* mid-infrared data, gives a slightly different picture, suggesting more of a continuum of properties (Allen et al. 2007). Using a variety of techniques, they catalogued protostellar candidates within a number of nearby molecular cloud complexes, and assessed the size distribution of the embedded clusters. Overall, they found that $\sim$65 % of the sources were found in large ($N > 100$) clusters, $\sim$15 % in clusters with between 30 and 100 detected sources and $\sim$5 % in groups with $10 - 30$ sources. This result echoes LL03 in placing the majority of protostars in relatively large embedded clusters, but some $20 - 25$ % of the sources appear to be relatively isolated, in groups consisting of fewer than 10 sources.

Allen et al. (2007) also noted that while the total number of stars in embedded clusters varies by over 2 orders of magnitude, the cluster size scales accordingly and the space densities vary by only a factor of a few between the sparsest and most populous clusters. The larger clusters often show elongation and, in some cases, hierarchical structure with multiple density peaks, clearly indicating that these systems are not dynamically relaxed. It remains unclear what fraction of even the most populous embedded clusters are actually gravitationally bound.

Bressert et al. (2010) have combined samples of young stellar objects (YSOs) from a number of surveys, and have considered the distribution as a function of surface density. Rather than characterising their results as a dichotomy between clustered and isolated sources, they find a reasonable match to a continuous distribution, close to log-normal in form, with the peak at a surface density of $\sim$22 YSOs pc^{-2}. They conclude that this places less emphasis on the role of embedded clusters in star production. However, it has to be noted that these sources typically have ages of 10–20 Myr, comparable with field OB associations and an order of magnitude older than the structures being studied within the molecular clouds. This clearly gives scope for some dynamical evolution to at least partially erase the initial spatial distribution.

20.6 Summary and Future Prospects

Over the past 40 years we have achieved a reasonable conceptual understanding of the progression of environments occupied by stars in the early years of evolution. However, uncertainties still persist in how we connect the various concentrations of protostellar sources detected within star-forming clouds to the open clusters and loose aggregations that emerge from the shadows. In particular, there continues to be a disconnect between the apparent dominance of highly-clustered sources among the obscured protostars, and observed spatial distribution of YSOs and the number of surviving open clusters. Do most embedded clusters dissipate rapidly after they shed their dusty cocoons, or are we still missing a significant population of isolated protostars in the cloud? Resolving those issues depends on measuring the mass function and kinematic state of multiple embedded clusters spanning a range of properties, and determining whether the stellar population is bound.

© Anglo-Australian Obs./Royal Obs. Edinburgh

Fig. 20.6 Ground-based optical (AAO; *left*) and *Spitzer* mid-infrared (*right*) images of the Large Magellanic Cloud

Until recently, observations of young clusters and star-forming regions have focussed almost exclusively on the Galaxy. However, while such systems have the advantage of proximity, there are also some complications by our internal perspective, with the potential for extraneous material along the line-of-sight to intrude on measurements. One option for mitigating such problems may be to undertake a more thorough investigation of star formation within our near neighbour, the Large Magellanic Cloud (LMC; see Fig. 20.6). The LMC has long been a target of optical observations, and extensive catalogues exist. Those are now supplemented by extensive observations at near-infrared wavelengths (2MASS and DENIS), as well as mid-infrared observations that extend to 24 μm obtained by the *Spitzer* and *Herschel* SAGE projects (Meixner et al. 2006; Kemper et al. 2010). Those surveys are already proving a powerful tool in mapping YSOs that have just escaped the parental cloud. In addition, high resolution HST imaging and spectroscopy are available for a number of clusters, notably the Tarantula survey of 30 Doradus (Sabbi et al. 2013), while JWST will be ideal for detailed follow-up observations of obscured sources at mid-infrared wavelengths. Perhaps our external view will permit a clearer view of how clustering evolves during the earliest phases of stellar development.

Addendum: Brown dwarf evolution
Brown dwarfs are objects that form through gravitational collapse, like stars. As the protostellar core collapses, potential energy is transformed to kinetic and the central temperature increase. Brown dwarfs have sufficiently low masses that they are unable to generate sufficient energy to increase the central temperature to above $\sim 3 \times 10^6$ K, the critical threshold for triggering hydrogen fusion. As a consequence, they lack a self-sustaining energy source and undergo cooling on a relatively rapid timescale, leading to decreasing luminosities and surface temperatures. It is important to bear these characteristics in mind when interpreting observations of cool dwarfs in star-forming regions.

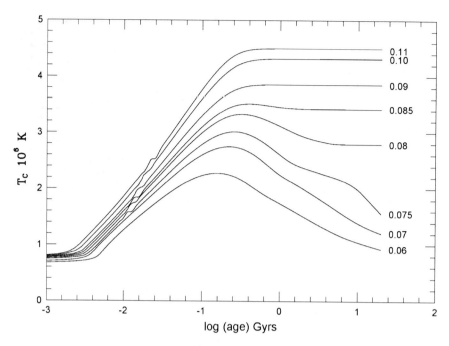

Fig. 20.7 The evolution of the central temperature in low-mass stars and brown dwarfs. Based on model calculations by Burrows et al. (1997)

Figures 20.7–20.9 provide a more quantitative picture of brown dwarf evolution using theoretical models computed by Burrows et al. (1997). In each case, the individual tracks are labelled by the mass of the model in solar units. Figure 20.7 shows the time evolution of the central temperature. For objects more massive than 0.085 $M_\odot$, the central temperature increases monotonically until it reaches a plateau marking hydrostatic equilibrium, where the energy generated by hydrogen fusion supports the star against further collapse i.e. the star is on the main-sequence. Note that the evolutionary timescale to reach this point is several hundred million years.

The 0.08 and 0.085 $M_\odot$ models also reach equilibrium, but the central temperature peaks with a subsequent decline before settling into the equilibrium configuration. This behaviour stems from the fact that these low-mass dwarfs become electron degenerate. As a result, energy is diverted from heating the core to overcoming degeneracy pressure. At lower masses, counteracting degeneracy depletes the available energy to the extent that the central temperature drops below the threshold for maintaining long-term fusion. The 0.075 $M_\odot$ model is a transition object, able to trigger fusion but unable to support activity beyond ∼10 Gyr; in lower mass dwarfs, the central temperature declines monotonically at ages beyond ∼200 Myr. Degeneracy also constrains the radius, with the consequence that all objects with masses below ∼0.1 $M_\odot$ have radii within 10 % of the radius of Jupiter.

Fig. 20.8 The evolution of the surface temperature in low-mass stars and brown dwarfs. Based on model calculations by Burrows et al. (1997)

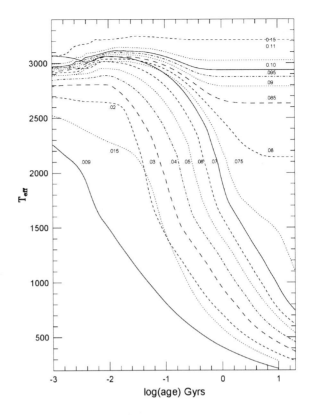

The formal threshold for activating hydrogen fusion, the star/brown dwarf boundary, lies at $\sim$0.073 M$_\odot$ for a solar-metallicity composition (lower metallicity objects have lower opacities, and the boundary moves to higher mass). Note that the maximum central temperature achieved by the 0.06 M$_\odot$ model is $\sim$2 $\times$ 10^6 K. This corresponds to the critical temperature for lithium fusion; lower mass dwarfs do not deplete lithium and, as a result, the detection of lithium absorption (6708 Å) in all save the youngest objects identifies low-mass brown dwarfs (Rebolo et al. 1992).

Figure 20.8 illustrates the behaviour predicted by Burrows et al. (1997) for the variation of surface temperature with time for low-mass stars and brown dwarfs with masses from 0.15 and 0.009 M$_\odot$. At the time of formation, all these objects span a relatively small range in temperature, between $\sim$2300 and 3100 K. For stars, the surface temperature follows the central temperature in peaking at an age of $\sim$100 Myr, with a subsequent decline before settling on the main-sequence. Higher mass brown dwarfs show similar initial behaviour, but the peak temperature is followed by a monotonic decline at a rate that increases with decreasing mass. Below $\sim$0.04 M$_\odot$, the initial temperature is almost constant, sustained by deuterium burning for masses above $\sim$0.013 M$_\odot$, and followed by a precipitous decline.

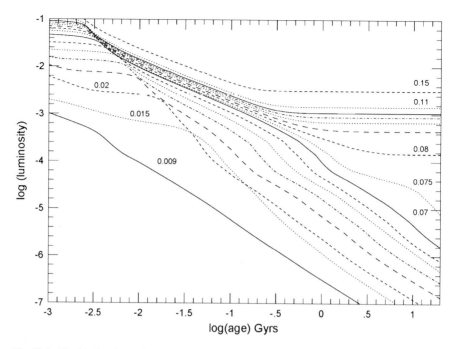

Fig. 20.9 The luminosity evolution of low-mass stars and brown dwarfs. Based on model calculations by Burrows et al. (1997)

The observational consequences are that brown dwarfs evolve in spectral type on timescales that are rapid by astronomical standards. A ~0.06 M⊙ brown dwarf has an initial spectral type of ~M5, and a spectrum dominated by TiO absorption at optical wavelengths and broad water bands in the near-infrared. The surface temperature peaks at an age of 20–30 Myr, and then declines; as the atmosphere cools, dust condenses within the atmosphere, weakening and then eliminating TiO and VO absorption, and the dwarf crosses the M/L boundary at ~2100 K at an age of ~800 Myr. The optical spectrum is now dominated by metal hydride bands and highly-broadened absorption by neutral sodium and potassium. The atmosphere continues to cool, and after ~5 Gyr, methane forms in the upper levels and the dwarf crosses the L/T boundary at a surface temperature of ~1300 K.

Figure 20.9 shows the predicted luminosity evolution for the same set of models. After an initial standstill, where the increasing temperature compensates for the decreasing radius, all objects show a monotonic decline, with stars settling onto the stable main-sequence configuration after ~1 Gyr and brown dwarfs fading into oblivion. Clearly, from Figs. 20.8 and 20.9, surveys of young (τ < 10 Myr) star-forming regions offer significant advantages in searching for brown dwarfs, since they are both hotter and significantly more luminous than older brown dwarfs in the field. Moreover, since age is a known quantity, masses can be estimated for individual objects, which is not the case for field brown dwarfs. For more extensive discussion of these phenomena see Reid and Hawley (2005).

Theoretical models are crucial to interpreting observations of low-mass members of star-forming regions, particularly in estimating their masses and constructing an IMF. The lowest mass brown dwarfs may well overlap in mass with massive planetary companions. Ideally, one would hope to distinguish these two objects based on their location—planets, forming in a disc, are most likely to be in orbit around a parent star. However, it is clear from dynamical models that planets can be ejected from their parent system, while brown dwarfs can also form as companions to stars. One can set a theoretical lower-mass limit to brown dwarf formation by invoking the minimum mass set by opacity limit for fragmentation, This represents the point where fragmentation within a collapsing cloud ceases because the fragment becomes optically thick, trapping the energy and leading to heating (Rees 1976; Low and Lynden-Bell 1976). This limit is generally estimated as between 0.003 and 0.01 $M_\odot$ depending on the composition, with lower abundances having a higher limit due to the lower opacities and the reduced ability to radiate away heat.

References

Allen, L., Megeath, S. T., Gutermuth, R., et al. 2007, Protostars and Planets V, ed. B. Riepurth, D. Jewitt & K. Keil, University of Arizona Press, 361

Aumann, H. H. 1985, PASP, 97, 885

Blaauw, A. 1964, ARA&A, 2, 213

Bressert, E., Bastian, N., Gutermuth, R., et al. 2010, MNRAS, 409, L54

Burrows, A., Marley, M., Hubbard, W. B., et al. 1997, ApJ, 491, 856

Chauvin, G., Lagrange, A.-M., Dumas, C., et al. 2004, A&A, 425, L29

Dahm, S. E. 2005, AJ, 130, 1805

de Zeeuw, P. T., Brown, A. G. A., de Bruijne, J. H. J., et al. 1997, in ESA Special Publication, Vol. 402, Hipparcos - Venice '97, ed. R. M. Bonnet, E. Høg, P. L. Bernacca, L. Emiliani, A. Blaauw, C. Turon, J. Kovalevsky, L. Lindegren, H. Hassan, M. Bouffard, B. Strim, D. Heger, M. A. C. Perryman, & L. Woltjer, 495

Gizis, J. E. 2002, ApJ, 575, 484

Golimowski, D. A., Ardila, D. R., Krist, J. E., et al. 2006, AJ, 131, 3109

Grasdalen, G. L., Strom, K. M., & Strom, S. E. 1973, ApJ, 184, L53

Hernández, J., Hartmann, L., Calvet, N., et al. 2008, ApJ, 686, 1195

Isobe, S. 1987, Ap&SS, 135, 237

Kemper, F., Woods, P. M., Antoniou, V., et al. 2010, PASP, 122, 683

Kenyon, S. J., Gómez, M., & Whitney, B. A. 2008, Handbook of Star Forming Regions, Volume I: The Northern Sky, ed. B. Reipurth, ASP Monograph Publications, 405

Kirk, H. & Myers, P. C. 2012, ApJ, 745, 131

Konopacky, Q. M., Ghez, A. M., Rice, E. L., & Duchêne, G. 2007, ApJ, 663, 394

Lada, C. J. & Lada, E. A. 2003, ARA&A, 41, 57

Lada, E. A., Depoy, D. L., Evans, II, N. J., & Gatley, I. 1991, ApJ, 371, 171

Lagrange, A.-M., Gratadour, D., Chauvin, G., et al. 2009, A&A, 493, L21

Lombardi, M., Lada, C. J., & Alves, J. 2010, A&A, 512, 67

Low, C. & Lynden-Bell, D. 1976, MNRAS, 176, 367

Luhman, K. L. 2004, ApJ, 617, 1216

Luhman, K. L., Stauffer, J. R., & Mamajek, E. E. 2005, ApJ, 628, L69

Luhman, K. L., Allen, P. R., Espaillat, C., Hartmann, L., & Calvet, N. 2010, ApJ, 186, 111

Meixner, M., Gordon, K. D., Indebetouw, R., et al. 2006, AJ, 132, 2268

Meyer, M. R., Flaherty, K., Levine, J. L., et al. 2008, Handbook of Star Forming Regions, Volume I: The Northern Sky, ed. B. Reipurth, ASP Monograph Publications, 662

Morgan, W. W., Sharpless, S., & Osterbrock, D. 1952, AJ, 57, 3

Neugebauer, G., Martz, D. E., & Leighton, R. B. 1965, ApJ, 142, 399

Pecaut, M. J., Mamajek, E. E., & Bubar, E. J. 2012, ApJ, 746, 154

Piskunov, A. E., Kharchenko, N. V., Röser, S., Schilbach, E., & Scholz, R.-D. 2006, A&A, 445, 545

Porras, A., Christopher, M., Allen, L., et al. 2003, AJ, 126, 1916

Rebolo, R., Martin, E. L., & Magazzu, A. 1992, ApJ, 389, L83

Rees, M. J. 1976, MNRAS, 176, 483

Reid, I. N. & Hawley, S. L. 2005, New Light on Dark Stars: Red Dwarfs, Low-Mass Stars, Brown Dwarfs, ed. Reid, I. N. and Hawley, S. L., Springer-Praxis

Sabbi, E., Anderson, J., Lennon, D. J., et al. 2013, AJ, 146, 53

Torres, C. A. O., Quast, G. R., Melo, C. H. F., & Sterzik, M. F. 2008, Handbook of Star Forming Regions, Volume II: The Southern Sky, ed. B. Reipurth, ASP Monograph Publications, 757

Wielen, R. 1971, Ap&SS, 13, 300

Zuckerman, B. & Song, I. 2004, ARA&A, 42, 685

Chapter 21
Where Was the Sun Born?

I. Neill Reid

21.1 Introduction

Most people want to know where they came from. Extensive genealogical databases now render it possible for many of us to work back through the generations, and TV programmes have become commonplace where celebrities trace their timelines and discover unexpected family members in unusual places. We would like to do the same for the Sun. In this case, as with some celebrities, we cannot expect to arrive at a definitive answer. However, we can take what we know about the intrinsic properties of the Sun and its attendant planetary system, synthesise that with our knowledge of the range of star-forming environments, and at least set constraints on its place of origin. This question was addressed in a recent review (Adams 2010), and we draw heavily from that work in the present chapter.

Considering its gross properties, the Sun is somewhat above average as a star. It is the tenth brightest main-sequence star among the 8 pc sample, and only 15 % of the stellar systems in that sample include a component that is or was more massive than the Sun. It is a single star, which places it in a minority, albeit a substantial minority, (∼30 %) among G dwarfs. Neither characteristic sets particular constraints on where the Sun might have formed.

In examining the Sun's more detailed intrinsic properties, we have the salient advantage we are close enough that we can make spatially resolved observations of its surface. The Sun is also exceedingly bright; the sheer number of photons arriving per square metre per second enables observational measurements at a level of precision that can scarcely be dreamt of for other stars. As a result, we can probe its properties to much greater precision than for any other object. Even so, significant uncertainties still remain.

I.N. Reid (✉)
Space Telescope Science Institute, Baltimore, MD, USA
e-mail: inr@stsci.edu

© Springer-Verlag Berlin Heidelberg 2015

C.P.M. Bell et al. (eds.), *Dynamics of Young Star Clusters and Associations*,
Saas-Fee Advanced Course 42, DOI 10.1007/978-3-662-47290-3_21

21.2 The Age of the Sun

The Sun is the only star whose age can be measured to four significant figures. This is because we can carry out laboratory experiments on primordial material, rather than relying solely on inference from long distance observations. Meteorites give us direct access to materials that coalesced with the Sun in the earliest stages of the formation of the Solar System.

Not all meteorites are suitable for this purpose. Iron meteorites originated from parent bodies that were large enough to have sufficient radioactive ^{26}Al to heat, melt and re-process material, producing a chemically differentiated asteroid. Iron meteorites are the cores of such bodies, while stony-iron meteorites are generally believed to originate in the core/mantle boundary. Stony meteorites composed of rocky materials account for the overwhelming majority, classed as chondrites or achondrites depending on whether they include small, spherical particles known as chondrules.

Chondrites are generally believed to have escaped inclusion in a larger body. Consequently, their materials have not been subject to extensive re-processing since formation in the early Solar System. In particular, carbonaceous chondrites are composed primarily of silicates, oxides and sulphides and can also include a significant fraction of water and organic compounds. The presence of those volatile materials demonstrates that these stony meteorites have not been subjected to significant heating since their formation in the early Solar System, suggesting that they have never been part of a larger body. Carbonaceous chondrites are sub-divided into groups based on their composition, with each group named after the archetype meteorite. They often contain small nodules of metals, known as refractory inclusions. Those inclusions, particularly calcium-aluminium inclusions (CAIs), can be extracted and subjected to radiometric dating to give a lower limit to the age of the oldest solids in the Solar System.

Radiometric dating requires the presence of a radioactive isotope with a half life, λ, comparable with the likely age of the material. A radioactive element decays such the number of atoms at present, $P(t)$, is given by $P_0 e^{-\lambda t}$, where P_0 is the original number and t, the age of the sample. The mathematical relation appropriate to radioactive decay is:

$$D = D_0 + P(t)(e^{\lambda t} - 1), \qquad (21.1)$$

where D is the number of atoms of the daughter isotope in a given sample and D_0 the original number of atoms of the daughter isotope. Generally, one can eliminate D_0 by taking the ratio between the number of daughter and parent atoms against a reference isotope for a number of samples; plotting those ratios for a number of samples should give a straight line, known as an isochron, whose slope gives the age in terms of the half-life.

A number of parent-daughter decay series are available for age-dating, including potassium to argon (^{40}K–^{40}Ar, $\lambda \sim 1.25$ Gyr), uranium to lead (^{238}U–^{206}Pb,

$\lambda \sim 4.5$ Gyr; ^{235}U–^{207}Pb, $\lambda \sim 704$ Myr) and rubidium to strontium (^{87}Rb–^{87}Sr, $\lambda \sim 49$ Gyr). Initial studies focussed on measuring lead in iron meteorites, matched against measurements of terrestrial basaltic rocks. The resultant age was derived as $\sim$4.55 Gyr, significantly older than the $\sim$3.5 Gyr measured for the oldest terrestrial rocks then available. Those studies, however, had to make assumptions regarding the evolution of the terrestrial samples. One of the earliest measurements based solely on meteoritic material was by Wasserburg et al. (1965), who used the Rb–Sr decay series to determine an age of 4.7 Gyr from samples of the Weekeroo Station meteorite.

More recent studies tend to focus on the uranium/lead decay series, which have half-lifes closer to the desired timescale. Analyses of CAIs from carbonaceous chondrites of the CV sub-class (named after the Vigarano meteorite) indicate that chondrules formed over a period of roughly 4 Myr, from 4.5672 to 4.5627 Gyr ago, with the uncertainties in the individual ages given as $\sim$0.5–1 Myr (Amelin et al. 2002; Amelin and Krot 2007). This sets a lower limit of 4.567 $\pm$ 0.001 Gyr for the age of the Sun. Variations in the primordial ^{238}U/^{235}U ratio may add complications at the level of a few Myr (Amelin et al. 2009), but this still represents a level of precision seldom (if ever) encountered elsewhere in astronomy. There is, however, a purely astronomical technique that can be used to probe stellar ages at the 1 % level: asteroseismology.

Asteroseismology had its origin in helioseismology which, in turn, came into being when Robert Leighton noticed periodic Doppler shifts as a function of position on the Sun's surface, indicating that the surface was oscillating with a period of $\sim$5 min (Leighton et al. 1962). Initially, those oscillations were regarded as a purely surface phenomenon, but, almost ten years after their discovery, Roger Ulrich (1970) demonstrated that they are a fundamental, global phenomenon, generated by and dependent on the interior properties of the Sun. Quoting from Ulrich's paper, '..the 5-minute oscillations are acoustic waves trapped below the solar photosphere and that power in the (k_h, ω)-diagram should be observed only along discrete lines.' Leibacher and Stein (1971) independently arrived at the same solution.

Analysis of the solar oscillations produced by these acoustic waves (known as p-mode oscillations) probes the internal sound speed, which depends on the mean molecular weight, that is, the abundance of helium and the heavy element composition. The sound speed within the convection zone is sensitive to the heavy element abundance; the sound speed within the core probes the relative abundance of hydrogen and helium in regions where nuclear fusion is changing that ratio as a function of time.

Detailed measurements of the solar oscillation frequency distribution (and more than 5000 modes have been measured to date) put strong constraints on solar models and enable derivation of an independent age estimate for the Sun. In the former context, helioseismology played an important role in resolving the solar neutrino problem (Bahcall et al. 1998). In the latter context, the most recent analyses give an age of $\tau_{Sun} = 4.60 \pm 0.04$ Gyr (Houdek and Gough 2011), encompassing, within 1σ, the more accurate age given by analyses of meteoritic data. There is, however, a complication, in that these very successful models require an overall heavy element abundance of $Z = 0.0155$, an issue noted in Chap. 15 and discussed further below.

21.3 The Sun's Chemical Composition

One might think that our proximity to the Sun would give us an advantage in determining its chemical composition. Such is not the case. We can conduct laboratory
experiments on meteorites or the Earth itself, but those materials have undergone
substantial processing and modification in terms of the relative abundance, particularly for volatile elements. We have not yet come to the point where we can pluck
the golden apples from the Sun; we are restricted to the same techniques that are
available for measuring stellar abundance high-resolution spectroscopy (albeit with
exceptional spectral resolution and signal-to-noise) coupled with theoretical modelling of the underlying photosphere.

We already touched on some of the issues regarding the solar heavy element abundance in Chaps. 15 and 17 in this series. Figure 21.1 shows the overall abundance
distribution: Hydrogen and helium are the two most abundant elements in the Sun,
contributing 73.9 and 24 %, respectively, by mass; Oxygen is the third most abundance element, contributing approximately 1 % by mass and therefore dominating
the overall abundance of heavy elements in the Sun; Carbon, next in line, contributes
less than 0.4 % by mass.

Oxygen may be abundant, but it offers only limited opportunities for spectroscopic
measurements. There are only a handful of permitted lines that can be used for
abundance analyses, with the most accessible being the O I 7772 Å triplet, which
requires correction for non-LTE effects before analysis. The best three available
forbidden [OI] lines (5577, 6300, and 6366 Å) are all blended with other features
(due to Ni I, Ca I and C_2, respectively), requiring significant corrections to the
measured equivalent widths. Finally, vibration-rotation and pure rotation lines of
OH can also be used to derive the oxygen abundance.

As noted in Chap. 15, estimates of the solar oxygen abundance have declined
considerably over the past two decades. Anders and Grevesse (1989) derived a
value of 8.93 ± 0.04 dex on a logarithmic scale where the hydrogen abundance

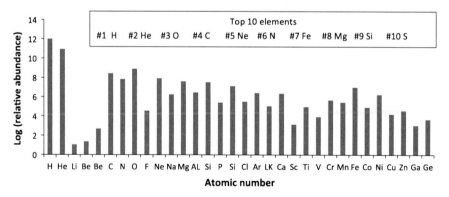

Fig. 21.1 The abundance distribution of elements in the Sun (based on data from Asplund et al.
2009)

is set to 12.0; based on revised atomic data for some species, Grevesse and Sauval (1998) reduced the estimate to 8.83 dex; Asplund (2005) applied more sophisticated three-dimensional models to the data analysis, and lowered the estimate even further to 8.66 dex, or almost a factor of two below the 1989 value; Asplund et al. (2009) adjusted the value slightly upwards, to 8.69 ± 0.05 dex. Consequently, the estimate for the overall abundance of heavy elements in the Sun was reduced from $Z = 0.0201$ (Anders and Grevesse 1989) to $Z = 0.0134$ (Asplund et al. 2009). This lower metal abundance leads to a serious conflict with helioseismology models, which require $Z \sim 0.0155$ in order to maintain sufficient opacity within the convective envelope to account for the inferred sound speeds, depth of the convection zone ($R = 0.722\,R_\odot$ vs. $0.713 \pm 0.001\,R_\odot$ from helioseismology) and the surface abundance of helium ($Y = 0.235$ vs. $Y = 0.248 \pm 0.004$).

A resolution to this conflict may be in sight. Caffau et al. (2011) have re-analysed the solar spectrum using the CO5BOLD three-dimensional model atmosphere. They derive somewhat higher oxygen, carbon and nitrogen abundances than Asplund et al. (2009; 8.76 vs. 8.69 dex, 8.50 vs. 8.43 dex and 7.86 vs. 7.83 dex respectively), and raise the total heavy metal content of the Sun to $Z = 0.0153$. They also note that diffusion may have reduced the abundance of metals in the photosphere by ∼0.04 dex over the Sun's lifetime, offering an even closer reconciliation with the requirements set by helioseismology.

Placing these results in a broader context, the latest value for the oxygen abundance in the Sun is consistent with direct measurements of HII regions and OB stars at the same radial distance from the Galactic Centre (see Fig. 21.2). This strongly mitigates one of the main arguments raised by Wielen et al. (1996) in support of past radial migration by the Sun (see Chap. 17).

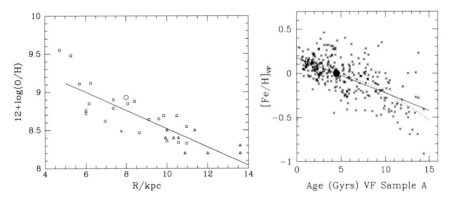

Fig. 21.2 *Left panel*: The oxygen abundance as a function of radius in the Galaxy, as measured from HII regions, with the *open circle* marking the (Anders and Grevesse 1989) value for the solar abundance (from Sellwood and Binney 2002). *Right panel*: The age-metallicity distribution for nearby stars based on data from Valenti and Fischer (2005); the *filled hexagon* marks the Sun's location in this diagram (from Reid et al. 2007)

In the opening chapter we noted that the Sun lies close to the mode of the metallicity distribution outlined by nearby stars, with approximately 40 % of the sample having higher metallicities. The Sun is also average in terms of its contemporaries. Figure 21.2 also shows the age-metallicity distribution of nearby stars from Reid et al. (2007). As discussed in Chap. 19, those data are taken from the analysis of high-resolution spectra of nearby solar-type stars (Valenti and Fischer 2005), with ages estimated by matching against theoretical tracks in the (L, T_{eff}) plane. With an age of 4.5 Gyr, the Sun is closely consistent with the average relation, $[\text{Fe}/\text{H}] = 0.177 - 0.04\tau$, with τ in Gyr before the present time.

21.4 The Characteristics of the Solar System

The major planetary bodies in the Solar System follow near-circular, closely planar orbits. This is in contrast with many of the extrasolar planetary systems discovered in recent years. The overall structure of the Solar System at the present day allows us to set limits on both its internal evolution and on the range in external environments since its formation. This section summarises the present-day characteristics that allow us to set constraints; in the following section, we use those constraints to examine the likely environment where the Sun formed.

Planets form through accretion within gaseous discs that are as a natural consequence of angular momentum conservation within the collapsing protostellar clump in a molecular cloud. This statement is now taken as commonplace fact, but it is worth bearing in mind that the possibility of planet formation within a protostellar discs was first suggested as a modification to the Laplace solar nebular hypothesis some 60 years ago (Kuiper 1951). Their existence was only confirmed through millimetre-wave observations as recently as the 1980s (Beckwith et al. 1986).

Under this hypothesis, planet formation and the subsequent evolution were taken as quiescent, almost static processes, bolstered by the regular, ordered nature of the present-day Solar System. Refractory grains of silicates form as the disc cools below 1700 K, and coagulate to form 1–10 km-sized planetismals which, in turn, accrete material to form planetary embryos and, eventually, terrestrial mass planets. Solar radiation limits the process at small radii, but super-embryos accrete icy condensates and gas-rich envelopes beyond the snow-line ($\sim$3 au for the Solar System), leading to the formation of the gas giants and ice giants at larger distances from the parent star.

The first exoplanet discoveries overturned this static picture. The detection of Jovian-mass planets on sub-Mercurian orbits, well interior to regions within the disc that can support their formation, has highlighted the important role played by orbital migration (Ida and Lin 2008). This migration can take place either within the protoplanetary disc, or subsequently, through planet-planet interactions. Even within the Solar System, dynamical models indicate that fairly substantial changes are possible. In particular, recent simulations based on the Nice [city in France and site of an important meeting] model (Tsiganis et al. 2005) show that mutual perturbations between planets, notably Saturn and Jupiter, can lead to significant changes in the semi-major axes. Such interactions could well account for the late heavy bombardment experienced by Earth some 400–600 Myr after its formation.

Internal orbital evolution is generally not accompanied by significant changes in orbital eccentricity. In contrast, many exoplanets are found to be on orbits of moderate or high eccentricity. Internal planetary interactions or external encounters with other stars are the most likely source of this phenomenon. Thus, the near-circularity of the orbit of major planets in the Solar System suggests that the Sun has not been subject to a relatively close encounter. As discussed in the following section, this limits the likely density of stars within its parent cluster.

In the same vein, the present extent of the Solar System sheds light on the likely extent of the parent protoplanetary disc, and perhaps on how that disc was shaped by external factors. Observations of young stars indicate that gas dissipates from the disc in less than ∼10 Myr (Hernández et al. 2008; Hillenbrand 2008), leaving a rocky debris disc that disperses progressively over time. This sets the critical timescale for forming planets and shaping the disc.

Neptune is the last major planet in the Solar System, with semi-major axis of ∼30 au; the Edgeworth-Kuiper belt lies beyond (Edgeworth 1943; Kuiper 1951), comprised of icy planetesimals ranging in size from dwarf planets like Pluto and Eris to fragments smaller than 1 km in size. Its total mass is estimated as 0.01–0.1 $M_\oplus$, concentrated primarily in ∼100 km objects although there may be a secondary peak at ∼10 km. There is evidence for a cut-off in the distribution of larger bodies at ∼50 au. This outer edge may have been set by external perturbations.

Beyond the Kuiper belt lies the Oort Cloud, a near-spherical distribution of cometary bodies extending from ∼2000 au to as much as 60,000 au (0.3 pc). The existence of this large cometary reservoir was hypothesised originally by Ernst Öpik (1932), but gained traction after Oort (1950) revived the idea as a means of explaining the relatively uniform sky-distribution of very-long period comets. The Oort cloud members are believed to have formed in the inner Solar System, and were ejected to large distances through interactions with major planets, notably Jupiter. The cloud is easily disrupted by close stellar passages, but dynamical models suggest that it was populated primarily some 800 Myr after the Sun's formation. Thus, the continued existence of the cloud sets limits on the number of close stellar passages since that time, but provides little information on the Sun's natal environment.

Individual members of the outer Solar System may also illuminate past dynamical interactions. In particular, the dwarf planet Sedna follows a particularly unusual orbit. It was discovered by Brown et al. (2005) as part of a larger survey for Trans-Neptunian (or Kuiper Belt) Objects using wide-field CCDs on the Palomar Schmidt. At its discovery, Sedna lay some 89.6 au from the Sun, making it the most distant member of the Solar System. Initially, its diameter was estimated as 10–40 % that of Pluto, but subsequent observations indicate that it has very red colours suggestive of a tholin-coated surface and a (larger) diameter closer to ∼1600 km, or ∼50 % that of Pluto. The mass is estimated as ∼2 × 10^{21} kg, or one-tenth that of Pluto.

Sedna is on a highly eccentric orbit, $e = 0.85$, with perihelion at ∼76 au and aphelion at ∼940 au. The perihelion distance lies well-beyond Neptune's orbit, and the dynamics might suggest that Sedna was a member of the Oort Cloud. Its mass, however, is substantially higher than comets typically associated with that body, while it seems very unlikely that the protoplanetary disc would have been substantial enough to permits its formation close to its current perihelion distance. An alternative possibility is that Sedna formed within the Kuiper belt and was perturbed into its present orbit by gravitational interactions with a passing star (Kenyon and Bromley 2004). Such an interaction would require a star passing within 400–800 au of the Sun, and might imply the existence of a family of Sedna-like objects in the outer Solar System.

Finally, meteorites not only provide us with an opportunity to probe the age of the first solids that formed in the Solar System, but also set constraints on the local environment. Analysis of some meteoritic material reveals the presence of decay products form short-lived radionucleides (Cameron et al. 1995; Kita et al. 2005). These species include ^{26}Mg, derived from ^{26}Al (half-life of $\sim$730,000 yr), ^{60}Ni from ^{60}Fe (half-life 1.5 Myr) and ^{53}Cr from ^{53}Mn (half-life 3.7 Myr). The parent radionucleides are generated by nucleosynthesis in Type II supernovae explosion, generated by core collapse within a massive star. The supernovae debris must have polluted the protoplanetary disc while it was still gas-rich, enabling their incorporation in solid grains and, later, meteoritic rocks (Williams and Gaidos 2007). This implies that the Sun was in relatively close proximity to a moderately massive star, $M \gtrsim 8 \, M_\odot$, which in turn suggests formation within a star cluster of at least modest number density.

21.5 The Sun's Natal Environment

Summarising the previous section, the present dynamical state of the Solar System suggests that the Sun was not subjected to particularly close encounters over its lifetime. However, the presence of supernovae detritus within meteorites indicates that it must have formed in a moderately-dense cluster that was capable of producing at least a few high-mass stars. Thus we have a tension between dynamical and chemical constraints that we aim to quantify in the present section. We should note that this topic is discussed extensively by Adams (2010) in his review.

Starting with the supernova ejecta, there are two lines of enquiry that we can follows: first, what constraints can we place on the progenitor mass, and what does that tell us about the parent cluster? Second, where would the Sun need to be relative to the supernova in order to accrete ejecta, but still survive the explosion? We should note that there have been suggestions that the protostellar nebula might have been polluted by more than one supernova, which would reduce the proximity requirement (Gounelle et al. 2009). However, there are issues both with regard to the probability of multiple, closely-spaced supernovae within a single molecular cloud, and the probability of those events occurring within a sufficiently short timespan to pollute the solar nebula. For the present juncture, we shall assume that one supernova was responsible for the pollution.

The mass of the supernova progenitor is constrained by the mass fractions of the daughter products in the meteoritic material. That distribution can be used to infer the mass fractions of the parent nucleides, and hence the overall yield from the supernova. Based on those data, Adams (2010) concludes that the progenitor must have had a mass exceeding $25 \, M_\odot$, while Williams and Gaidos (2007) set an upper limit of $75 \, M_\odot$. The lower the mass, the higher the efficiency required in dispersing processed materials in the ejecta.

Stellar evolution sets another constraint on the likely progenitor mass. If we assume coeval star formation within the cluster, the evolutionary lifetime of the supernova progenitor must be shorter than the lifetime of the Sun's protoplanetary disc. Protoplanetary discs rapidly become gas-poor, with fewer than 5 % of stars aged $\sim 10 \, \mathrm{Myr}$ showing evidence for residual gas (Hernández et al. 2008). In comparison, a $15 \, M_\odot$ solar-metallicity star has a main-sequence lifetime of $\sim 11.1 \, \mathrm{Myr}$, and undergoes core collapse after $\sim 12.1 \, \mathrm{Myr}$; a $25 \, M_\odot$ solar-metallicity star leaves the main-sequence after $\sim 6.7 \, \mathrm{Myr}$ and explodes after $\sim 7.54 \, \mathrm{Myr}$. Thus, unless the formation of the massive stars preceded the formation of the Sun, these evolutionary calculations suggest that it is unlikely that a supernova progenitor less massive than $\sim 20 \, M_\odot$ could be responsible for the radio nucleotides in the solar nebula.

If we assume that the parent cluster had the standard form of the IMF, then we can estimate its likely size by computing the total number of stars that would be required to give a 50 % chance of forming a progenitor with the critical mass. Characterising that parameter as N_{50} and the crucial mass as m_c, Adams (2010) estimates that:

$$N_{50} \sim 7 m_c^{3/2}. \tag{21.2}$$

Thus, we require that $N_{50} \sim 825$, 2500 and 10700 for a 25, 50 or $75 \, M_\odot$ progenitor. This implies that the Sun had its origin in a sizeable open cluster.

Supernovae are extremely energetic events ($\sim 10^{51}$ erg) and have the potential to disrupt a protoplanetary disc. Two avenues are available: ram pressure from supernova ejecta; and momentum transfer from the blast wave. In either case, the interaction strips material from the disc where the incident energy exceeds the gravitational binding energy, with the net result that the disc is truncated. Assuming spherical symmetry in the supernovae ejecta, the final radius of the disc scales linearly with the distance of the supernova. A supernova at 0.1 pc can reduce a protoplanetary disc to a radius of $\sim 100 \, \mathrm{au}$ through ram pressure and to less than 30 au from momentum transfer. Chevalier (2000) has concluded that viable planet-forming discs do not survive a supernova event within 0.2 pc.

Turning the coin, the supernova has to occur close enough to the Sun to allow the protoplanetary nebula to accrete sufficient quantities of ejecta to materially affect the overall abundances. Again assuming spherically symmetric ejection and a disc radius of 30 au, a supernova with a $25 \, M_\odot$ progenitor would need to occur at a distance between 0.1 and 0.3 pc from the Sun (Adams 2010). Higher-mass supernovae are effective at larger distances, and an asymmetric ejection could also mitigate the distance requirements. In any event, these calculations suggest that the Sun may have been in a relatively unusual (lucky!) location during a crucial stage in the formation of its planetary system.

Gravitational encounters with other stars within the Sun's natal cluster will also modify the properties of the protoplanetary disc. In particular, the current architecture suggests a transition in disc properties at $\sim$30 au and a truncation at $\sim$50 au, the edge of the Kuiper belt. A truncation in disc properties at a radial distance, b, generally implies a gravitational encounter at distance $d_b \sim 3b$. This suggests that the Sun was subject to a close encounter at $d_b \sim 150$ au during its first few Myr, modifying the extent of the gas-rich disc. One can compute the expected encounter rate for a star cluster with a given space density. For a typical embedded cluster with density $\sim$100 stars pc^{-3} and velocity dispersion $\sigma \sim 1$ km s^{-1}, the characteristic distance for an encounter is a few hundred au.

Similarly, the low orbital eccentricity exhibited by the major planets in the Solar System implies that the Sun has not been subject to a close stellar encounter since their formation (i.e. after the Sun's protoplanetary disc has become gas-poor). This sets an effective lower limit of 200–225 au for such encounters. The characteristic timescale for such encounters is $\sim$2.5 Gyr for stellar densities of $\sim$100 stars pc^{-3} and $\sim$250 Myr for $\sim$1000 stars pc^{-3}.

The gas-rich protoplanetary disc is also vulnerable to modification by photoionisation and ablation by UV irradiation from nearby high-mass stars. Such processes are clearly evident among the proplyds in the Orion Nebula Cluster. The driving force is extreme-UV radiation, $\lambda < 940$ Å, and the effects are substantial only when in closer proximity to very-high-mass stars. Such stars are likely confined closely to the cluster core, and most stars will be subject to significant irradiation during relatively short periods as their orbits take them close to the core regions.

Finally, Sedna's unique orbit likely requires an event of comparable rarity. As noted in the previous section, the orbital parameters can be induced through an encounter with a Sun-like star at a separation of 400–800 au. A lower-mass star would need to encounter the Sun at a correspondingly closer separation. Clearly, that encounter must take place after Sedna's formation, and therefore postdates the gas-rich stage of protoplanetary disc evolution. The encounter could have taken place at any time since then, either while the Sun was still within its parent cluster or in subsequent years, after the clusters dispersal. Indeed, Kaib et al. (2011) have argued that a particularly propitious period for such an encounter would be during a radial migration episode, if the Sun underwent such an event. There are, however, a wide range of potential environments for such an event.

Pulling these various constraints together, the Sun probably formed in a cluster that was sufficiently populous to produce a (fortuitously placed) massive supernova progenitor, but without sufficient high-mass stars to ionise a substantial fraction of the solar nebula. The cluster also had to be sparse enough that the Sun did not undergo an encounter closer than $\sim$225 au following the formation of the major planets. Adams (2010) has combined these constraints to estimate the Sun's parent cluster had a total membership of $\sim$2,000–3,000 stars. This is smaller than the Orion cluster, but perhaps comparable with the Pleiades.

21.6 Can We Find the Sun's Companions?

If the Sun formed in a cluster, might we be able to locate some of its compatriots? A number of astronomers have espoused this cause. In particular, Portegies Zwart et al. (2009) used dynamical simulations to model the evolution of a compact (1–3 pc radius) massive (500–3,000 $M_\odot$) cluster. He has argued that, depending on when the cluster dissipated, up to 50–60 solar siblings might reside within 100 pc of the Sun.

Given that incentive, astronomers have searched for stellar systems that have photometric properties consistent with an age of 4.6 Gyr, near-solar metallicites and space motions consistent with a past close encounter with the Sun (Bobylev et al. 2011). Several candidates have been identified, notably HD 28676 (Brown et al. 2010), HD 83423 and HD 175740 (Batista and Fernandes 2012) and 162826 (Bobylev et al. 2011). However, on closer inspection, none meet all the criteria: HD 28676 and 162826 have velocities inconsistent with a common point of origin; HD 162826 is also more than twice the metallicity of the Sun; and HD 83423 and HD 175740 both have metallicities that are significantly sub-solar.

In the same vein, the cluster M67 has been suggested as the Sun's point of origin. The cluster has a metal abundance within 0.04 dex of solar, an age estimated as 4.57 Gyr and lies at a distance of ~900 pc. However, Pichardo et al. (2012) have modelled the relative orbits of the Sun and M67, and their simulations fail to find any encounters where the relative velocity is less than 20 km s^{-1}. This would require that the Sun was ejected from the cluster, which would demand a stellar encounter well within the 225 au limit set by the planet's current orbits. Overall, Pichardo et al. (2012) conclude that the probability of the Sun originated in M67 is less than 1 part in 10^7, a level of improbability that gives even astronomers pause.

Moreover, detailed simulations by Mishurov and Acharova (2011) have raised significant doubts as to whether one would even expect to see solar siblings in the local neighbourhood. As they point out, the original analysis by Portegies Zwart et al. (2009) did not take into account the effect of density waves in spiral arms, including interactions that lead to dispersion of cluster members through the Galaxy. If the Sun's parent cluster included 1,000 stars, they estimate that none are likely to lie within 100 pc; indeed, unless the parent cluster had more than 10,000 stars, Mishurov and Acharova (2011) assert that the task of finding solar siblings '...becomes less than hopeless'.

Finally, what about solar migration through the Galactic disc? The original impetus for considering this phenomenon was the apparent mismatch between the Sun's oxygen abundance and the abundance in the local ISM. It is no longer clear whether that mismatch exists—recent analyses are moving towards reconciling the spectroscopic abundance analyses of the Sun's photosphere with the heavy metal content required within the convection zone by helioseismology. That being said, radial migration clearly occurs within the disc, and likely accounts for many moving groups in the Solar Neighbourhood. *Gaia* will provide more insight on these matters, but even *Gaia* is unlikely to be able to allow us to reconstruct the exact behaviour of an individual star that happens to host our planet.

21.7 Summary and Conclusions

A key question, perhaps THE key question, underlying investigations of the Sun's birthplace is 'How likely is the formation of a habitable planetary system?'. As results from the *Kepler* satellite rolled in, it is clear that forming planets is not a difficult process (Batalha et al. 2013). The initial radial velocity surveys were effective at picking up Jovian-mass companions, and indicated that 7–10 % of solar-type stars had planetary companions in that mass range. *Kepler* has pushed the detection limit to Neptunes, super-Earths and even Earths. Those results strongly indicate that almost every star, irrespective of its mass, has a planetary system. Many, probably most, do not have Earth-like planets within their habitable zone, but the statistics have accumulated to the point where we can assert that $\eta_{Earth} > 0.1$ (Petigura et al. 2013) and start making appropriate plans to find and characterise the nearest habitable system. What about solar analogues? Adams (2010) has compiled a table of Solar System properties with a set of associated probabilities: mass, metallicity, multiplicity, structure of the planetary system, influence by supernovae and gravitational encounters. If one assumes that all these factors are independent, then one arrives at a total probability of 4.6×10^{-7} for the existence of the Solar System. Given $\sim 10^{10}$ stars in total and $\sim 10^{9}$ solar-type stars, this would imply that there are a few hundred planetary systems nearly identical to ours distributed throughout the Milky Way.

However, this calculation is only relevant in a broader sense if we believe that there is some causal connection between the specific properties of the Sun and one or more critical events that enabled life to develop on Earth. Do Sedna's properties or supernova proximity actually influence the emergence of sentience? We can list many solar characteristics, and catalogue past events, such as stellar encounters and proximity to supernovae, but we cannot point with any certainty to which factors as played a crucial role in enabling (or even not hindering) the development of life on Earth.

In essence, we are still trying to solve the Drake equation, but recast as a set of astronomical observables combined with a set of biological quantities. Following Reid and Hawley (2005), as adapted by M. Mountain, we can rewrite the classical equation as follows:

$$N_{l,t} = (N_{*,t} P_p n_e P_w)(P_a P_e P_s),   \tag{21.3}$$

where $N_{l,t}$ is the number of life-bearing planets at time t. The first four quantities on the right hand side of the revised equation are astronomical: $N_{*,t}$ is the number of stars at time t, given by the convolution of the star formation history, the initial mass function and the stellar death rate; P_p is the probability of forming a planetary system; n_e is the number of terrestrial planets; and P_w is the probability of finding liquid water. The latter two quantities could be combined as η_e, the number of terrestrial planets within the habitable zone, with the loss of systems such as Europa; *Kepler*'s latest results point to a value of $\eta_e > 0.1$ for Sun-like stars. The last three quantities are biological: P_a, the probability of abiogenesis; P_e, the probability of evolution; and

P_s, the probability of survival. We now have observational constraints on all the astronomical quantities, while laboratory and field analyses are starting to constrain the biological factors. It may be, however, that the real breakthrough will only come with the discovery of life elsewhere, when we can look for systematic patterns and stop arguing from the statistics of one.

References

Adams, F. C. 2010, ARA&A, 48, 47

Amelin, Y. & Krot, A. 2007, Meteoritics and Planetary Science, 42, 1321

Amelin, Y., Krot, A. N., Hutcheon, I. D., & Ulyanov, A. A. 2002, Science, 297, 1678

Amelin, Y., Connelly, J., Zartman, R. E., et al. 2009, Geochim. Cosmochim. Acta., 73, 5212

Anders, E. & Grevesse, N. 1989, Geochim. Cosmochim. Acta., 53, 197

Asplund, M. 2005, ARA&A, 43, 481

Asplund, M., Grevesse, N., Sauval, A. J., & Scott, P. 2009, ARA&A, 47, 481

Bahcall, J. N., Basu, S., & Pinsonneault, M. H. 1998, Physics Letters B, 433, 1

Batalha, N. M., Rowe, J. F., Bryson, S. T., et al. 2013, ApJS, 204, 24

Batista, S. F. A. & Fernandes, J. 2012, New Astron., 17, 514

Beckwith, S., Sargent, A. I., Scoville, N. Z., et al. 1986, ApJ, 309, 755

Bobylev, V. V., Bajkova, A. T., Mylläri, A., & Valtonen, M. 2011, Astronomy Letters, 37, 550

Brown, A. G. A., Portegies Zwart, S. F., & Bean, J. 2010, MNRAS, 407, 458

Brown, M. E., Trujillo, C. A., & Rabinowitz, D. L. 2005, ApJ, 635, L97

Caffau, E., Ludwig, H.-G., Steffen, M., Freytag, B., & Bonifacio, P. 2011, Sol. Phys., 268, 255

Cameron, A. G. W., Hoeflich, P., Myers, P. C., & Clayton, D. D. 1995, ApJ, 447, L53

Chevalier, R. A. 2000, ApJ, 538, L151

Edgeworth, K. E. 1943, Journal of the British Astronomical Association, 53, 181

Gounelle, M., Meibom, A., Hennebelle, P., & Inutsuka, S.-I. 2009, in Lunar and Planetary Science Conference, Vol. 40, Lunar and Planetary Science Conference, 1624

Grevesse, N. & Sauval, A. J. 1998, Sci. Rev., 85, 161

Hernández, J., Hartmann, L., Calvet, N., et al. 2008, ApJ, 686, 1195

Hillenbrand, L. A. 2008, Physica Scripta Volume T, 130, 014024

Houdek, G. & Gough, D. O. 2011, MNRAS, 418, 1217

Ida, S. & Lin, D. N. C. 2008, ApJ, 673, 487

Kaib, N. A., Roškar, R., & Quinn, T. 2011, Icarus, 215, 491

Kenyon, S. J. & Bromley, B. C. 2004, Nature, 432, 598

Kita, N. T., Huss, G. R., Tachibana, S., et al. 2005, in Astronomical Society of the Pacific Conference Series, Vol. 341, Chondrites and the Protoplanetary Disk, ed. A. N. Krot, E. R. D. Scott, & B. Reipurth, 558

Kuiper, G. P. 1951, Proceedings of the National Academy of Science, 37, 1

Leibacher, J. W. & Stein, R. F. 1971, Astrophys. Lett., 7, 191

Leighton, R. B., Noyes, R. W., & Simon, G. W. 1962, ApJ, 135, 474

Mishurov, Y. N. & Acharova, I. A. 2011, MNRAS, 412, 1771

Oort, J. H. 1950, Astron. Inst. Netherlands, 11, 91

Öpik, E. J. 1932, Proceedings of the American Academy of Arts and Sciences, 67, 169

Petigura, E. A., Howard, A. W., & Marcy, G. W. 2013, Proceedings of the National Academy of Science, 110, 19273

Pichardo, B., Moreno, E., Allen, C., et al. 2012, AJ, 143, 73

Portegies Zwart, S. F. 2009, ApJ, 696, L13

Reid, I. N. & Hawley, S. L. 2005, New Light on Dark Stars: Red Dwarfs, Low-Mass Stars, Brown Dwarfs, ed. Reid, I. N. and Hawley, S. L., Springer-Praxis

Reid, I. N., Turner, E. L., Turnbull, M. C., Mountain, M., & Valenti, J. A. 2007, ApJ, 665, 767
Sellwood, J. A. & Binney, J. J. 2002, MNRAS, 336, 785
Tsiganis, K., Gomes, R., Morbidelli, A., & Levison, H. F. 2005, Nature, 435, 459
Ulrich, R. K. 1970, ApJ, 162, 993
Valenti, J. A. & Fischer, D. A. 2005, ApJS, 159, 141
Wasserburg, G. J., Burnett, D. S., & Frondel, C. 1965, Science, 150, 1814
Wielen, R., Fuchs, B., & Dettbarn, C. 1996, A&A, 314, 438
Williams, J. P. & Gaidos, E. 2007, ApJ, 663, L33